THE CHEMICAL FORMULARY

The
Chemical Formulary

*A Collection of Valuable, Timely, Practical,
Commercial Formulae and Recipes for
Making Thousands of Products in
Many Fields of Industry*

VOLUME XII

Editor-in-Chief
H. BENNETT, F.A.I.C.

*Director, B. R. Laboratory
Miami Beach, Florida*

1965
CHEMICAL PUBLISHING CO., INC.
212 FIFTH AVENUE   NEW YORK, N. Y.

11,977

PRINTED IN THE UNITED STATES OF AMERICA

PREFACE TO VOLUME XII

This new volume of the CHEMICAL FORMULARY series is a collection of new, up-to-date formulae. The only repetitious material is the introduction (Chapter I) which is used in every volume for the benefit of those who may have bought only one volume and who have no educational background or experience in chemical compounding. The simple basic formulae and compounding methods given in the introduction will serve as a guide for beginners and students. It is suggested that they read the introduction carefully and even make a few preparations described there before compounding the more intricate formulae included in the later chapters.

The list of chemicals and their suppliers has been enlarged with new trade-mark chemicals, so that buying the required ingredients will present no problem.

Grateful acknowledgement is made to the Board of Editors for their valuable suggestions and contributions.

H. BENNETT

NOTE: All the formulae in volumes I, II, III, IV, V, VI, VII, VIII, IX, X, XI, XII (except in the introduction) are different. Thus, if you do not find what you are looking for in this volume, you may find it in one of the others.

NOTE: This book is the result of cooperation of many chemists and engineers who have given freely of their time and knowledge. It is their business to act as consultants and to give advice on technical matters for a fee. As publishers, we do not maintain a laboratory or consulting service to compete with them. Therefore, please do not ask *us* for advice or opinions, but confer with a chemist.

Formulae for which patent numbers are listed can be manufactured only after obtaining a license from the patentees.

PREFACE

Chemistry, as taught in our schools and colleges, concerns chiefly synthesis, analysis, and engineering — and properly so. It is part of the right foundation for the education of the chemist.

Many a chemist entering an industry soon finds that most of the products manufactured by his concern are not synthetic or definite chemical compounds, but are mixtures, blends, or highly complex compounds of which he knows little or nothing. The literature in this field, if any, may be meager, scattered, or obsolete.

Even chemists with years of experience in one or more industries spend considerable time and effort in acquainting themselves with any new field which they may enter. Consulting chemists similarly have to solve problems brought to them from industries foreign to them. There was a definite need for an up-to-date compilation of formulae for chemical compounding and treatment. Since the fields to be covered are many and varied, an editorial board of chemists and engineers engaged in many industries was formed.

Many publications, laboratories, manufacturing firms, and individuals have been consulted to obtain the latest and best information. It is felt that the formulae given in this volume will save chemists and allied workers much time and effort.

Manufacturers and sellers of chemicals will find, in these formulae, new uses for their products. Nonchemical executives, professional men, and interested laymen will make through this volume a "speaking acquaintance" with products which they may be using, trying, or selling.

It often happens that two individuals using the same ingredients in the same formula get different results. This may be due to slight deviations in the raw materials or unfamiliarity with the intricacies of a new technique. Accordingly, repeated experiments may be necessary to get the best results. Although many of the formulae given are being used commercially, many have been taken from the literature and may be subject to various errors and omissions. This should be taken into consideration. Wherever possible, it is advisable to consult with other chemists or technical workers regarding commercial production. This will save time and money and help avoid trouble.

A formula will seldom give exactly the results which one requires. Formulae are useful as starting points from which to work out one's own ideas. Also, formulae very often give us ideas which may help us in our specific problems. In a compilation of this kind, errors of omission, commission, and printing may occur. I shall be glad to receive any constructive criticism.

H. BENNETT

CONTENTS

ABBREVIATIONS

amp	ampere
amp /dm$_2$	amperes per square decimeter
amp /sq ft	amperes per square foot
anhydr	anhydrous
avoir	avoirdupois
bbl	barrel
Bé	Baumé
B.P.	boiling point
°C	degrees Centigrade
cc	cubic centimeter
c d	current density
cm	centimeter
cm$_3$	cubic centimeter
conc	concentrated
c.p.	chemically pure
cp	centipoise
cu ft	cubic foot
cu in.	cubic inch
cwt	hundredweight
d	density
dil	dilute
dm	decimeter
dm^2	square decimeter
dr	dram
E	Engler
°F	degrees Fahrenheit
f f c	free from chlorine
f f p a	free from prussic acid
fl dr	fluid dram
fl oz	fluid ounce
fl pt	flash point
F.P.	freezing point
ft	foot
ft^2	square foot
g	gram

gal ....gallon
gr ....grain
hl ....hectoliter
hr ....hour
in................inch
kg ....kilogram
l.................liter
lb ....pound
liq ....liquid
m ....meter
min ....minim, minute
ml ....milliliter (cubic centimeter)
mm ....millimeter
M.P...............melting point
N.................Normal
N F ....National Formulary
oz ....ounce
pH...............hydrogen-ion concentration
p p m ....parts per million
pt ....pint
pwt ....pennyweight
q.s...............a quantity sufficient to make
qt ....quart
r p m ....revolutions per minute
sec ....second
sp ....spirits
Sp. Gr............specific gravity
sq dm ....square decimeter
tech ....technical
tinc ....tincture
tr ....tincture
Tw ....Twaddell
U S P ....United States Pharmacopeia
v ....volt
visc ....viscosity
vol ....volume
wt ....weight

CHAPTER I

INTRODUCTION

The following introductory matter has been included at the suggestion of teachers of chemistry and home economics.

This section will enable anyone, with or without technical education or experience, to start making simple products without any complicated or expensive machinery. For commercial production, however, suitable equipment is necessary.

Chemical specialties are composed of pigments, gums, resins, solvents, oils, greases, fats, waxes, emulsifying agents, dyestuffs, perfumes, water, and chemicals of great diversity. To compound certain of these with some of the others requires definite and well-studied procedures, any departure from which will inevitably result in failure. The steps for successful compounding are given with the formulae. Follow them rigorously. If the directions require that (a) is added to (b), carry this out literally, and do not reverse the order. The preparation of an emulsion is often quite as tricky as the making of mayonnaise. In making mayonnaise, you add the oil to the egg, slowly, with constant and even stirring. If you do it correctly, you get mayonnaise. If you depart from any of these details: If you add the egg to the oil, or pour the oil in too quickly, or fail to stir regularly, the result is a complete disappointment. The same disappointment may be expected if the prescribed procedure of any other formulation is violated.

The point next in importance is the scrupulous use of the proper ingredients. Substitutions are sure to result in inferior quality, if not in complete failure. Use what the formula calls for. If a cheaper product is desired, do not prepare it by substituting a cheaper ingredient for the one prescribed: use a different formula. Not infrequently, a formula will call for an ingredient which is difficult to obtain. In such cases, either reject the formula or substitute a similar substance only after a preliminary experiment demonstrates its usability. There is a limit to which this rule may reasonably be extended. In some cases, substitution of an equivalent ingredient may be made legitimately. For example, when the formula calls for white wax (beeswax), yellow wax can be used, if the color of the finished product is a matter of secondary importance. Yellow beeswax can often replace white beeswax, making due allowance for color, but paraffin wax will not replace beeswax, even though its light color seems to place it above yellow beeswax.

And this leads to the third point: the use of good-quality ingredients, and ingredients of the correct quality. Ordinary lanolin is not the same thing as anhydrous lanolin. The replacement of one with the other, weight for weight, will give discouragingly different results. Use exactly what the formula calls for: if you are not acquainted with the substance and you are in doubt as to just what is meant, discard the formula and use one you understand. Buy your chemicals from reliable sources. Many ingredients are obtainable in a number of different grades: if the formula does not designate the grade, it is understood that the best grade is to be used. Remember that a formula and the directions can tell you only part of the story. Some skill is often required to attain success. Practice with a small batch in such cases until you are sure of your technique. Many examples can be cited. If the formula calls for steeping quince seed for 30 minutes in cold water, steeping for 1 hour may yield a mucilage of too thin a consistency. The originator of the formula may have used a fresher grade of seed, or his conception of what "cold" water means may be different from yours. You should have a feeling for the right degree of mucilaginousness, and if steeping the seed for 30 minutes fails to produce it, steep them longer until you get the right kind of mucilage. If you do not know what the right kind is, you will have to experiment until you find out. This is the reason for the recommendation to make small experimental batches until successful results are obtained. Another case is the use of dyestuffs for coloring lotions and the like. Dyes vary in strength; they are all very powerful in tinting value; it is not always easy to state in quantitative terms how much to use. You must establish the quantity by carefully adding minute quantities until you have the desired tint. Gum tragacanth is one of those products which can give much trouble. It varies widely in solubility and bodying power; the quantity listed in the formula may be entirely unsuitable for your grade of tragacanth. Therefore, correction is necessary, which can be made only after experiments with the available gum.

In short, if you are completely inexperienced, you can profit greatly by experimenting. Such products as mouth washes, hair tonics, and astringent lotions need little or no experience, because they are, as a rule, merely mixtures of simple liquid and solid ingredients, which dissolve without difficulty and the end product is a clear solution that is ready for use when mixed. However, face creams, tooth pastes, lubricating greases, wax polishes, etc., whose formulation requires relatively elaborate procedure and which must have a definite final viscosity, need some skill and not infrequently some experience.

Figuring

Some prefer proportions expressed by weight or volume, others use percentages. In different industries and foreign countries different systems of weights and measures are used. For this reason, no one set of units could be satisfactory for everyone. Thus divers formulae appear with different units, in accordance with their sources of origin. In some cases, parts are given instead of percentage or weight or volume. On the pages preceding the index, conversion tables of weights and measures are listed. These are used for changing from one system to another. The following examples illustrate typical units:

EXAMPLE No. 1

Ink for Marking Glass

Glycerin	40	Ammonium Sulfate	10
Barium Sulfate	15	Oxalic Acid	8
Ammonium Bifluoride	15	Water	12

Here no units are mentioned. In this case, it is standard practice to use parts by weight throughout. Thus here we may use ounces, grams, pounds, or kilograms as desired. But if ounces are used for one item, the ounce must be the unit for all the other items in the formula.

EXAMPLE No. 2

Flexible Glue

Powdered Glue	30.90%	Glycerin	5.15%
Sorbitol (85%)	15.45%	Water	48.50%

Where no units of weight or volume, but percentages are given, forget the percentages and use the same method as given in Example No. 1.

EXAMPLE No. 3

Antiseptic Ointment

Petrolatum	16 parts	Benzoic Acid	1 part
Coconut Oil	12 parts	Chlorothymol	1 part
Salicylic Acid	1 part		

The instructions given for Example No. 1 also apply to Example No. 3. In many cases, it is not wise to make up too large a quantity of a product before making a number of small batches to first master the necessary technique and also to see whether the product is suitable for the particular purpose for which it is intended. Since, in many cases, a formula may be given in proportions as made up on a factory scale, it is advisable to reduce the quantities proportionately.

EXAMPLE No. 4

Neutral Cleansing Cream

Mineral Oil	80 lb	Water	90 lb
Spermaceti	30 lb	Glycerin	10 lb
Glyceryl Monostearate	24 lb	Perfume	To suit

Here, instead of pounds, ounces or even grams may be used. This formula would then read:

Mineral Oil	80 g	Water	90 g
Spermaceti	30 g	Glycerin	10 g
Glyceryl Monostearate	24 g	Perfume	To suit

Reduction in bulk may also be obtained by taking the same fractional part or portion of each ingredient in a formula. Thus in the following formula:

EXAMPLE No. 5

Vinegar Face Lotion

Acetic Acid (80%)	20	Alcohol	440
Glycerin	20	Water	500
Perfume	20		

We can divide each amount by ten and then the finished bulk will be only one tenth of the original formula. Thus it becomes:

Acetic Acid (80%)	2	Alcohol	44
Glycerin	2	Water	50
Perfume	2		

Apparatus

For most preparations, pots, pans, china, and glassware, which are used in every household, will be satisfactory. For making fine mixtures and emulsions, a malted-milk mixer or egg beater is necessary. For weighing, a small, low-priced scale should be purchased from a laboratory-supply house. For measuring fluids, glass graduates or measuring glasses may be purchased from your local druggist. Where a thermometer is necessary, a chemical thermometer should be obtained from a druggist or chemical-supply firm.

Methods

To understand better the products which you intend to make, it is advisable that you read the complete section covering such products. You may learn different methods that may be used and also to avoid errors which many beginners are prone to make.

Containers for Compounding

Where discoloration or contamination is to be avoided, as in light-colored, or food and drug products, it is best to use enameled or earthenware vessels. Aluminum is also highly desirable in such cases, but it should not be used with alkalis as these dissolve and corrode aluminum.

Heating

To avoid overheating, it is advisable to use a double boiler when temperatures below 212°F (temperature of boiling water) will suffice. If a double boiler is not at hand, any pot may be filled with water and the vessel containing the ingredients to be heated placed in the water. The pot may then be heated by any flame without fear of overheating. The water in the pot, however, should be replenished from time to time; it must not be allowed to "go dry." To get uniform higher temperatures, oil, grease, or wax is used in the outer container in place of water. Here, of course, care must be taken to stop heating when thick fumes are given off as these are inflammable. When higher uniform temperatures are necessary, molten lead may be used as a heating medium. Of course, with chemicals which melt uniformly and are nonexplosive, direct heating over an open flame is permissible, with stirring, if necessary.

Where instructions indicate working at a certain temperature, it is important to attain the proper temperature not by guesswork, but by the use of a thermometer. Deviations from indicated temperatures will usually result in spoiled preparations.

Temperature Measurement

In the United States and in Great Britain, the Fahrenheit scale of temperature is used. The temperature of boiling water is 212° Fahrenheit (212°F); the temperature of melting ice is 32° Fahrenheit (32°F).

In scientific work, and in most foreign countries, the Centigrade scale is used, on which the temperature of boiling water is 100 °Centigrade (100°C) and the temperature of melting ice is 0° Centigrade (0°C).

The temperature of liquids is measured by a glass thermometer. This is inserted as deeply as possible in the liquid and is moved about until the temperature reading remains steady. It takes a short time for the glass of the thermometer to reach the temperature of the liquid. The thermometer should not be placed against the bottom or side of the container, but near the center of the liquid in the vessel. Since the glass of the thermometer bulb is very thin, it breaks easily when striking it against any hard surface. A cold thermometer should be warmed gradually (by holding it over the surface of a hot liquid) before immersion. Similarly the hot thermometer when taken out of the liquid should not be put into cold water suddenly. A sharp change in temperature will often crack the glass.

Mixing and Dissolving

Ordinary dissolution (e.g., that of sugar in water) is hastened by stirring and warming. Where the ingredients are not corrosive, a clean stick, a fork, or spoon may be used as a stirring rod. These may also be used for mixing thick creams or pastes. In cases where very thorough stirring is necessary (e.g., in making mayonnaise, milky polishes, etc.), an egg beater or a malted-milk mixer is necessary.

Filtering and Clarification

When dirt or undissolved particles are present in a liquid, they are removed by settling or filtering. In the first procedure, the solution is allowed to stand and if the particles are heavier than the liquid they will gradually sink to the bottom. The liquid may be poured or siphoned off carefully and, in some cases, it is then sufficiently clear for use. If, however, the particles do not settle out, then they must be filtered off. If the particles are coarse they may be filtered or strained through muslin or other cloth. If they are very small, filter paper is used. Filter papers may be obtained in various degrees of fineness. Coarse filter paper filters rapidly but will not retain extremely fine particles. For fine particles, a very fine grade of filter paper should be used. In extreme cases, even this paper may not be fine enough. Then, it will be necessary to add to the liquid 1 to 3% infusorial earth or magnesium carbonate. These are filter aids that clog up the pores of the filter paper and thus reduce their size and hold back undissolved material of extreme fineness. In all such filtering, it is advisable to take the first portions of the filtered liquid and pour them through the filter again as they may develop cloudiness on standing.

Decolorizing

The most commonly used decolorizer is decolorizing carbon. This is added to the liquid to the extent of 1 to 5% and the liquid is heated, with stirring, for $\frac{1}{2}$ hour to as high a temperature as is feasible. The mixture is then allowed to stand for a while and filtered. In some cases, bleaching must be resorted to.

Pulverizing and Grinding

Large masses or lumps are first broken up by wrapping in a clean cloth, placing between two boards, and pounding with a hammer. The smaller pieces are then pounded again to reduce their size. Finer grinding is done in a mortar with a pestle.

Spoilage and Loss

All containers should be closed when not in use to prevent evaporation or contamination by dust; also because, in some cases, air affects the product adversely. Many chemicals attack or corrode the metal containers in which they are kept. This is particularly true of liquids. Therefore, liquids should be transferred into glass bottles which should be as full as possible. Corks should be covered with aluminum foil (or dipped in melted paraffin wax when alkalis are present).

Glue, gums, olive oil, or other vegetable or animal products may ferment or become rancid. This produces discoloration or unpleasant odors. To avoid this, suitable antiseptics or preservatives must be used. Cleanliness is of utmost importance. All containers must be cleaned thoroughly before use to avoid various complications.

Weighing and Measuring

Since, in most cases, small quantities are to be weighed, it is necessary to get a light scale. Heavy scales should not be used for weighing small amounts as they are not accurate enough for this type of weighing.

For measuring volumes of liquids, measuring glasses or cylinders (graduates) should be used. Since this glassware cracks when heated or cooled suddenly, it should not be subjected to sudden changes of temperature.

Caution

Some chemicals are corrosive and poisonous. In many cases, they are labeled as such. As a precautionary measure, it is advised not to inhale them and, if smelling is absolutely necessary, only to sniff a few inches from the cork or stopper. Always work in a well-ventilated room when handling poisonous or unknown chemicals. If anything is spilled, it should be wiped off and washed away at once.

Where to Buy Chemicals and Apparatus

Many chemicals and most glassware can be purchased from your druggist. A list of suppliers of all products is at the end of this book.

Advice

This book is the result of cooperation of many chemists and engineers who have given freely of their time and knowledge. It is their business to act as consultants and to give advice on technical matters for a fee. As publishers, we do not maintain a laboratory or consulting service to compete with them.

Please, therefore, do not ask us for advice or opinions, but confer with a chemist in your vicinity.

Extra Reading

Keep up with new developments of materials and methods by reading technical magazines. Many technical publications are listed under references in the back of this book.

Calculating Costs

Raw materials, purchased in small quantities, are naturally higher in price than when bought in large quantities. Commercial prices, as given in the trade papers and catalogs of manufacturers, are for large quantities such as barrels, drums, or sacks. For example, 1 lb. epsom salts, bought at retail, may cost 10 or 15 cents. In barrel lots its price is about 2 or 3 cents per pound.

Typical Costing Calculation
Formula for Beer- or Milk-Pipe Cleaner

Soda Ash	25 lb	@ $0.02½ per lb	= $ 0.63
Sodium Perborate	75 lb	@ 0.16 per lb	= 12.00
Total	100 lb		Total $12.63

If 100 lb cost $12.63, 1 lb will cost $12.63 divided by 100 or about $0.126, assuming no loss.

Always weigh the amount of finished product and use this weight in calculating costs. Most compounding results in some loss of material because of spillage, sticking to apparatus, evaporation, etc. Costs of making experimental lots are always high and should not be used for figuring costs. To meet competition, it is necessary to buy in large quantities and manufacturing costs should be based on these.

Elementary Preparations

The simple recipes that follow have been selected because of their importance and because they are easy to make.

The succeeding chapters go into greater detail and give many different types and modifications of these and other recipes for home and commercial use.

Cleansing Creams

Cleansing creams, as the name implies, serve as skin cleaners. Their basic ingredients are oils and waxes which are rubbed into the skin. When wiped off, they carry off dirt and dead skin. The liquefying type cleansing cream contains no matter and melts or liquefies when rubbed on the skin. To suit different climates and likes and dislikes harder or softer products can be made.

Cleansing Cream (Liquefying)

Liquid Petrolatum	5.5
Paraffin Wax	2.5
Petrolatum	2.0

Melt the ingredients together, with stirrings, in an aluminum or enamelled dish and allow to cool. Then stir in a perfume oil. Allow to stand until it becomes hazy and then pour into jars, which should be allowed to stand undisturbed overnight.

Cold Creams

The most important facial cream is the cold cream. This type of cream contains mineral oil and wax which are emulsified in water with a small amount of borax or glycosterin. The function of a cold cream is to form a film that takes up dirt and waste tissue, which are removed when the skin is wiped thoroughly. Many modifications of this basic cream are encountered in stores. They vary in color, odor, and in claims, but, essentially, they are not more useful than this simple cream. The latest type of cold cream is the nongreasy cold cream which is of particular interest because it is nonalkaline and, therefore, nonirritating for sensitive skins.

Cold Cream

Liquid Petrolatum	52 g
White Beeswax	14 g

Heat this in an aluminum or enamelled double boiler. (The water in the outer pot should be brought to a boil.) In a separate aluminum or enamelled pot dissolve:

Borax	1 g
Water	33 cc

and bring this to a boil. Add this in a thin stream to the melted wax, while stirring vigorously in one direction only. When the temperature drops to 140°F, add 0.5 cc perfume oil and continue stirring until the temperature drops to 120°F. At this point, pour into jars, where the cream will set after a while. If a harder cream is desired, reduce the amount of liquid petrolatum. If a softer cream is wanted, increase it.

Nongreasy Cold Cream

White Paraffin Wax	1.25
Petrolatum	1.50
Glycosterin or Glyceryl Monostearate	2.25
Liquid Petrolatum	3.00

Heat this mixture in an aluminum or enamelled double boiler. (The water in the outer pot should be boiling.) Stir until clear. To this slowly add, while stirring vigorously:

Boiling Water	10

Continue stirring until smooth and then add, with stirring, perfume oil. Pour into jars at 110 to 130°F and cover the jars as soon as possible.

Vanishing Creams

Vanishing creams are nongreasy soapy creams which have a cleansing effect. They are also used as a powder base.

Vanishing Cream

Stearic Acid	18 oz

Melt this in an aluminum or enamelled double boiler. (The water in the outer pot must be boiling.) Add, in a thin stream, while stirring vigorously, the following boiling solution made in an aluminum or enamelled pot:

Potassium Carbonate	¼ oz
Glycerin	6½ oz
Water	5 lb

Continue stirring until the temperature falls to 135°F, then mix in a perfume oil and stir from time to time until cold. Allow to stand overnight and stir again the next day. Pack into jars and close these tightly.

Hand Lotions

Hand lotions are usually clear or milky liquids or salves which are useful in protecting the skin from roughness and redness because of exposure to cold, hot water, soap, and other agents. Chapped hands are common. The use of a good hand lotion keeps the skin smooth, soft, and in a healthy condition. The lotion is best applied at night, rather freely, and cotton gloves may be worn to prevent soiling. During the day, it should be put on sparingly and the excess wiped off.

Hand Lotion
(Salve)

| Boric Acid | 1 |
| Glycerin | 6 |

Warm these in an aluminum or enamelled dish and stir until dissolved (clear). Then allow to cool and work this liquid into the following mixture, adding only a little at a time.

| Lanolin | 6 |
| Petrolatum | 8 |

To impart a pleasant odor a little perfume may be added and worked in.

Hand Lotion
(Milky liquid)

Lanolin	¼ tsp	
Glycosterin or Glyceryl Monostearate	1	oz
Tincture of Benzoin	2	oz
Witch Hazel	25	oz

Melt the first two items together in an aluminum or enamelled double boiler. If no double boiler is at hand, improvise one by placing a dish in a small pot containing boiling water. When the mixture becomes clear, remove from the double boiler and add slowly, while stirring vigorously, the tincture of benzoin and then the witch hazel. Continue stirring until cool and then put into one or two large bottles and shake vigorously. The finished lotion is a milky liquid comparable to the best hand lotions on the market sold at high prices.

Brushless Shaving Creams

Brushless or latherless shaving creams are soapy in nature and do not require lathering or water. The formula given here is of the latest type being free from alkali and nonirritating. It should be borne in mind, however, that certain beards are not softened by this type of cream and require the old-fashioned lathering shaving cream.

Brushless Shaving Cream

White Mineral Oil	10
Glycosterin or Glyceryl Monostearate	10
Water	50

Heat the first two ingredients together in a Pyrex or enamelled dish to 150°F and run in slowly, while stirring, the water which has been heated to boiling. Allow to cool to 150°F and, while stirring, add a few drops of perfume oil. Continue stirring until cold.

Mouth Washes

Mouth washes and oral antiseptics are of practically negligible value. However, they are used because of their refreshing taste and slight deodorizing effect.

Mouth Wash

Benzoic Acid	5⁄8
Tincture of Rhatany	3
Alcohol	20
Peppermint Oil	1⁄8

Mix together in a dry bottle until the benzoic acid is dissolved. One teaspoonful is used to a small-wine-glassful of water.

Tooth Powders

The cleansing action of tooth powders depends on their contents of soap and mild abrasives, such as precipitated chalk and magnesium carbonate. The antiseptic present is practically of no value. The flavoring ingredients mask the taste of the soap and give the mouth a pleasant aftertaste.

Tooth Powder

Magnesium Carbonate	420 g
Precipitated Chalk	565 g
Sodium Perborate	55 g
Sodium Bicarbonate	45 g
Powdered White Soap	50 g
Powdered Sugar	90 g
Wintergreen Oil	8 cc
Cinnamon Oil	2 cc
Menthol	1 g

Dissolve the last three ingredients together and then rub well into the sugar. Add the soap and perborate, mixing well. Add the chalk, with good mixing, and then the sodium bicarbonate and magnesium carbonate. Mix thoroughly and sift through a fine wire screen. Keep dry.

Foot Powders

Foot powders consist of talc or starch with or without an antiseptic or deodorizer. In the following formula the perborates liberate oxygen, when in contact with perspiration, which tends to destroy unpleasant odors. The talc acts as a lubricant and prevents friction and chafing.

Foot Powder

Sodium Perborate	3
Zinc Peroxide	2
Talc	15

Mix thoroughly in a dry container until uniform. This powder must be kept dry or it will spoil.

Liniments

Liniments usually consist of an oil and an irritant, such as methyl salicylate or turpentine. The oil acts as a solvent and tempering agent for the irritant. The irritant produces a rush of blood and warmth which is often slightly helpful.

Sore-Muscle Liniment

Olive Oil	6 fl oz
Methyl Salicylate	3 fl oz

Mix together and keep in a well-stoppered bottle. Apply externally, but do not use on chafed or cut skin.

Chest Rubs

In spite of the fact that chest rubs are practically useless, countless sufferers use them. Their action is similar to that of liniments and they differ only in that they are in the form of a salve.

Chest-Rub Salve

Yellow Petro-		
latum	1	lb
Paraffin Wax	1	oz
Eucalyptus Oil	2	fl oz
Menthol	½	oz
Cassia Oil	⅛	fl oz
Turpentine	½	fl oz

Melt the petrolatum and paraffin wax together in a double boiler and then add the menthol. Remove from the heat, stir, and cool a little; then mix in the oils, and turpentine. When it begins to thicken, pour into tins and cover.

Insect Repellents

Preparations of this type may irritate sensitive skins and they will not always work.

Mosquito-Repelling Oil

Cedar Oil	2 fl oz
Citronella Oil	4 fl oz
Spirits of Camphor	8 fl oz

Mix in a dry bottle and the oil is ready for use. This preparation may be smeared on the skin as often as is necessary.

Fly Sprays

Fly sprays usually consist of deodorized kerosene, perfume, and an active insecticide. In some cases, they merely stun the flies who may later recover and begin buzzing again.

Fly Spray

Deodorized	
Kerosene	80 fl oz
Methyl Salicylate	1 fl oz
Pyrethrum Powder	10 oz

Mix thoroughly by stirring from time to time; allow to stand covered overnight and then filter through muslin.

This spray is inflammable and should not be used near open flames.

Deodorant Spray

(For public buildings, sick rooms, lavatories, etc.)

Pine-Needle Oil	2
Formaldehyde	2
Acetone	6

* Isopropyl Alcohol 20

One ounce of this mixture is diluted with 1 pt. water for spraying.

Cresol Disinfectant

† Caustic Soda 25.5 g
Water 140.0 cc

Dissolve in a Pyrex or enamelled dish and warm. To this, add slowly the following warmed mixture:

‡ Cresylic Acid 500.0 cc
Rosin 170.0 g

Stir until dissolved and add water to make 1,000 cc.

Ant Poison

Sugar 1 lb
Water 1 qt
‡ Arsenate of Soda 125 g

Boil and stir until uniform; strain through muslin and add 1 spoonful honey.

Bedbug Exterminator

* Kerosene 90 fl oz
Clove Oil 5 fl oz
‡ Cresol 1 fl oz
Pine Oil 4 fl oz

Simply mix and bottle.

Nonstaining Mothproofing Fluid

Sodium Aluminum
Silicofluoride 0.50
Water 98.00

Glycerin 0.50
"Sulfatate" (Wetting
Agent) 0.25

Stir until dissolved.

Fly Paper

Rosin 32
Rosin Oil 20
Castor Oil 8

Heat this mixture in an aluminum or enamelled pot on a gas stove, with stirring, until all the rosin has melted and dissolved. While hot, pour on firm paper sheets of suitable size which have been brushed with soap water just before coating. Smooth out the coating with a long knife or piece of thin flat wood and allow to cool. If a heavier coating is desired, increase the amount of rosin. Similarly, a thinner coating results by reducing the amount of rosin. The finished paper should be laid flat and not exposed to undue heat.

Baking Powder

Bicarbonate of Soda 28
Monocalcium Phosphate 35
Corn Starch 27

Mix these powders thoroughly

---

* Inflammable.
† Do not get this on the skin as it is corrosive.
‡ Poison.

in a dry can by shaking and rolling for ½ hour. Pack into dry airtight tins as moisture will cause lumping.

Malted-Milk Powder

Powdered Malt Extract	5
Powdered Skim Milk	2
Powdered Sugar	3

Mix thoroughly by shaking and rolling in a dry can. Pack in an airtight container.

Cocoa-Malt Powder

Corn Sugar	55
Powdered Malt	19
Powdered Skim Milk	12½
Cocoa	13
Vanillin	⅛
Powdered Salt	⅜

Mix thoroughly and then run through a fine wire sieve.

Sweet Cocoa Powder

Cocoa	17½ oz
Powdered Sugar	32½ oz
Vanillin	¾ g

Mix thoroughly and sift.

Pure Lemon Extract

Lemon Oil	
U.S.P.	6½ fl oz
Alcohol	121½ fl oz

Shake together in 1-gal jug until dissolved.

Artificial Vanilla Flavor

Vanillin	¾ oz
Coumarin	¼ oz
Alcohol	2 pt

Stir the ingredients in a glass or china pitcher until dissolved. Then mix into the following solution:

Sugar	12 oz
Water	5¼ pt
Glycerin	1 pt

Color brown by adding sufficient burnt-sugar coloring.

Canary Food

Dried and Chopped Egg Yolk	2
Poppy Heads (Coarse Powder)	1
Cuttlefish Bone (Coarse Powder)	1
Granulated Sugar	2
Powdered Soda Crackers	8

Mix well together.

Blue-Black Writing Ink

Naphthol Blue Black	1 oz
Powdered Gum Arabic	½ oz
Carbolic Acid	¼ oz
Water	1 gal

Stir together in a glass or enamelled vessel until dissolved.

Indelible Laundry-Marking Ink

a Soda Ash   1 oz
  Powdered Gum
    Arabic   1 oz
  Water   10 fl oz
  Stir until dissolved.

b Silver Nitrate   4 oz
  Powdered Gum
    Arabic   4 oz
  Lampblack   2 oz
  Water   40 fl oz

Stir this in a glass or porcelain dish until dissolved. Do not expose the mixture to strong light or it will spoil. Then pour into a brown glass bottle. In using these solutions, wet the cloth with solution a and allow to dry. Then write on it with solution b using a quill pen.

Green Marking Crayon

Ceresin   8
Carnauba Wax   7
Paraffin Wax   4
Beeswax   1
Talc   10
Chrome Green   3

Melt the first four ingredients in a container and then add the last two slowly, while stirring. Remove from the heat and continue stirring until thickening begins. Then pour into molds. If other-color crayons are desired, other pigments may be used. For example, for black, use carbon black or bone black; for blue, Prussian blue; for red, orange chrome yellow.

Antique Coloring for Copper

Copper Nitrate   4 oz
Acetic Acid   1 oz
Water   2 oz

Dissolve by stirring together in a glass or porcelain vessel. Pack into glass bottles.

Wet the copper to be colored and apply the coloring solution hot.

Blue-Black Finish on Steel

a Place the object in molten sodium nitrate at 700 to 800°F for 2 to 3 minutes. Remove and allow to cool somewhat, wash in hot water, dry, and oil with mineral or linseed oil.

b Then put the object in the following solution for 15 minutes:

Copper Sulfate   ½ oz
Iron Chloride   1 lb
Hydrochloric Acid 4 oz
Nitric Acid   ½ oz
Water   1 gal

Allow to dry for several hours. Place in a solution again for 15 minutes, remove and dry for 10 hours. Place in boiling water for ½ hour, dry, and scratch-brush very lightly. Oil with mineral or linseed oil and wipe dry.

Rust-Prevention Compound

Lanolin	1
* Naphtha	2

Mix until dissolved.

The metal to be protected is cleaned with a dry cloth and then coated with the composition.

Metal Polish

Naphtha	62	oz
Oleic Acid	⅓	oz
Abrasive	7	oz
Triethanolamine Oleate	⅓	oz
Ammonia (26°)	1	oz
Water	1	gal

In one container mix together the naphtha and oleic acid to a clear solution. Dissolve the triethanolamine oleate in the water separately, stir in the abrasive, and then add the naphtha solution. Stir the resulting mixture at a high speed until a uniform creamy emulsion results. Then add the ammonia and mix well, but do not agitate so vigorously as before.

Glass-Etching Fluid

Hot Water	12
† Ammonium Bifluoride	15
Oxalic Acid	8
Ammonium Sulfate	10

---

* Inflammable — keep away from flames.
† Corrosive.

Glycerin	40
Barium Sulfate	15

Warm the washed glass slightly before writing on it with this fluid. Allow the fluid to act on the glass for about 2 minutes.

Leather Preservative

Cold-Pressed Neatsfoot Oil	10
Castor Oil	10

Mix.

This is an excellent preservative for leather book bindings, luggage, and other leather goods.

White-Shoe Dressing

Lithopone	19	oz
Titanium Dioxide	1	oz
Bleached Shellac	3	oz
Ammonium Hydroxide	¼	fl oz
Water	25	fl oz
Alcohol	25	fl oz
Glycerin	1	oz

Dissolve the last four ingredients by mixing in a porcelain vessel. When dissolved, stir in the first two pigments. Keep in stoppered bottles and shake before using.

Waterproofing for Shoes

Wool Grease	8
Dark Petrolatum	4
Paraffin Wax	4

Melt together in any container.

Polishes

Polishes are generally used to restore the original luster and finish of a smooth surface. They are also expected to clean the surface and to prevent corrosion or deterioration. There is no one polish which will give good results on all surfaces.

Most polishes contain oil or wax for their lustering or polishing properties. Oil polishes are easy to apply, but the surfaces on which they are used attract dust and show finger marks. Wax polishes are more difficult to apply, but are more lasting.

Oil or wax polishes are of two types: waterless and aqueous. The former are clear or translucent, the latter are milky in appearance.

For use on metals, abrasives of various kinds, such as tripoli, silica dust, or infusorial earth, are incorporated to grind away oxide films or corrosion products.

Black Shoe Polish

Carnauba Wax   5½ oz
Crude Montan Wax   5½ oz

Melt together in a double boiler. (The water in the outer container should be boiling.) Then stir in the following melted and dissolved mixture:

Stearic Acid   2 oz
Nigrosine Base   1 oz

Then stir in
Ceresin   15 oz
Remove all flames and run in slowly, while stirring,
Turpentine   90 fl oz
Allow the mixture to cool to 105°F. and pour into airtight tins which should stand undisturbed overnight.

Clear Oil-Type Auto Polish

Paraffin Oil   5   pt
Raw Linseed Oil   2   pt
China-Wood Oil   ½ pt
* Benzol   ¼ pt
* Kerosene   ¼ pt
Amyl Acetate   1   tbsp

Mix together in a glass jar and keep it stoppered.

Paste-Type Auto and Floor Wax

Yellow Beeswax   1   oz
Ceresin   2½ oz
Carnauba Wax   4½ oz
Montan Wax   1¼ oz
* Naphtha or
   Mineral Spirits   1   pt
* Turpentine   2   oz
Pine Oil   ½ oz

Melt the waxes together in a double boiler. Turn off the heat and run in the last three ingredients in a thin stream, with stirring. Pour into cans, cover, and allow to stand undisturbed overnight.

* Inflammable — keep away from flames.

Oil-and-Wax Type Furniture
Polish

Paraffin Oil	1 pt
Powdered Carnauba Wax	¼ oz
Ceresin Wax	⅛ oz

Heat together until all of the wax is melted. Allow to cool and pour into bottles before the mixture turns cloudy.

Liquid Polishing Wax

Yellow Beeswax	1 oz
Ceresin Wax	4 oz

Melt together and then cool to 130°F.; turn off all flames and stir in slowly:

*Turpentine	17 fl oz
Pine Oil	½ fl oz

Pour into cans or bottles which are closed tightly to prevent evaporation.

Floor Oil

Mineral Oil	46 fl oz
Beeswax	½ oz
Carnauba Wax	1 oz

Heat together in double boiler until dissolved (clear). Turn off the flame and stir in

*Turpentine	3 fl oz

Lubricants

Lubricants, in the form of oils or greases, are used to prevent friction and wearing of parts which are rubbed together. Lu-

---

* Inflammable.

bricants must be chosen to fit specific uses. They consist of oils and fats often compounded with soaps and other unctuous substances. For heavy duty, heavy oils or greases are used and light oils are suitable for light duty.

Gum Lubricant

White Petrolatum	15 oz
Acid-Free Bone Oil	5 oz

Warm gently and mix together.

Graphite Grease

Ceresin	7 oz
Tallow	7 oz

Warm together and gradually work in with a stick:

Graphite	3 oz

Stir until uniform and pack in tins when thickening begins.

Penetrating Oil
(For loosening rusted bolts, screws, etc.)

Kerosene	2 oz
Thin Mineral Oil	7 oz
Secondary Butyl Alcohol	1 oz

Mix and keep in a stoppered bottle.

Molding Compound

White Glue	13 lb
Rosin	13 lb
Raw Linseed Oil	⅓ qt
Glycerin	1 qt
Whiting	19 lb

Heat the white glue until it

melts. Then cook separately the rosin and raw linseed oil until the rosin is dissolved. Add the rosin, oil, and glycerin to the glue, stirring in the whiting until the mass reaches the consistency of a putty. Keep the mixture hot.

Press this mass into the die firmly and allow it to cool slightly before removing. The finished product is ready to use within a few hours after removal. Suitable pigments may be added to secure brown, red, black, or other color.

In applying ornaments made of this composition to a wood surface, they are first steamed to make them flexible; in this condition, they will adhere to the wood easily and securely. They can be bent to any shape, and no nails are required for applying them.

Grafting Wax

Wool Grease	11
Rosin	22
Paraffin Wax	6
Beeswax	4
Japan Wax	1
Rosin Oil	9
Pine Oil	1

Melt together until clear and pour into tins. This composition can be made thinner by increasing the amount of rosin oil and thicker by decreasing it.

Candles

Paraffin Wax	30.0
Stearic Acid	17.5
Beeswax	2.5

Melt together and stir until clear. If colored candles are desired, add a very small amount of any oil-soluble dye. Pour into vertical molds in which wicks are hung.

Adhesives

Adhesives are sticky substances used to unite two surfaces. Adhesives are specifically called glues, pastes, cements, mucilages, lutes, etc. For different uses different types are required.

Wall-Patching Plaster

Plaster of Paris	32
Dextrin	4
Pumice Powder	4

Mix thoroughly by shaking and rolling in a dry container. Keep away from moisture.

Cement-Floor Hardener

Magnesium Fluosilicate	1 lb
Water	15 pt

Mix until dissolved.

The cement should first be washed with clean water and then drenched with this solution.

Paperhanger's Paste

White or Fish Glue	4 oz
Cold Water	8 oz

Venice Turpentine   2 fl oz
Rye Flour   1 lb
Cold Water   16 fl oz
Boiling Water   64 fl oz

Soak the glue in the first amount of cold water for 4 hours. Dissolve on a water-bath (glue-pot) and while hot stir in the Venice turpentine. Use a cheap grade of rye or wheat flour, mix thoroughly with the second amount of cold water to about the consistency of dough or a little thinner, being careful to remove all lumps. Stir in 1 tbsp of powdered alum to 1 qt flour, then pour in the boiling water, stirring rapidly until the flour is thoroughly cooked. Let this cool and finally add the glue solution. This makes a very strong paste which will also adhere to a painted surface, owing to the Venice turpentine content.

Aquarium Cement

Litharge	10
Plaster of Paris	10
Powdered Rosin	1
Dry White Sand	10
Boiled Linseed Oil	Sufficient

Mix all together in the dry state, and make a stiff putty with the oil just before use.

Do not fill the aquarium for 3 days after cementing. This cement hardens under water, and will stick to wood, stone, metal, or glass and as it resists the action of sea water, it is useful for marine *aquaria*.

Wood-Dough Plastic

* Collodion	86
Powdered Ester Gum	9
Wood Flour	30

Allow the first two ingredients to stand until dissolved, stirring from time to time. Then, while stirring, add the wood flour, a little at a time, until uniform. This product can be made softer by adding more collodion.

Putty

Whiting	80
Raw Linseed Oil	16

Rub together until smooth. Keep in a closed container.

Wood-Flour Bleach

Sodium Metasilicate	90
Sodium Perborate	10

Mix thoroughly and keep dry in a closed can. Use 1 lb to 1 gal boiling water. Mop or brush on the floor, allow to stand ½ hour, then rub off and rinse well with water.

---

* Paint Remover

Benzol	5	pt
Ethyl Acetate	3	pt
Butyl Acetate	2	pt
Paraffin Wax	½	lb

* Inflammable.

Stir together until dissolved.

Soaps and Cleaners

Soaps are made from a fat or fatty acid and an alkali. They lather and produce a foam which entraps dirt and grease. There are many kinds of soaps.

Cleaners contain a solvent, such as naphtha, with or without a soap. Abrasive cleaners are soap pastes containing powdered pumice, stone, silica, etc.

Concentrated Liquid Soap

Water	11
* Solid Caustic Potash	1
Glycerin	4
Red Oil (Oleic Acid)	4

Dissolve the caustic soda in water, add the glycerin, and bring to a boil in an enamelled pot. Remove from the heat, add the red oil slowly, while stirring. If a more neutral soap is wanted, use more red oil.

Saddle Soap

Beeswax	5.0
* Caustic Potash	0.8
Water	8.0

Boil for 5 minutes, while stirring. In another vessel heat:

Castile Soap	1.6
Water	8.0

Mix the two solutions with

* Do not get on the skin as it is corrosive.

good stirring; remove from the heat and add, while stirring:

Turpentine	12

Mechanics' Hand-Soap Paste

Water	1.8 qt
White Soap Chips	1.5 lb
Glycerin	2.4 oz
Borax	6.0 oz
Dry Sodium Carbonate	3.0 oz
Coarse Pumice Powder	2.2 lb
Safrol	To suit

Dissolve the soap in two thirds of the water by heat. Dissolve the last three ingredients in the rest of the water. Pour the two solutions together and stir well. When it begins to thicken, sift in the pumice, stirring constantly till thick, then pour into cans. Vary the amount of water, for heavier or softer paste. Water cannot be added to the finished soap.

Dry-Cleaning Fluid

Glycol Oleate	2 fl oz
Carbon Tetrachloride	60 fl oz
Naphtha	20 fl oz
Benzine	18 fl oz

This is an excellent cleaner that will not injure the finest fabrics.

Wall-Paper Cleaner

Whiting	10 lb
Calcined Magnesia	2 lb

Fuller's Earth   2 lb
Powdered Pumice   12 oz
Lemenone or
    Citronella Oil   4 oz
Mix well together.

Household Cleaner

Soap Powder   2
Soda Ash   3
Trisodium Phosphate   40
Finely Ground Silica   55
Mix well and pack in the usual containers.

Window Cleanser

Castile Soap   2
Water   5
Chalk   4
French Chalk   3
Tripoli Powder   2
Petroleum Spirits   5
Mix well and pack in tight containers.

Straw-Hat Cleaner

Sponge the hat with a solution of:

Sodium Hyposulfite   10 oz
Glycerin   5 oz
Alcohol   10 oz
Water   75 oz

Lay the hat aside in a damp place for 24 hours and then apply a mixture of:

Citric Acid   2 oz
Alcohol   10 oz
Water   90 oz

Press with a moderately hot iorn after stiffening with gum water, if necessary.

Grease, Oil, Paint, and Lacquer Spot Remover

Alcohol   1
Ethyl Acetate   2
Butyl Acetate   2
Toluol   2
Carbon Tetrachloride   3

Place the garment with the spot over a piece of clean paper or cloth and wet the spot with this fluid; rub with a clean cloth toward the center of the spot. Use a clean section of cloth for rubbing and clean paper or cloth for each application of the fluid. This cleaner is inflammable and should be kept away from flames. Cleaners of this type should be used out of doors or in well-ventilated rooms as the fumes are toxic.

Paint-Brush Cleaner

a   Kerosene   2.00
    Oleic Acid   1.00
b   Strong Liquid
      Ammonia (28%)   0.25
    Denatured Alcohol   0.25

Slowly stir b into a until a smooth mixture results. To clean brushes, pour into a can and leave the brushes in it overnight. In the morning, wash out with warm water.

Rust and Ink Remover

Immerse the part of the fabric with the rust or ink spot alternately in solutions a and b, rinsing with water after each immersion.

a Ammonium Sulfide
   Solution   1
   Water   19
b * Oxalic Acid   1
   Water   19

Javelle Water
(Laundry bleach)

Bleaching Powder .2 oz
Soda Ash   2 oz.
Water   5 gal
Mix well until the reaction is completed. Allow to settle overnight and siphon off the clear liquid.

Liquid Laundry Blue

Prussian Blue   1
Distilled Water   32
* Oxalic Acid   ¼
Dissolve by mixing in a crock or wooden tub.

Glassine Paper

Paper is coated with or dipped in the following solution and then hung up to dry.
Copal Gum   10 oz
Alcohol   30 fl oz
Castor Oil   1 fl oz
Dissolve by letting stand overnight in a covered jar and stirring the next day.

* Poisonous.

Waterproofing Paper and Fiberboard

The following composition and method of application will make uncalendered paper, fiberboard, and similar porous material waterproof.

Paraffin (M.P.
   about 130°F.)   22.5
Trihydroxyethylamine
   Stearate   3.0
Water   74.5
The paraffin wax is melted and the stearate added to it. The water is then heated to nearly the boiling point and vigorously agitated with a suitable mechanical stirring device while the mixture of melted wax and emulsifier is being slowly added. This mixture is cooled while it is stirred.

The paper or fiberboard is coated on the side which is to be in contact with water. This method works most effectively on paper-pulp molded containers and has the advantage of being much cheaper than dipping in melted paraffin as only about one tenth as much paraffin is needed. In addition, the outside of the container is not greasy and can be printed on after treatment which is not the case when treating with melted wax.

*Waterproofing Liquid

Paraffin Wax	⅖ oz
Gum Dammar	1⅕ oz
Pure Rubber	⅛ oz
Benzol	13 oz
Carbon Tetrachloride	
To make 1 gal	

Dissolve the rubber in the benzol, add the other ingredients, and allow to dissolve.

This liquid is suitable for wearing apparel and wood. It is applied by brushing on two or more coats, allowing each to dry before applying another coat. Apply outdoors as vapors are inflammable and toxic.

Waterproofing Heavy Canvas

Raw Linseed Oil	1 gal
Crude Beeswax	13 oz
White Lead	1 lb
Rosin	12 oz

Heat, while stirring, until all lumps are removed and apply warm to the upper side of the canvas, wetting it with a sponge on the underside before application.

Waterproofing Cement

China-Wood Oil

Fatty Acids	10	oz
Paraffin Wax	10	oz
Kerosene	2½	gal

* Inflammable.

Stir until dissolved. Paint or spray on cement walls, which must be dry.

Oil- and Greaseproofing Paper and Fiberboard

This solution, applied by brush, spray, or dipping, will leave a thin film which is impervious to oil and grease. Applied to paper or fiber containers, it will enable them to retain oils and greases.

Starch	6.6
Caustic Soda	0.1
Glycerin	2.0
Sugar	0.6
Water	90.5
Sodium Salicylate	0.2

The caustic soda is dissolved in the water. Then the starch is made into a thick paste by adding a portion of this solution. The paste is then added to the water. The resulting mixture is placed on a water bath and heated to about 85°C, until all the starch granules have broken. The temperature is maintained about ½ hour longer at 85°C. The other substances are then added and thoroughly mixed. The composition is now ready for application. Less water may be used if applied hot and then a thicker coating will result.

Fireproof Paper

Ammonium Sulfate	8.00
Boric Acid	3.00
Borax	1.75
Water	100.00

The ingredients are mixed together in a gallon jug by shaking until dissolved.

The paper to be treated is dipped into this solution in a pan, until uniformly saturated. It is then taken out and hung up to dry. Wrinkles can be prevented by drying between cloths in a press.

Fireproofing Canvas

Ammonium Phosphate	1	lb
Ammonium Chloride	2	lb
Water	½	gal

Impregnate with the solution; squeeze out the excess, and dry. Washing or exposure to rain will remove fireproofing salts.

Fireproofing Light Fabrics

Borax	10 oz
Boric Acid	8 oz
Water	1 gal

Impregnate, squeeze, and dry. Fabrics so impregnated must be treated again after washing or exposure to rain as the fireproofing salts wash out easily.

Dry Fire Extinguisher

Ammonium Sulfate	15
Sodium Bicarbonate	9
Ammonium Phosphate	1
Red Ochre	2
"Silex"	23

Use powdered substances only. Mix well and pass through a fine sieve. Pack in tight containers to prevent lumping.

Fire-Extinguishing Liquid

Carbon Tetrachloride	95
Solvent Naphtha	5

The naphtha minimizes the development of toxic fumes when extinguishing fires.

Fire Kindler

Rosin or Pitch	10
Sawdust	10 or more

Melt, mix, and cast in forms.

Solidified Gasoline

* Gasoline	½	gal
Fine-Shaved White Soap	12	oz
Water	1	pt
Ammonia	5	oz

Heat the water, add the soap, mix and, when cool, add the ammonia. Then slowly work in the gasoline to form a semisolid mass.

* Inflammable.

Boiler Compound

Soda Ash	87
Trisodium Phosphate	10
Starch	1
Tannic Acid	2

Use powders, mix well, and then pass through a fine sieve.

Noncorrosive Soldering Flux

Powdered Rosin	1
Denatured Alcohol	4

Soak overnight and mix well.

Photographic Solutions

Developing Solution

Stock Solution *a*

Pyro	4 oz
Pure Sodium Bisulfite	280 gr
Potassium Bromide	32 gr
Distilled Water	64 oz

Dissolve in a glass or enamel dish.

Stock Solution *b*

Pure Sodium Sulfite	7 oz

Pure Sodium

Carbonate	5 oz
Distilled Water	64 oz

Dissolve separately in a glass or enamel dish.

Use the following proportions:

Stock Solution *a*	2
Stock Solution *b*	2
Distilled Water	16

At 65°F., this developer requires about 8 minutes.

Acid-Hardening Fixing Bath

a	Sodium Hyposulfite	32
	Distilled Water	8

Stir until dissolved and then add the following chemicals in the order given, stirring each until dissolved:

b	Warm Distilled Water	2½
	Pure Sodium Sulfite	½
	Pure Acetic Acid (28%)	1½
	Potassium Alum Powder	½

Add *b* to *a* and store in dark bottles away from light.

CHAPTER II

ADHESIVES

Hot Melt Adhesives

Formula No. 1

"Oronite" Poly-	
butene	25-30
Paraffin Wax	
(130°F)	30-50
Hydrogenated	
Resin-Glycerol	
Ester	25-30
Rubber Base	10-20

The rubber base can be one of the following elastomers or a combination of them: pale crepe, GR-S, GR-I or solid polyisobutylene. In general, the higher the molecular weight of the elastomer or the greater the rubber base content, the greater the adhesive strength of a given formulated mass.

The best procedure is to melt the "Oronite" Polybu-tene and resin together and work the molten mixture into the rubber base on a mill. In this way, the extended rubber will dissolve more readily in the wax. "Oronite" Polybu-tenes Nos. 32, 122, and 128 are suggested.

This basic formula offers a very wide range of variation in control of initial tack, "quick grab," "legs," adhesive strength, softness, removal characteristics, and hot melt viscosity. By the required control of these properties, adhesive masses can be developed for many types of paper-lamination jobs. Moreover, fillers such as zinc oxide can be incorporated. In special applications where rubber substitutes can be tolerated, factice and mineral rubber may be used.

In the formulation of surgical tape masses for calendar application, for example, white factice is suggested in concentration of 10 percent or less on the weight of the formula along with 60 to 70% of zinc oxide filler.

The solution type of adhesive is recommended when milling or Banbury equipment is not available, or for other special conditions of production. A basic formulation for a starting point follows:

No. 2

"Oronite"	
Polybutene	50-100
Hydrogenated	
Resin-Glycerol	
Ester	25-75
Rubber Base	75-125
Solvent	
(Petroleum	
Thinner)	750-1500

This formulation offers wide adaptability in the manufacture of cements for leather in the fabrication and repair of shoes and the laminating of many types of paper foils and fibers. Fillers can be incorporated and the amount of petroleum thinner can be regulated to produce the consistency required. In applications, this consistency is calculated at about five times the weight of the ingredients other than fillers, or about ten times the weight of the rubber base. "Oronite" Polybutenes Nos. 32, 122, and 128 are suggested.

Another solution-type of adhesive which deposits a clear pressure-sensitive film that can be used for tapes and labels can be formulated as follows:

No. 3

"Oronite" Polybutene	3
Terpene Hydrocarbon	
Resin	5
Petroleum Oil (as	
SAE 20)	7
Rubber Base	10
Solvent	To desired
(Petroleum	consist-
Thinner)	ency

One method of preparation is to dissolve the rubber base in about 60 parts of solvent thinner between 150° and 250°F, the boiling range. At the same time, work in the molten mixture of the other ingredients, it should be added slowly and under moderate agitation to avoid loss of solvent. The solution can be adjusted to lower solids content by adding more solvent to the consistency required.

The preparation of solution adhesives is greatly facilitated by mastication of the rubber base on a mill. In this way, it is possible to put in resins, fillers, and moderate amounts of solvent so that the worked mass can be more readily taken into the remaining solvent. The temperature in the working mill must be kept low enough to avoid excessive loss of solvent.

The use of fillers is recommended for application by doctor blade or spreader. Here, however, dispersal of the filler in the rubber base and incorporation of as much solvent as can be tolerated on the mill is suggested as the first step. The milled mass can be dissolved in the rest of the solvent under agitation or churning.

Because of the ease with which "Oronite" Polybutenes of all viscosities can be emulsified, their use by themselves or in conjunction with emulsions of resins and elastomers is suggested.

No. 4

Half-Second Cellulose Acetate Butyrate	35.00
"Aroclor" 5460	30.00
Dioctyl Phthalate	15.00
"Newport," V-40	19.89
"Santonox"	0.10
"Synfleur #6"	0.01

This coating can be applied at about 350°F; ventilation should be provided.

No. 5

Half-Second Cellulose Acetate Butyrate	35.00
"Aroclor" 5460	30.00
Dioctyl Phthalate	15.00
"Newport," V-40	19.89
"Santonox"	0.10
"Synfleur #6" (Odorant)	0.01

This coating can be applied at about 350°F; ventilation should be provided.

No. 6

Ethyl Cellulose, 50 cpr	24
"Aroclor" 5460	7
"Lopor" No. 45 Mineral Oil	57

"Bakers" No. 15

Castor Oil	5
Epoxy Soybean Oil	3
Paraffin Wax (m.p. 135°F)	3
"Santonox"	1

No. 7

"Elvax" 150	35
Paraffin Wax (145 AMP)	40
Microcrystalline Wax (140 AMP)	15
Stabilized Rosin (Laminating Grade)	10

The tack after activation can be increased by raising the rosin content. The blocking temperature can be raised by increasing the amount of paraffin wax or using a paraffin of higher melting point. Coatings based on formulations similar to the one shown above can be applied by hot-melt gravure processes, or by casting from hot toluene solution.

No. 8

"Elvax"	30
Butyl Rubber	10
Microcrystalline Wax (Laminating Grade)	30
Rosin	30

No. 9

"Elvax"	40
Microcrystalline Wax (Laminating Grade)	30
Rosin or Rosin-Derived Resin	30

	No. 10		No. 11	No. 12
	Application Temperature 165°-175°C		Low-Melting	Alkali-Dispersible
Polyvinyl Acetate	95	100	50	38
"Opalon" 505 PVC Resin	5	—	—	—
Modified Rosin	—	—	25	—
Coumarone-Indene Resin (Soft)	—	—	—	28.3
"Aroclor" 1254	—	55	12	27
"Aroclor" 1260	50-75	—	—	—
Dibutyl Phthalate	—	—	13	7
"Santicizer" 160 (or Dibutyl Phthalate)	50-75	30	—	—

	No. 10	No. 11	No. 12
Clay	—	25	—	—
Fluxing Resin ("Filtros" Resin WW)	—	75	—	—
"Acrawax C"	—	—	—	1
Sodium Benzoate	—	—	—	0.76
Viscosity (centipoises)			2000	3500
			(120°C)	(175°C)
			Non-Blocking	Non-Blocking

No. 13

"Versamid" 950	57.6
"Versamid" 100	38.5
Dicapryl Phthalate	3.9

This formulation retains its flexibility at 0°F.

No. 14

This formulation is especially useful as a heat-seal coating where retention at low temperature is required.

"Versamid" 940	85
"Staybelite" Ester #10	5
Tributyl Phosphate or Dibutyl Phthalate	10

This adhesive composition can be applied either by techniques for hot melt or solvent.

No. 15

| "Versamid" 940 | 100 |

130°F Paraffin	4
"Staybelite" Ester #10	10
"Santicizer" #8	10

The rosin derivative gives increased tack, the plasticizer aids flexibility at low temperature, and the paraffin offsets the tendency to blocking caused by the other modifiers.

No. 16

"Parlon" (125 Centipoise Type)	20
"Aroclor" 1254	6
"Aroclor" 5460	6
Toluene	68

No. 17

"Versamid" 930	17.5
"Versamid" 950	17.5
Isopropanol	40.0
"Skellysolve" B	25.0

No. 18

(Aluminum Foil to Wood)

Ethyl Cellulose	
(Hercules Type N-7)	3
"Nevillac" 10°	5
"Nevinol"	14
R-14 "Neville" Resin	30

The ethyl cellulose, "Nevillac" 10°, and "Nevinol" are heated and mixed until a uniform mixture is obtained. The R-14 "Neville" Resin is added with further mixing. Where color is unimportant, it is suggested that the 1-D Heavy Oil be substituted for the "Nevinol" and R-16-A "Neville" Resin for the R-14 "Neville" Resin.

No. 19

(Glass to Polyvinyl Chloride)
Japanese Patent 856 (1961)

Chlorinated	
Diphenyl	100
Chromium Chloride	6
Carbon Tetrachloride	100
Cobalt Naphthenate	0.5
Zirconium Octoate	0.5

This solution is applied to the glass heated to 100°C and then the plastic is pressed to it.

No. 20

(Metal to Metal)

"Versamid" 100	60
"Versamid" 900	40

This combination provides an excellent metal-to-metal thermoplastic adhesive and caulking compound. It is extremely tough and resistant to impact. In addition to its use as a metal-to-metal adhesive it effectively joins plastics, foils, and papers. It is applied as a hot-melt adhesive.

No. 21

(For Metals:* Heat Cure)

a "Versamid" 115	50	
"Gen Epoxy" 190		
or M180	50	

High tensile strength, moderate peel strength, moderate temperature-resistance.

b "Versamid" 140 35 to 40 Good temperature-re-

"Gen Epoxy" 190 sistance, high tensile

 or M180 65 to 60 strength, moderate peel

  strength.

c "Versamid" 115

 or 140 60 Improved peel strength,

"Gen Epoxy" 190 high tensile strength,

 or M180 40 moderate temperature-

Glass Cloth resistance.

No. 22

(For Metals:* Cure at Room Temperature)

a "Versamid" 125 40 Short hardening time,

"Gen Epoxy" 190 60 lower strength.

"DMP-30" 8

b "Versamid" 125 60 Improved peel strength.

"Gen Epoxy" 190 40

Glass Cloth

* Fillers can be added. Aluminum flakes are especially suitable.

No. 23		Automotive Sealer	
(Cloth to Metal)		"Opalon" 410 PVC	
"R-16" or "R-16-A"		Resin	96
"Neville" Resin	70	"Vinylite" VMCH	
Polystyrene ("Kop-		Resin	4
pers" KTPL-3		"Aroclor" 1260	50
Coating Grade)	20	Diisodecyl Phthalate	100
"Nevinol" or "1-D		"Santicizer" 160	28
Heavy Oil"	10	Calcium Carbonate	125
		Epoxy Soybean Oil	5
		Dibasic Lead	
The resin and polystyrene		Phosphite	3
are melted and mixed to-			
gether. The "Nevinol" or			
"1-D Heavy Oil" is then		For Granulated Cork	
added.		"Geon" 121	50

"Geon" 202	20
Union Carbide	
"Stabilizer A-5"	2
Dioctyl Phthalate	100
"Neville" LX-685,	
180	40

The polyvinyl chloride, the vinylidene chloride, and "Stabilizer A-5" are added to the dioctyl phthalate under agitation. The "Neville" LX-685, 180 previously powdered, is added last. The entire mixture is then passed through a three-roll paint mill twice. At this point after the grinding operation, the LX-685, 180 is a finely dispersed powder in the mixture and remains stable for hours. After sitting overnight, the mixture should be agitated a little before use. When mixed with granulated cork, the mixture is molded under low compression at 330°F, and the resulting product is flexible and tough.

Abrasives to Cloth

Zein	22
"Nevillac" Hard	7
"Nevillac" 10°	4
Methyl "Cellosolve"	54
Isopropanol	15

The zein is dissolved in the methyl "Cellosolve" while the "Neville" Resins are dissolved in the isopropanol. The two solutions are then combined.

For Non-Slip Sandpaper

Smoked Sheets*	63.0
"Dixie Clay"	78.0
"Calcene" TM	15.5
Zinc Oxide	3.0
Stearic Acid	0.6
"Butyl Eight"	7.0
Sulfur	2.5
"Neville"	
LX-685, 125	17.0
Toluene	746.0

* Previously masticated for 10 minutes on a cool, tight mill and aged for 16 hours.

The Smoked Sheets are banded on a cool, tight mill and masticated for 5 minutes. The zinc oxide, stearic acid, "Dixie Clay," and calcium carbonate are all added together. The sulfur is added last. The total milling time should be no longer than ten minutes.

This mixture plus "Neville" LX-685, 125 is then added to the toluene under agitation.

The "Butyl Eight" is added last.

NOTE: The "Butyl Eight" should always be added to the mixture just before use as this compound will set in 12–36 hours at room temperature when the "Butyl Eight" is present.

For Food Packaging

R-1 "Nevindene"	20
"Staybelite" Resin A-1	6
Ethyl Cellulose Type N-22	8
Dibutyl Phthalate	6
Toluene	48
Isopropanol (Anh.)	12

The ethyl cellulose is dissolved in the toluene-dibutyl phthalate mixture. The R-1 "Nevindene" Resin and "Staybelite" Resin are then added. The anhydrous isopropanol is added last.

For Linoleum

R-16-A "Neville" Resin, 70% Solution in "Hi-Flash" Solvent	30
Asphalt (Solid)	20
Filler (Slate, Slate Flour, or Whiting)	50

Add Mineral Spirits to attain desired viscosity or consistency.

A uniform mixture is made of the R-16-A "Neville Resin" Solution and asphalt. The filler is then added along with the mineral spirits. Water may be added to adjust the troweling properties and to lower the cost.

Aluminum Foil to Fiber Glass

Flat Bark (Natural Rubber)	8.5
Zinc Oxide	4.0
R-29 "Neville" Resin, 10°	15.0
"Nebony" 100	5.0
Toluene	37.0
"Nevsolv" T	60.0
"Nevastain" B	0.1

The flat bark is banded on cool tight rolls (0.020 inch) and the zinc oxide and "Nevastain" B are added together. Milling is continued for about ten minutes. The batch is removed from the mill, cut into small pieces, and dissolved in the toluene.

The "Nebony" 100 and R-29 "Neville" Resin 10° are dissolved in the "Nevsolv T." The rubber solution and resin

solution are then mixed together.

Aluminum Metal to Glass

GR-S 50	5.5
Calcium Silicate	1.0
"Neville" LX-685, 135	35.0
"Bentone" 34	2.5
Calcium Carbonate	35.0
"Nevsolv" T or Toluene	20.0

The GR-S is added to cool
tight rolls and masticated for
20 minutes. The calcium silicate is then added and dispersed.

The batch is cut from the
mill and dissolved in approximately half the amount of
"Nevsolv" T shown.

The "Neville" LX-685, 135
is dissolved in the remainder
of "Nevsolv" T. The "Ben-
tone" 34 is added to the resin
solution and mixed well. The
rubber solution is then mixed
into the resin solution. Finally
the calcium carbonate is
added.

Aluminum Foil to Cork

Coating Grade Poly-styrene (Dow PS-3)	10
R-17 "Neville" Resin	60
"Nevinol"	30

The R-17 "Neville" Resin
and polystyrene are melted
and mixed. The "Nevinol" is
added after a uniform mixture
has been obtained. This adhesive could also be dissolved
in a suitable aromatic solvent,
such as xylene, and applied
from solution.

"Butyl Eight"	12
Zinc Oxide	3
Stearic Acid	1

The natural rubber is milled
along with the other ingredients with the exception of the
"Neville" Resin. The R-6
"Nevindene" is dissolved in
100 parts of toluene. To this
solution is added the milled
rubber compound. Additional
toluene can be added for viscosity adjustment.

Rayon to Rubber

Mix

Sodium Hydroxide (10% solution)	3.0
Resorcinol	3.4
Formaldehyde (40% solution)	7.2

Water 86.4

Add

Sodium Hydroxide
(10% solution) 3

Water 47

Concentrated Latex
(60% total solids) 50

Allow reaction to continue 6 hours at 75°-80°F.

For Sanding Cloth

"Neville"
LX-685, 135 75

China Wood Oil 25

Mineral Spirits 100

The resin and oil are cooked at 500°F until a sample droplet remains clear at room temperature. The batch is then cooled and thinned with an equal amount of mineral spirits or to any desired range of viscosity. The addition of a trace of manganese naphthenate is suggested to speed the drying time.

For Coated Cloth

Natural Rubber 100

R-6 "Nevindene" 30

Sulfur 4

"Nevastain" B 1

Pressure-Sensitive Adhesives

Formula No. 1

For Sanding Cloth

R-16-A "Neville"
Resin 80

Polystyrene, Medium
Molecular
Weight 10 to 20

"LX-77 Oil" 0 to 10

The polystyrene is melted and mixed hot with half of the "R-16-A" until the mixture is uniform. The remaining resin and the "LX-77 Oil" are then added. An aromatic type of solvent can also be added to obtain the desired viscosity for application. "The LX-77 Oil" is suggested as a plasticizer, if one is needed.

No. 2

Plastisol-Coated Tape,
to Cloth Backing

Smoked Sheets 160.0

"Calcene" TM 30.0

R-28 "Neville"
Resin, 35° 30.0

"Nevastain" B 1.5

Toluene 700.0

The rubber is broken down on successive days on a cool

mill. The filler is added on the second day and the rubber is cut from the mill and added to the toluene under agitation. The R-28 "Neville" Resin-35° and "Nevastain" B are then added to the rubber solution.

No. 3

R-27 "Neville" Resin, 10°	60
"901 Compound"	40

The R-27 "Neville" Resin-10° and "901 Compound" are melted and mixed together. This mixture may also be dissolved in a suitable aromatic solvent for solution application.

No. 4

Polyvinyl Acetate ("Vinylite" AYAT)	14
"Nevillac" 10°	30
R-29 "Neville" Resin, 10°	55

The polyvinyl acetate and "Nevillac"-10° are melted and mixed until a uniform mixture is obtained. The R-29 "Neville" Resin-10° is then slowly added. This mixture

can also be dissolved in a suitable aromatic solvent for solution application.

No. 5

Ethyl Cellulose (Hercules Type N-7)	10
R-17 "Neville" Resin	50
"Nevillac" 10°	20

The "Nevillac" 10° is heated to approximately 180°C. The ethyl cellulose is slowly added under vigorous agitation. When the mixture is uniform, the R-17 "Neville" Resin is melted and added.

No. 6

Ethyl Cellulose (Hercules Type N-7)	5
"Paradene" No. 35	45
"No. 1-D Heavy Oil"	5

A slurry is made of the ethyl cellulose and a small amount of the "Paradene" No. 35. When this mixture is uniform, the remainder of the "Paradene" No. 35 can be added. Finally, the No. "1-D Heavy Oil" is added.

No. 7

GR-1-15	35
Calcium Silicate	13
Mineral Rubber	7
"Nebony" 100	3
"X-743 Plasticizing Oil"	19
Hard Clay	22

Procedure: Roll Opening,
0.030 inch
Roll Temperature,
240°-250°F.

NOTE: The fact that butyl rubber tends to stick on a cool mill necessitates a high milling temperature. The consistency of the stock, coupled with the high processing temperature, leaves the stock with very little uncured tensile strength. Therefore, it cannot be taken from the mill by conventional methods, such as cutting and stripping. When a scraper blade is pressed against the roll, the stock leaves the roll with a clean break and can be fed into a tuber without any difficulty. The temperature of the die on the head of the tuber can be much lower than 250°F.

This particular formulation will stay on the fast roll at this temperature. The butyl rubber is added to the mill and masticated for three minutes. The calcium silicate is added and dispersed well before adding the mineral rubber, "Nebony" 100, "X-743 Plasticizing Oil," and hard clay. The compound is taken from the mill by using a scraper.

No. 8

"Vistanex" MM L-100	100
"Vistanex" LM-MH	25
"Amberol" ST-137X	125
Solvent Naphtha or Heptane	310

No. 9

U.S. Patent 2,965,592

Polyvinyl Butyl Ether	100
Terpene Phenolic Resin	15
Antioxidant	$1\frac{1}{2}$
Heptane	800

	No. 10	No. 11	No.12
PVP/VA 1-335	2	2	2
Nonyl Phenol	2	—	2
"Igepal" CO-210	—	2	—
"Piccolyte" S-70	—	2	3
"Picco" 25	2	—	—
Toluene	6	6	7

No. 13

"Elvax" 150	30
"Aroclor" 5460	45
"Aroclor" 1254	25

No. 14

Friction Tape, Electrical

Tire Carcass	
Reclaim	22.0
Roll Brown Crepe	64.0
Carbon Black	1.0
Whiting	46.5
Barytes	8.5
Rosin Oil	2.5
R-28 "Neville"	
Resin, 35°	34.0

The roll brown crepe is milled on successive days with the reclaim; fillers are added in the second mixing cycle along with the R-28 "Neville" Resin-35°, and rosin oil. This adhesive can be applied by using calender rolls or from a solvent system.

No. 15

Paper to Glass

"Neoprene" Type GN	90
Magnesium Oxide	4
Zinc Oxide	3
Polyvinyl Acetate	
Resin	11

| "Nevillac" Hard | 31 |
| "Nevillac" 10° | 59 |

The "Neoprene," magnesium oxide, and zinc oxide are milled on a cool tight mill for approximately fifteen minutes. The batch is removed from the mill, cut into small pieces, and added to a toluene solution of the polyvinyl acetate, "Nevillac" Hard and "Nevillac" 10°.

No. 16

Felt to Metal

Butyl Tube Reclaim	
(Pequanoc Reclaim	
No. 5950)	100
Smoked Sheets	61
FF Wood Rosin	36
Zinc Oxide	45
R-28 "Neville"	
Resin, 35°	14
Toluene	512

The smoked sheets and reclaim are placed in a tight mill. The temperature of both rolls is 200°F. After a band is formed, the rolls are opened to permit a rolling bank, the zinc oxide is added, and the batch is cut from the mill. The rubber is then added to the toluene under agitation. After the rubber is in solution, the R-28 "Neville" Resin, 35° and the wood rosin are added.

Block-Resistant Adhesive

"Versamid" 940	85
Wax Ester 60	
("Petrolite")	15

This combination has exceptional blocking resistance and a low sealing temperature. The wax makes the coating more economical. It may be prepared by hot melt blending or solution blending.

Rubber Tape Sealant

No. 1

GR-1-15	25.0
Butyl Reclaim	9.0
Calcium Silicate	10.0
Mineral Rubber	14.0
"Nebony" 100	2.5
"X-743 Plasticizing Oil"	17.0
R-29 "Neville" Resin, 25°	13.0
Hard Clay	9.5
Roll Opening, 0.030 inch	
Roll Temperature, 240°-250°F.*	

* See NOTE under *Pressure-Sensitive Adhesives*, No. 7.

The GR-1-15 and butyl reclaim are banded on the mill rolls. The calcium silicate, hard clay, R-29 "Neville" Resin-25°, "X-743 Plasticizing Oil," mineral rubber and "Nebony" 100 are added in that order.

No. 2

R-16-A "Neville" Resin	80
Rubber Reclaim	10 to 20
"LX-77 Oil"	0 to 10

The reclaim is milled on a heated two-roll mill. The resin is added to the reclaim along with the "LX-77 Oil."

For Surgical Tape

"Vistanex" MM L-100	100
Zinc Oxide	50
Hydrated Alumina	50
Paraffin Oil	50
"Amberol" ST-137X	70
Solvent Naphtha or Heptane	600

Release Coating for Adhesive Tape

Zein G200	100
"Santicizer" 9	25
Stearic Acid	25

Morpholine	7.5
Proprietary	
Alcohol	500
Water	25

This is prepared by stirring the stearic acid and "Santicizer" 9 in the alcohol until dissolved. The morpholine and water are added slowly with stirring. When a clear solution is obtained the Zein G200 is added with stirring until solution is complete. Although stearic acid alone has limited solubility in proprietary alcohol at room temperature, the morpholine soap is readily dispersed in this solvent system. When formulation has aged at room temperature for one to two weeks, a layer of fatty acid soap may be apparent at the bottom of the container. Gentle warming to 35° to 40°C redisperses this soap.

Adhesive-Tape Remover

Carbon Tetrachloride	80
Mineral Oil	To make 100

For Attaching Soles to Shoes

"Neoprene" AC	100
Magnesium Oxide	4

Age Rite Powder	1
"Silene" EF	5
Zinc Oxide	5
"Nevillac" Hard	23
Toluene	460
Butyl Acetate	92
Stearic Acid	1

The "Neoprene" AC is banded on a cool tight mill. The magnesium oxide is added first, then the stearic acid, antioxidant, and "Silene" EF, and the zinc oxide is added last. The total milling time should be no longer than 5 minutes. (Additional time of mastication will decrease viscosity of the adhesive and lower the initial adhesive strength.) This mixture is added to the toluene under agitation.

The "Nevillac" Hard is added to the butyl acetate under agitation. The two solutions are then mixed.

For Shoes: Rubber Canvas Soles (Curing)

Smoked Sheets, No. 3 or No. 4	100.0
Whiting (Precipitated)	75.0
Soft Clay	25.0

Zinc Oxide	10.0
R-29 "Neville" Resin, 25°	8.0
Stearic Acid	1.0
"Nevastain" B	1.0
Sulfur	2.3
"Altax"	0.8
"Captax"	0.8

The smoked sheets are milled on a cool tight mill for approximately one-half hour. The "Nevastain" B, zinc oxide, stearic acid, soft clay, whiting, "R-29 Neville" Resin-25°, "Altax," "Captax," and sulfur are added in that order. The compound can then be dissolved in a rubber solvent. The adhesive film, after the solvent evaporates, can then be press-cured from eight to twelve minutes at 315°F.

For Shoes, Solvent
Cement for "Neoprene"
Soles (Non-curing)

"Neoprene" AC	50
"Neoprene" W	50
Zinc Oxide	5
Magnesium Oxide	4
"Silene" EF	10
"Nevillac" Hard	25
Toluene	432

Both types of "Neoprene" are banded on a cool, tight mill. The magnesium oxide is added first, followed by the "Silene" EF, and the zinc oxide is added last. The rubber is then dissolved in toluene, after which the "Nevillac" Hard is added.

For Arch Supports
in Shoes

R-7 "Neville" Resin, 108-112	27
Polystyrene (Clear, Molding Grade)	14
"Sun 91" Oil	57

Equal amounts of the polystyrene and "R-7 Neville Resin" are melted and mixed together. When this mixture is uniform, the remaining "R-7 Neville" Resin can then be added along with the plasticizing oil.

Acid- and Flame-
Resistant Adhesive

Chlorinated Rubber (125-cp type)	20
"Aroclor" 1254	6
"Aroclor" 5460	6
Toluene	68

Bonding of Polyethylene
or Polypropylene

1. Wash in acetone and dry.
2. Using a solution of 10 parts sulfuric acid, 1 part sodium dichromate and 30 parts of water (by weight) at 140-160°F, wash the internal surface for 10 minutes.
3. Rinse in clear water and then in dry air. Apply the coating just as soon as possible after the surface treatment. Then seal with heat.

Polyethylene Adhesive

Formula No. 1

| "Versamid" 950 | 40 |
| "Versamid" 100 | 40 |

| Dicapryl Hexaphthalate | 13.3 |
| "Santicizer" 9 | 6.7 |

This combination shows excellent adhesion to polyethylene and retains the necessary flexibility.

No. 2

PVP K-30	20
Nonyl Phenol	40
PVM/MA (10% in ethanol)	92
Triethylamine	8

Dissolve PVP in the mixture of PVM/MA and triethylamine and add nonyl Phenol without heat.

Adhesives for Plastics

Mixture		Used for
1. "Versamid" 125	65	Polyethylene (flame-treated)
"Gen Epoxy" 190	35	
2. "Versamid" 125	60	Rigid polyvinyl chloride
"Gen Epoxy" 190	40	
3. "Versamid" 125	50	Rigid polyvinyl chloride
"Gen Epoxy" M180	50	
4. "Versamid" 125	50	Nylon (pretreated)
"Gen Epoxy" 190	50	
5. "Versamid" 125	65	Butadiene-styrene rubber
"Gen Epoxy" 190	35	

Mixture		*Used for*
6. "Versamid" 125	50	Natural rubber
"Gen Epoxy" M180	50	(pretreated)
7. "Versamid" 125	60-55	"Teflon" (pretreated)
"Gen Epoxy" 190	40-45	
8. "Versamid" 140	80	Polyester resins
"Gen Epoxy" M180	20	

Rubber to Polyvinyl Chloride

"Neoprene" AC,	
Soft	100
Magnesium Oxide	4
"Nevastain" A	1
"Silene" EF	5
Zinc Oxide	5
"Nevillac" Hard	25
Toluene	552

The "Neoprene" and "Nevillac" Hard can be dissolved in the solvent under agitation with the pigments and "Nevastain" A being added slowly after the "Neoprene" is in solution. Ball-milling would result in better dispersion of the pigment.

Asphalt to Vinyl Coated Objects

Polyvinyl Acetate	
("Vinylite AYAT or AYAF)	17
"Nevillac" 10°	49
R-5 "Nevindene"	34

The "Nevillac" 10° and polyvinyl acetate are mixed at 140° to 150°C until the mixture is clear and uniform. The R-5 "Nevindene" is then added with continued agitation.

Nylon to Nylon
(Self-Curing, 2-3 Days)

"Neoprene" AC	50
"Neoprene" CG	50
Zinc Oxide	5
Magnesium Oxide	4
Stearic Acid	1
Litharge	20
Sulfur ("Blackbird")	4
DuPont "Neozone" A	2
DuPont "Accelerator" 833*	4
R-12 "Neville" Resin	20
Toluene	480

* To be withheld until cement is ready to be used; "Accelerator" 808 can be substituted for 833.

Milling:
Mill Roll
Temperature 90-100°F

"Neoprene"
mastication
time   6 minutes
"Neozone" A   3 minutes
Magnesium
Oxide and
Stearic Acid   5 minutes
Litharge,
Sulfur and
Zinc Oxide   5 minutes

Mixing:

All components with the exception of the R-12 "Neville Resin" are added to the toluene under agitation. The R-12 "Neville" Resin is then added after the rubber compound is in solution.

Plastic Foam to Metal

Polyvinyl Acetate
  ("Vinylite" AYAT)   5
"Nevillac" 10°   14
R-5 "Nevindene"   5
"Durez" Resin
  No. 12987   4
Methyl Ethyl Ketone   50

The polyvinyl acetate is added to the methyl ethyl ketone in an open container under agitation. After the polyvinyl acetate is dissolved, the "Nevillac" 10° and R-5 "Nevindene" are added together. The "Durez" Resin No. 12987 should be added last. The adhesive solution is applied to the metal and allowed to dry. The adhesive is later activated and cured by heat.

For Polyurethane
Flocking

I. "Multranil" FLD
    Urethane Resin   100
    "Mondur" C
      Isocyanate   5
    "Aroclor" 1254   20
II. "Multranil" FLD   100
    "Mondur" C   5-10

I is applied to the fabric by knife coating and allowed to dry thoroughly. The fabric is then coated with II, and the material is flocked immediately.

For Glass-Foamed
Plastics

Japanese Patent 875 ('61)
Chlorinated Paraffin
  Wax   1.0
Chlorinated
  Biphenyl   2.0
Titanium Hydroxy
  Stearate   0.4

Melted mixture is applied to the glass heated to 100°C.

"Hycar" Rubber Cement

"Hycar" 1001 x 225	15
Methyl Ethyl Ketone	250
"Carbopol" 934	6
Di(2-ethylhexyl)amine	6
Water	35

Dissolve the rubber in the MEK by standard procedures. Add the "Carbopol" 934 with good stirring, followed by addition of the amine. No thickening occurs but the cement may appear somewhat lumpy. Lastly, add the water with vigorous mixing, and the cement will immediately thicken to a smooth, easily spread compound. Viscosity is 3500 cps (Brookfield, 20 rpm) and Brookfield yield value is 1100.

This cement is characterized by clarity and stable viscosity, with no change in color or adhesion. Even though the rubber solids are only 5%, complete control of the flow properties of the cement are possible by using "Carbopol" 934. Higher solids formulations are also possible using the same procedure.

Sprayable Rubber Cement

"Hycar" 1042	10.0
Methyl Ethyl Ketone	79.0
Water	11.0
"Carbopol" 934	0.5
Di(2-ethylhexyl)- amine	0.5

All the ingredients, except the water, are placed in a can and rolled on a roller mill until completely dissolved and dispersed. This mixture is then stirred vigorously while the water is rapidly added. The cement immediately thickens to a viscosity of 3000 cps (Brookfield, 20 rpm).

Dipping & Coating Cement

"Estane" 5740 x 1	15.0
Tetrahydrofuran	42.5
Dimethyl Formamide	42.5
"Carbopol" 934	1.7
Di(2-ethylhexyl)- amine	1.7
Water	4.2

Dissolve the "Estane" 5740 x 1 in the THF and DMF. Slowly add the "Carbopol" 934 with vigorous agitation. When the "Carbopol" 934 is completely dispersed, carefully blend in the amine. At this point, the cement will still be relatively fluid. Lastly, add the water rapidly with as much agitation as possible.

Vinyl Resin Cement

"Geon" 427	150.0
"Geon" 443	150.0
Methyl Ethyl Ketone	900.0
Titanium Dioxide	396.0
"Carbopol" 934	26.1
Di(2-ethylhexyl)-amine	45.0
Water	44.5
Methanol	71.2

Dissolve the resins in the MEK. Next add the titanium dioxide and amine and ball mill to the desired pigment fineness. Using good agitation, carefully disperse the "Carbopol" 934 in the resin-dispersion. Lastly, combine the water and methanol and add this blend with vigorous mixing. The mixture will immediately thicken.

To Improve Bondability of "Teflon"

Treat for 1-5 minutes in a bath of sodium naphthalene and tetrahydrofuran.

Paper Cement

"Vistanex" MM L-100	100
Solvent Naphtha	900

For Paper Board

"Gelva," V-2½	50
"Gelva," V-15	50
Rosin WW	50
Dibutyl Phthalate	10

Melt dibutyl phthalate, Rosin WW, and "Gelva" V-2½ at 130°C. Add "Gelva" V-15 and melt at 170°C. When a melt is uniform adjust to desired operating temperature.

Paper Adhesive, Aerosol (Spray-on, pressure-sensitive)

PVP K-30	2.5
Nonyl Phenol	2.5
Isopropanol (Anhydrous)	20.0
Propellent 11/12 (50/50)	75.0

For Glassine and
General Purpose Paper

| "Versamid" 940 | 50 |
| "Versamid" 950 | 50 |

This combination is especially useful for heat-sealing glassine and general applications where low heat-sealing temperature and a low rate of water vapor transmission is desired. The solid resins may be easily blended by melting them together, or by using a solvent.

For Laminating Glassine

Microcrystalline Wax	50
Hydrogenated Rosin	40
Polyethylene (Ac-6)	10

For Bookbinding

Polyvinyl Acetate ("Vinylite" AYAA or AYAT)	60
"Arochlor" 1242	20
"Nevillac" Hard	20

Ingredients are dissolved in a suitable aromatic or ketone solvent.

Paper to Tin

R-28 "Neville" Resin, 35°	60
Crepe Rubber	40
"Nevastain" A	1
Mineral Spirits	360

The crepe rubber is milled and dissolved in the mineral spirits at approximately 10% solids. The "Nevastain" A and R-28 "Neville" Resin-35° are then dissolved in the rubber solution.

Adhesive Vinyl Coatings
for Glass, Aluminum, Nylon

The following formulations show excellent adhesion to substrates typified by glass, aluminum, and nylon.

Plastisol

PVP/VA 1-535	20
"Exon" 654 (Polyvinyl Chloride)	60
Dioctyl Phthalate	40

Dissolve PVP/VA 1-535 in dioctyl phthalate and then add "Exon" 654. Coat on substrate and cure at 170°C for 15 minutes.

Organosol

PVP/VA 1-535	20
"Vinylite" VYHH	20
Methyl Ethyl Ketone	40
Toluene	40
Dioctyl Phthalate	3

Dissolve "Vinylite" VYHH in a 1:1 mixture of toluene and methyl ethyl ketone. Add PVP/VA 1-535 in dioctyl phthalate. Coat on substrate and air-dry at room temperature.

Construction Sealants (Flexible)

"Vistanex" MM L-80	100
Petroleum-Base Oxidizable Oil*	220

Talc	100
Soft Clay	100
Barytes	200
Short-Fibered Asbestos	200
Lithopone	110
"Pentalyn" B-25	25
Mineral Spirits	115

* CTLA Polymer, Enjay Chemical Co.

Thixotropic Patching Compound

"Versamid" 125	40
Modified Epoxy Resin	60
Atomized Aluminum Powder #101	50
"Santocel" 54 or "Cab-O-Sil"	9

Thixotropic Gel Coat

	Formula No. 1	No. 2
"Versamid" 125	40	40
Modified Epoxy Resin	60	60
"Santocel" 54 or "Cab-O-Sil"	9	9
Atomized Aluminum Powder #101	10	0

A ⅛-inch thick film of No. 2 gelled in five hours at room temperature.

ADHESIVES

For Cracks in Wood & Masonry

"Parlon" (20-cp)	100
"Aroclor" 1254	30
"Aroclor" 5460	49
Asbestos	160
Titanium Dioxide (Rutile)	25
Toluene	300

Although this mastic formulation adheres well to either wood or masonry, a primer is suggested before applying it to steel. For better fire-retardance, substitution of 33 parts of the asbestos by 47 parts of antimony oxide is suggested as a starting formulation.

Gun Grade Caulking Compounds

	No. 1	No. 2	No. 3	No. 4
Vehicle or Binder:		Wt	%	
Blown Soybean Oil (Z-4)	17.1	14.2	17.0	11.3
Raw Soybean Oil	—	—	10.0	—
"Oronite" Polybutene No. 24	7.7	6.4	—	5.0
"Oronite" Polybutene No. 32	—	—	15.4	—
Soya Fatty Acids	1.1	1.1	1.3	1.1
Soya Lecithin	—	—	—	0.3
6% Cobalt Naphthenate	0.3	0.3	0.3	0.1
Mineral Spirits	3.8	7.7	—	2.7
3% "Methocel" Solution in Water	—	—	—	10.2
Total Vehicle	30.0	29.7	44.0	30.7
Filler or Pigment:				
"Atomite"	56.0	68.6	16.0	69.3
Talc	14.0	—	—	—
7M Asbestos	—	—	40.0	—
"Thixcin"	—	1.7	—	—
Total Filler	70.0	70.3	56.0	69.3

Caulking Compound

	No. 1 (White)	No. 2 (Aluminum)
50% Butyl 035 in Mineral Spirits	388	388
60% "Piccopale" 150 Mineral Spirits	74	74
"Hercolyn"	8.5	8.5
Denatured Alcohol	1.3	1.3
"Bentone" 38	25	25
"Atomite"	400	400
International Fiber	100	100
"TiPure" R-510	50	—
Aluminum Paste (73.5% NV)	—	34
Mineral Spirits	57	50

Ingredients are mixed in a Baker Perkins or similar sigma-bladed kneader. The "Piccopale," "Hercolyn," and "Bentone" 38, and alcohol are blended or preblended into a uniform heavy paste. Half the butyl solution is added, followed by the "TiPure," "Atomite" and Fiber. More of the butyl solution is added as needed to aid addition of pigment; however, a stiff mixture should be maintained for best dispersion. Add carefully, in increments, the rest of the butyl and the mineral spirits. If aluminum paste is used, add it near the end to get a shiny color, and mix only enough for good dispersion.

No. 3

"Vistanex" MM L-80	100
Petroleum-Base Oxidizable Oil*	220
Talc	100
Soft Clay	100
Barytes	200
Short-Fibered Asbestos	200
Lithopone	110
"Pentalyn" B-25	25
Mineral Spirits	115

* CTLA Polymer, Enjay Chemical Co.

No. 4

Calcium Carbonate	800.0
Fibrous Talc	200.0

Blown Soya	
Oil (Z_2)	246.0
Polybutene Resin	112.0
Soya Fatty Acids	16.0
Mineral Spirits	60.0
Cobalt Naph-	
thenate (6%)	4.0
"Bentone" 38	7.0
Methanol/Water	
(95/5)	2.2

For Core (Construction)
Base

Formula No. 1

"Nuba" No. 1	25
Dixie Clay	55
"1-D Heavy Oil"	15
Xylene	10
Water (As Required)	

A uniform mixture is made of the "Nuba" No. 1, "1-D Heavy Oil," and xylene. The Dixie Clay is then mixed into this solution. Depending on the type of clay employed and subject to the costs of raw material, a considerable amount of water may be added to this formulation to adjust troweling properties.

No. 2

"Neville"	
LX-685, 135	32
"1-D Heavy Oil"	14
Dixie Clay	50
VM & P Naphtha	14
Water (As Required)	

The "Neville" LX-685, 135 is added to the mineral spirits under agitation. The "1-D Heavy Oil" is then added followed by the Dixie Clay. Water may be added to adjust the troweling properties and to reduce the costs of raw materials.

Plastic Putty

"Versamid" 115	50
Modified Epoxy	
Resin	50
Atomized Aluminum	
Powder #101	20
Hammermilled Glass	
Fiber	10

It is often desired to formulate an adhesive or sealant which is resistant to sag or flow. Flow-control agents, such as "Cab-O-Sil" and "Santocel" 54, have been found useful for giving thixotropic properties. The amount of agent necessary for flow-control will vary from 3 to 12% depending on the agent and the degree of thixotropy desired.

Nonfreezing Putty

Dutch Patent 67,940

Litharge	32.0
Whiting	20.3
Linseed Oil	11.4
Drier	1.3
Pipe Clay	22.8
Water	11.0
Ethyl Lactate	1.2

Mastic-Type Adhesive

A simple mastic-type adhesive is prepared utilizing "Carbopol" 934 as a thickener. Even at its high viscosity, the compound spreads and trowels easily.

Mastic Adhesive

Water	180
SBR 2002 Latex (48% T.S.)	352
Whiting	336
Titanium Dioxide	5
"Carbopol" 934	5
Ammonium Hydroxide	To mucilage, pH 9.5

Disperse the "Carbopol" 934 in the water with moderate mixing and neutralize to pH 9.5 with ammonium hydroxide. Slowly add the latex to this preneutralized "Carbopol" 934 mucilage with moderate mixing, such as with a Lightnin' propeller type mixer. Add the pigment, very slowly, with moderate agitation and stir until the product is smooth.

Lanolin Modified Putty

Kaolin	150
Lanolin	25
Raw Linseed Oil	27

Results obtained to date show that this type of putty has excellent storage properties, showing no tendency to separation of oil. Skinning is also less pronounced.

Outside results show that when used on wooden frames the putty does not harden and become brittle as quickly as linseed oil putty and has good adhesion. Steel sash putties can also be made by using a little more kaolin. This putty gives added anti-corrosive protection to steel due to the lanolin content.

Sash Putty

Calcium Carbonate	1640.0
"Bentone" 34	4.0
Raw Linseed Oil	185.0
Mineral Spirits	9.0
Liquid Drier	8.0
Methanol-Water (95-5)	1.3

Asphalt Tile Cement

German Patent 1,023,166

Bitumen	8
Rosin	7
Solvent	15
Pigment	6-8
Filler, to make	100

For Asphalt Tile and Linoleum

"Durez" 209 Resin-powdered	9.0 lb
Ethyl Cellulose N type	1.0 lb
Isopropanol	13.7 gal

Mix all ingredients with agitation; allow mixture to stand to dissolve the ethyl cellulose and resin.

—OR—

Make separate solution of ethyl cellulose in alcohol at 10% solids. Allow this to stand overnight and mix. Dissolve "Durez" 209 separately, then blend the solutions with agitation.

PVA Emulsion Adhesive

This is a stable adhesive which is compatible with polyvinyl acetate emulsions.

PVP K-90 (20% soln)	50
"Igepal" CO-210 (GAF)	10-20
"Alipal" CO-436 (GAF)	1-1.5
Water	20

Binder for Drug Tablets

No. 1

Zein G200	100
Propylene Glycol	10
Stearic Acid	10
Ethanol, 90%	200

The stearic acid is dissolved in the alcohol at 35-40°C, as it is not soluble in 90% aqueous ethanol at room temperatures. The propylene glycol is added and the solution cooled to 30°C. The Zein G200 is added slowly with good agitation, and stirred until dissolved. If desired, the formulation may be further diluted with 90% aqueous ethanol.

A suitable alcohol-soluble dye may be added for colored tablets.

No. 2

A 5% aqueous solution of polyvinyl alcohol gives a firmer tablet than does starch or dextrin; 20 grains is used for 100 of product.

Dental Cement

U.S. Patent 2,937,099

Zinc Oxide	100.000
Calcium Acid Phosphate	3.500
Mineral Oil	0.025
Magnesium Stearate	0.025

Mix with eugenol to form a paste for use.

Dental (Root) Cement

German Patent 1,027,848

Barium Sulfate	65.0 g
Glycerol, Anhydrous	35.0 g

Methyl p-Hydroxybenzoate	0.3 g
Tyrothricin	1.0 mg

Carbon Cement

U.S. Patent 2,710,812

Carbon, Powdered	45-55
Sugar Syrup	45-55
Ethylene Glycol	5-20

Nonskid Rug Backing

"Elvax" 250	50
"Chlorowax" LV or "Aroclor" 1254	50

This mixture can be applied from a 20-25% solids solution in toluene; or from a hot-melt system with addition of heat-stable chlorinated biphenyls.

Antislip Belt Dressing

PVP K-30	1.0
Nonyl Phenol	0.25-0.5
Isopropanol (Anhydrous)	3.0

For an aerosol system, 25 parts of this mixture may be used with 75 parts propellents 11/12 (50/50 or 40/60).

Hot-Press Blood Adhesives

These are basic adhesive mixtures, intended primarily for hot-press application which can be substantially changed for other specific uses. Dry blood and soy flour are interchangeable, pound for pound. Blood will increase the water-resistance and degree of adhesion, depending upon the adherents.

Premix Formula:

Soy Four (52% Protein)	60
Adhesive-Grade Dry Blood	40
Disodium Phosphate	2
Sodium Fluoride	1
Pine Oil	2
Trisodium Phosphate	3

Compounding of Formulas:

1	2	3	4	5	6	To
100	100	100	100	100	100	Parts Premix, add
230	200	315	200	200	200	Parts of Water at 60°F
						Mix 5 minutes or until smooth, add
—	—	—	135	135	135	Parts Water at 60°F
						Mix 5 minutes or until smooth, add
30	36	12	10	6.7	12	Parts of Lime as slurry in
60	72	24	20	13.3	24	Parts of Water
						Mix 5 minutes, add
3.6	4	6	6	4	6	Parts Caustic dissolved in
7.3	8	12	12	8	12	Parts of Water
						Mix 5 minutes, add
25	30	25	25	15	25	Parts of Sodium Silicate 41° Be.
						Mix 5 minutes, add

1	2	3	4	5	6	To
1.5	2.5	1.8	2.0	1.1	2	Parts of 75-25 mixture of CS_2 and CCl_4.
						Mix 5 minutes. Allow to stand for 15 minutes, then apply
457.3	452.3	472.3	510	383	516	Parts total weight of formula

Typical Hot-Press Cycle:

15 minutes, closed assembly, under light pressure. Pressing temperature,   240-280°F, at 120-160 psi for 1.5 minutes.

CHAPTER III

CERAMICS AND GLASS

Glass Electrode Seal

It is often difficult to find a suitable wax or cement for sealing silver electrodes into glass tubing. A satisfactory seal can be achieved by melting narrow strips of polyethylene tubing in a small flame and working the melt into place around the silver-glass union. After cooling, any excess polyethylene may be trimmed.

Foamed Glass

Czechoslovakian Patent 88,037

Ground glass is mixed with 0.1% naphthalene and heated to 820°C. The naphthalene creates a reducing atmosphere and stable foam.

Coating for Optical Lenses
(Colored Solution for Temporary Identification of Lenses)

PVP K-30	2.0
Alcohol/Water-soluble Dye	0.25-0.5
"Alipal" CO-436	0.1
Ethanol/Water	q.s.

This coating solution should be kept below pH 10 as higher values may cause some insolubilization of the film after storage. PVP also has been insolubilized by certain azo compounds under exposure to light; therefore, it is suggested that azo-type dyestuffs be avoided. The coating may be removed easily from the glass with a slightly alkaline spray, before further processing of the lenses.

Coating for Optical Lens

German Patent

Thorium Fluoride	33
Magnesium Fluoride	67

Dissolve in water, filter, coat lens, dry in vacuum.

Typical Compositions of Glass Fluxes
(% Oxide Compositions)

Formula	No. 1	No. 2	No. 3	No. 4	No. 5	No. 6	No. 7	No. 8	No. 9	No. 10
PbO	81.8	78.8	66.6	60.4	59.5	55.5	45.5	53.1	49.4	48.6
B_2O_3	10.6	9.6	18.6	8.2	3.0	3.0	2.2	2.0	5.2	6.9
SiO_2	7.6	11.6	14.8	21.4	32.0	29.6	36.3	31.7	26.4	25.0
Al_2O_3	—	—	—	3.0	—	—	—	—	—	—
CdO	—	—	—	7.0	4.0	3.5	4.3	3.8	3.2	3.6
Na_2O	—	—	—	—	1.5	4.4	4.8	3.3	4.0	3.2
TiO_2	—	—	—	—	—	4.0	4.9	4.3	3.6	2.4
Li_2O	—	—	—	—	—	—	2.0	1.8	1.5	—
ZrO_2	—	—	—	—	—	—	—	—	6.7	6.7
Na_2SiF_6	—	—	—	—	—	—	—	—	—	3.6
Firing Temperature (°F)	850	950	1050	1100	1120	1120	1100	1050	1100	1120
(°C)	454	510	567	593	604	604	593	567	593	604
Coefficient of Expansion ($\times 10^{-7}$)	104	85	76	73	78	89	97	98	95	85
Lead Solubility (parts per million)	cs	cs	cs	cs	500	40	2	60	100	100
Alkali Resistance	—	—	—	Fair	Good	Good	Good	Good	Exc.	Exc.

Note: cs indicates complete solubility.

Tableware and Illuminating Glass
(Percent)

	Tableware	Tableware	Tableware	Opal Illuminating Glass
PbO	17.0	7.1	14.8	3.0
SiO_2	67.0	73.1	67.2	59.0
Al_2O_3	0.4	—	—	8.9
CaO	—	2.0	0.9	4.6
MgO	—	—	—	2.0
Na_2O	6.0	12.7	9.5	7.5
K_2O	9.6	4.6	7.1	—
As_2O_3	Tr.	0.5	0.5	—
F_2	—	—	—	5.0
ZnO	—	—	—	10.0

Approximate Melting Range   2500-2550°F
                    1371-1399°C
                    (24-28 Hours)

Optical Glass
(Percent)

	Light Barium Flint	Light Flint	Dense Flint	Dense Flint	Very Heavy Silicate Flint	Heaviest Silicate Flint
PbO	16.3	36.5	47.0	51.5	71.0	80.0
SiO_2	47.4	52.8	46.3	40.5	27.3	20.0
Al_2O_3	0.2	0.2	0.2	0.2	—	—
CaO	0.3	0.3	0.3	0.2	—	—
BaO	15.3	—	—	—	—	—
ZnO	8.3	—	—	—	—	—
Na_2O	3.0	—	5.0	—	0.2	—
K_2O	9.1	10.1	1.1	7.5	1.5	—
As_2O_3	0.1	0.1	0.1	0.1	—	—

Approximate Melting Range   2450-2500°F   2200°F
                    1343-1371°C   1204°C
                    (24-30 hours)

Lead-Containing Glazes

GLAZES	BATCH WEIGHTS (percent)	MOLECULAR FORMULA		
Cone 012-08 Color: Red-orange				
White Lead	89.0	1.00 PbO	0.101 Al_2O_3	0.203 SiO_2
China Clay	9.0		0.038 Cr_2O_3	
Chrome Oxide	2.0			
Cone 06-05 Color: Robin's egg blue				
White Lead	47.8			
Whiting	4.4	0.68 PbO		2.15 SiO_2
Buckingham Feldspar	15.2	0.16 CaO	0.15 Al_2O_3	0.10 SnO_2
Zinc Oxide	1.3	0.10 K_2O		
China Clay	3.5	0.06 ZnO		
Flint	23.7			
Tin Oxide	4.1			
Mill Addition:				
Copper Oxide	5.0			
Iron Oxide	1.3			
Cone 04 Clear raw lead bright* glaze				
White Lead	52.5	0.6 PbO		
Whiting	10.2	0.3 CaO	0.25 Al_2O_3	1.7 SiO_2
Buckingham Feldspar	20.2	0.1 K_2O		
China Clay	1.2			
Flint	15.9			

* For matte glaze substitute BaO for CaO

Lead-Containing Glazes (Continued)

GLAZES	BATCH WEIGHTS (Percent)	MOLECULAR FORMULA	
Cone 08-1 Medium opaque light blue bright glaze			
Lead Monosilicate	9.88	0.448 PbO	
Frit No. 5, Table 3-2	69.10	0.174 Na$_2$O	2.137 SiO$_2$
Florida Kaolin	9.87	0.058 K$_2$O	0.163 ZrO$_2$
Zirconium Spinel*	9.87	0.186 CaO	0.265 Al$_2$O$_3$
Copper Oxide	1.28	0.080 ZrO	0.420 B$_2$O$_3$
		0.054 CuO	
Cone 01 Wall tile glaze			
Lead Monosilicate	20.0		
Frit	21.0		
Feldspar	20.0	0.194 PbO	
Zinc Oxide	12.0	0.187 Na$_2$O	1.895 SiO$_2$
Barium Carbonate	4.0	0.192 CaO	0.235 Al$_2$O$_3$
Clay	8.0	0.051 BaO	0.179 B$_2$O$_3$
Calcined Clay	5.0	0.376 ZnO	
Flint	10.0		

Plus coloring and opacifier.

* Zirconium Spinel

ZrO$_2$	40.3	ZnO	19.6
SiO$_2$	20.4	Al$_2$O$_3$	19.7

Lead-Containing Glazes (*Continued*)

GLAZES	BATCH WEIGHTS (percent)	MOLECULAR FORMULA		
Cone 5 Sanitary ware glaze				
Lead Bisilicate	1.0	0.010 PbO		
Keystone Spar	20.8	0.058 Na$_2$O		
Flint	22.5	0.015 K$_2$O		2.770 SiO$_2$
Exeter Clay	8.7	0.607 CaO	0.140 Al$_2$O$_3$	0.067 SnO$_2$
Whiting	17.4	0.270 ZnO		0.325 ZrO$_2$
Zinc Oxide	5.3	0.028 BaO		
Barium Carbonate	1.6	0.012 MgO		
Talc	0.3			
Zirconium Spinel*	4.5			
Superpax	15.0			
Tin Oxide	2.9			

* See Cone 08-1

Lead-Containing Glazes (Continued)

GLAZES	BATCH WEIGHTS (Percent)	MOLECULAR FORMULA

Cone 5 Semi-vitreous dinnerware glaze
Lead Frit No. 6, Table 3-2

Ingredient	Percent	Molecular Formula
"Mayer" Frit*	25.0	0.26 PbO
Feldspar	50.0	0.43 CaO
Whiting	3.0	0.12 K₂O — 0.27 Al₂O₃, 0.31 B₂O₃ — 2.60 SiO₂
Zinc Oxide	1.0	0.06 Na₂O
Clay	3.0	0.13 ZnO
Flint	6.0	
	12.0	

Molecular formula:
$0.26\ PbO,\ 0.43\ CaO,\ 0.12\ K_2O,\ 0.06\ Na_2O,\ 0.13\ ZnO \mid 0.27\ Al_2O_3,\ 0.31\ B_2O_3 \mid 2.60\ SiO_2$

Cone 7 Crystalline glaze

Ingredient	Percent	Molecular Formula
Red Lead	34.6	0.50 PbO
Feldspar	16.9	0.20 CaO
Flint	25.4	0.09 Na₂O — 0.11 Al₂O₃ — 2.04 SiO₂, 0.50 TiO₂
Rutile	12.1	0.01 K₂O
Whiting	6.0	0.20 ZnO
Zinc Oxide	5.0	

Molecular formula:
$0.50\ PbO,\ 0.20\ CaO,\ 0.09\ Na_2O,\ 0.01\ K_2O,\ 0.20\ ZnO \mid 0.11\ Al_2O_3 \mid 2.04\ SiO_2,\ 0.50\ TiO_2$

---

* "Mayer" Frit

	Melted Weight Composition (%)			Molecular Formula	
K₂O	0.7	Al₂O₃	10.0	K₂O 0.02	Al₂O₃ 0.27
Na₂O	6.5	B₂O₃	14.4	Na₂O 0.29	B₂O₃ 0.57
CaO	14.0	SiO₂	54.4	CaO 0.69	SiO₂ 2.49

Solder Glass
(Percent)

PbO	80.0	SiO_2	5.0
Al_2O_3	5.0	B_2O_3	10.0

Coefficient of Linear Thermal Expansion — 93.4×10^{-7} per °C.

Softening Point — 750°F / 399°C

Typical Flux for Glass Color
(Percent)

PbO	58.6	ZrO_2	3.5
SiO_2	26.6	Na_2O	5.1
B_2O_3	2.5	TiO_2	3.7

Pigment to be added.

Dry Process Ground Coat Enamels for Cast Iron
Batch Weights
(Percent)

	FRIT A	FRIT B
Lead Bisilicate	3.0	—
Litharge	—	4.0
Dehydrated Borax	18.6	21.0
Feldspar	65.1	45.0
Quartz	12.7	30.0
Manganese Dioxide	0.6	—

	PARTS	PARTS
Mill Additions:		
Frit A and B	100	100
Clay	5	15
Iron Oxide	1	—

	PARTS	PARTS
Flint	—	20
Zircon	½	—
Soda Ash	—	10
Borax	2	—

Firing Temperature for Frits A and B: $\begin{cases} 1650\text{-}1700°F \\ 899\text{-}927°C \end{cases}$

Dry Process Cover Coat Enamels for Cast Iron
Batch Weights
(Percent)

	REGULAR ENAMELS		ACID RESISTING ENAMELS	
	FRIT A	FRIT B	FRIT C	FRIT D
Litharge	19.0	—	12.5	6.4
Lead Bisilicate	—	7.2	—	—
Hydrated Borax	6.3	26.8	8.4	18.6
Quartz	34.3	1.3	36.0	33.7
Soda Ash	17.6	—	14.9	13.0
Sodium Nitrate	5.0	3.3	7.4	4.9
Fluorspar	2.0	7.0	—	2.0
Sodium Antimonate	11.0	10.0	8.7	9.6
Calcium Carbonate	2.8	1.2	4.1	4.4
Sodium Silico Fluoride	1.0	—	2.0	—
Bone Ash	1.0	—	—	—
Cryolite	—	1.2	—	2.0
Feldspar	—	27.7	—	—
Zinc Oxide	—	9.3	—	—
Barium Carbonate	—	5.0	—	—
Titanium Dioxide	—	—	6.0	5.4

Firing Temperature for Frits A, B, C and D: $\begin{cases} 1550\text{-}1600°F \\ 843\text{-}871°C \end{cases}$

Batches for Matte Glazes
Cone 06-1

"Pemco" Pb-740	88.0
Milled Alumina	3.0
Clay	9.0
	100.0
*Opacifier	15.0

Cone 05-2

Pb-83	40.8	44.5
Pb-63	40.8	27.4
Feldspar	2.8	14.5
Clay	6.1	9.9
Whiting	2.4	—
Zinc Oxide	7.1	3.7
	100.0	100.0
*Opacifier	17.0	25.0

Cone 05-2

P-54	11.9
Pb-316	37.0
Pb-63	34.5
Clay	11.9
Zinc Oxide	4.7
	100.0
*Opacifier	20.0

Cone 04-2

Pb-1279	60.0
Feldspar	18.5
Zinc Oxide	2.5
Whiting	3.5
Barium Carbonate	1.8
Flint	3.7
Clay	10.0
	100.0
*Opacifier ·	8-15.0%

Cone 04-1

P-238	40.0
P-283	15.5
Flint	28.4
Clay	5.1
Whiting	2.0
Zinc Oxide	5.0
Barium Carbonate	4.0
	100.0
*Opacifier	22.0

* Zirconium Silicate Type

Cone 02-1

P-626	30.2
Feldspar	33.2
Alumina Hydrate	7.1
Zinc Oxide	9.3
Clay	9.1
Whiting	7.1
Flint	4.0
	100.0
*Opacifier	10.0

Cone 01-3

Pb-1279	50.0
P-64	15.0
Flint	17.0
Clay	9.0
Alumina Hydrate	4.0
Barium Carbonate	5.0
	100.0
*Opacifier	10.0

Cone 01-3

P-991	74.0
Zinc Oxide	10.0
Clay	8.0
Flint	8.0
	100.0
*Opacifier	10.0

Cone 01-1

Pb-197	60.0
Pb-113	5.0
Talc	20.0
Milled Alumina	5.0
Clay	10.0
	100.0
*Opacifier	10.0

Cone 01-1

Pb-704	70.0
Talc	10.0
Zinc Oxide	5.0
Milled Alumina	5.0
Clay	10.0
	100.0
*Opacifier	10.0

Cone 01-1

P-1679	70.0
Flint	9.0
Zinc Oxide	9.0
Milled Alumina	2.0
Clay	10.0
	100.0
*Opacifier	10.0

* Zirconium Silicate Type

Cone 01-1

P-926	30.0
Feldspar	32.0
Clay	9.0
Zinc Oxide	22.0
Whiting	4.0
Flint	3.0
	100.0
*Opacifier	10.0

Cone 01-3

	Satin	Matte	Frost Matte
P-54	34.2	32.6	28.7
Feldspar	41.2	38.6	34.0
Flint	—	—	7.4
Clay	3.5	8.2	11.7
Whiting	4.3	—	—
Barium Carbonate	8.2	7.6	6.7
Fluorspar	3.2	3.0	2.6
Talc	5.4	10.0	8.9
	100.0	100.0	100.0
* Opacifier	22.0	20.0	18.0

Cone 3-5

Pb-349	12.4
P-1090	36.7
Nepheline Syenite	6.4
Dolomite	12.0
Talc	6.3
Milled Alumina	4.0
Clay	9.8
Flint	12.4
	100.0
*Opacifier	8-15.0

Cone 3-5

Pb-1279	60.0
Feldspar	13.5
Zinc Oxide	2.5
Whiting	3.5
Barium Carbonate	1.8
Flint	3.7
Clay	10.0
Milled Alumina	5.0
	100.0
*Opacifier	8-15.0

Wollastonite Satin and Matte Glazes

Wollastonite type satin and matte glazes, without opacifier, generally produce good transparent glazes for use over underglazes. In the following formulas, it may be necessary to adjust the frit-wollastonite ratio to obtain the desired texture.

* Zirconium Silicate Type

Cone 06-04

Pb-63	30
Pb-742, Pb-63 or Pb-1114	15
Pb-716	15
Wollastonite	30
Clay	10
	100
*Opacifier	0-15

Cone 04-1

Pb-349	70
Wollastonite	20
Clay	10
	100
*Opacifier	0-15

Cone 03-5

P-238	60
Wollastonite	30
Clay	10
	100
*Opacifier	0-15

Cone 06-04

P-54	15
Pb-742	40
Wollastonite	30
Tin Oxide	1
Clay	14
	100
*Opacifier	0-15

* Zirconium Silicate Type

Cone 02-2

Pb-349	35
P-1225	35
Wollastonite	20
Clay	10
	100
*Opacifier	0-10

Cone 01-5

P-311	60
Wollastonite	30
Clay	10
	100
*Opacifier	0-15

Cone 3-5

P-786	65-70
Wollastonite	25-20
Clay	10
	100
*Opacifier	0-15

Cone 01-5

P-926	60
Wollastonite	30
Clay	10
	100
*Opacifier	0-15

Cone 3-5

Pb-349	59
Wollastonite	20
Clay	10
Flint	11
	100

*Opacifier   0-15

Single-Application Textured Glazes*

	Cone 06-03	Cone 06-03	Cone 06-03	Cone 04-02	Cone 03-1
P-1733**	52.0	55.0	30.0	29.5	27.0
P-926	—	—	23.0	—	22.5
P-1836	—	—	—	20.0	—
Kaolin	7.5	8.0	9.0	9.0	9.0
Zinc Oxide	10.0	14.5	15.0	10.8	16.2
Fluorspar	7.5	5.5	9.5	6.7	6.3
Opacifier***	12.0	12.0	13.5	13.5	13.5
Silica	2.5	5.0	—	5.5	—
Alumina	2.5	—	—	—	5.5
White Lead	3.7	—	—	—	—
Titania	2.3	—	—	—	—
Whiting	—	—	—	5.0	—
	100.0	100.0	100.0	100.0	100.0

* Recommended thickness of application is 0.030"–0.040" for maximum shade and uniform texture.

** The substitution of P-1805 for P-1733 has produced more satisfactory results in some plants.

*** Zirconium Silicate Type

	Cone 01-2	Cone 01-2	Cone 01-2	Cone 01-2	Cone 01-2
P-1733	49.5	27.2	26.5	29.2	—
P-926	—	20.8	—	—	—
Pb-1K82	—	—	29.0	—	39.4
P-54	—	—	—	28.3	15.6
Kaolin	9.0	8.5	7.5	10.0	7.5
Zinc Oxide	10.8	14.0	9.5	14.5	10.0
Fluorspar	11.7	9.5	12.5	5.0	12.5
Opacifier*	13.5	20.0	15.0	13.0	15.0
Alumina	5.5	—	—	—	—
	100.0	100.0	100.0	100.0	100.0

	Cone 01-2	Cone 01-2	Cone 01-2	Cone 01-2	Cone 2-3
Pb-1K82	39.4	39.4	—	—	—
B-683	15.6	—	—	—	—
P-1A88	—	15.6	—	—	—
Pb-1307	—	—	46.0	—	—
P-1F07	—	—	—	53.0	—
P-1733	—	—	—	—	49.5
Kaolin	7.5	7.5	5.0	11.0	9.0
Zinc Oxide	9.6	9.6	9.0	11.0	16.2
Fluorspar	12.5	12.5	12.0	11.0	6.3
Opacifier*	15.0	15.0	18.0	14.0	13.5
Silica	—	—	10.0	—	—
Alumina	—	—	—	—	5.5
	100.0	100.0	100.0	100.0	100.0

* Zirconium Silicate Type

	Cone 4-5	Cone 6-8	Cone 6-8	Cone 6-8	Cone 6-9
P-1733	40.5	40.0	21.4	25.2	20.0
P-786	—	—	—	—	26.5
P-926	—	—	16.3	19.3	—
Kaolin	9.0	9.0	6.7	7.8	8.0
Opacifier	13.5	17.0	22.0	26.0	16.0
Fluorspar	11.7	6.0	9.9	8.7	5.0
Zinc Oxide	10.8	16.6	11.1	13.0	16.5
Alumina	5.5	6.0	—	—	3.0
Silica	9.0	6.0	—	—	3.0
Feldspar	—	—	12.6	—	2.0
	100.0	100.0	100.0	100.0	100.0

Two-Coat Textured
Glazes

Cone 1-4
Base Coat Glazes**

	No. 1	No. 2
P-54	27.3	18.4
Feldspar	32.4	38.2
Kaolin	2.7	—
Whiting	6.9	9.2
Barium Carb.	6.4	6.1
Fluorspar	2.5	2.4
Titanium Oxide	—	5.0
* Opacifier	21.8	20.7
	100.0	100.0

	No. 3	No. 4
P-239	30.3	37.6
Feldspar	—	15.7
Neph. Syenite	18.3	—
Flint	5.3	5.4
Whiting	11.7	9.5
Barium Carb.	6.1	5.6
Fluorspar	2.5	2.2
Titanium Oxide	5.0	4.7
* Opacifier	20.8	19.3
	100.0	100.0

* Zirconium Silicate Type
** Two-coat application of textured glazes consists of the base coat applied about twice as heavy as a regular glaze. The overspray glaze is then applied over the wet base glaze at about one-tenth the weight of a regular glaze.

Overspray Glazes**

No. 1

P-688	90.0
Kaolin	10.0
	100.0

No. 2

P-1733	30.0
P-926	23.0
Kaolin	9.3
* Opacifier	13.5
Fluorspar	9.2
Zinc Oxide	15.0
	100.0

* Zirconium Silicate Type
** Two-coat application of textured glazes consists of the base coat applied about twice as heavy as a regular glaze. The overspray glaze is then applied over the wet base glaze at about one-tenth the weight of a regular glaze.

Batches for Gloss Glazes

The popularity of all-fritted glazes is growing rapidly because of the general trend toward automation. Many preliminary steps are eliminated or reduced—buying and storing of batch materials, control checking of raw materials, handling, mixing, and milling. The suggested glaze batches that follow are, for the most part, all-fritted glazes. They have been arranged according to firing range of the glaze.

Most of the suggested glazes do not include opacifier. They produce clear, transparent glazes for dinnerware and artware. When these glazes are to be used for colored ware or wall tile, the addition of zirconium silicate opacifiers should be up to 5% for intense colors and 5-15% for pastel colors. Colored glazes made with tin-bearing stains are often intensified by additions of 2-5% tin oxide.

Glazes containing zinc oxide are not compatible with all glaze stains. Where zinc oxide is detrimental to the color, other fluxes may be used instead of the zinc oxide mill addition.

The glaze formulas suggested are to be used as a guide for your preliminary trials. You may find that adjustments are necessary to suit your special plant conditions.

Batches for Gloss Glazes
Cone 012-08

P-760 or P-1N55	75
Pb-41	24
Bentonite	1
	100

Cone 012-06

Pb-461	71
Pb-41	18
Clay	11
	100

Cone 08-04

Pb-1K75	90
Clay	10
	100

Cone 08-02

P-64	90 - 92
Clay	10 - 8
	100

Cone 08-4

Pb-742	46
Pb-349	46
Clay	8
	100

Cone 06-02

Pb-1038	90
Clay	10
	100

Cone 012-08

Pb-545	75
Clay	10
Zirconium Silicate Opacifier	15
	100

Cone 010-06

P-1701	95
Ball Clay	5
	100

Cone 08-04

Pb-1114	90
Clay	10
	100

Cone 08-1

Pb-742 or Pb-63	90 - 92
Clay	10 - 8
	100

Cone 06-02

P-626	89
Clay	11
	100

Cone 06-02

P-930	90 - 92
Clay	10 - 8
	100

Cone 06-03

Pb-716	20
Pb-1421	70
Clay	10
	100

Cone 05-1

Pb-1241	90
Clay	10
	100

Cone 04-1

P-238	26.3
Feldspar	29.5
Wollastonite	13.7
Flint	12.6
Zinc Oxide	9.5
Clay	8.4
	100.0

Cone 04-1

P-1403	90	75	55
Pb-1421	—	15	25
Clay	10	10	10
Flint	—	—	10
	100	100	100

Cone 03-4

P-1836	90
Clay	10
	100
Zirconium Silicate Opacifier	0-15

Cone 05-02

P-586	74.0
P-626	12.7
Flint	7.1
Clay	6.2
	100.0

Cone 05-1

Pb-1421	90
Clay	10
	100

Cone 04-6

Pb-197	92
Clay	8
	100

Cone 04-2

Pb-1492	90	60	45
Clay	10	10	10
Feldspar	—	25	25
Wollastonite	—	5	10
Zinc Oxide	—	—	5
Flint	—	—	5
	100	100	100

Cone 03-1

P-1341	92
Clay	8
	100

Cone 02-1

P-609	28.0
Pb-349 or P-786	18.0
Nepheline Syenite or Feldspar	28.0
Wollastonite	8.5
Whiting	5.5
Clay	12.0
	100.0

Cone 02-2

P-991	90 - 96
Clay	10 - 4
	100

Cone 02-4

Pb-1041	90
Clay	10
	100

Cone 02-2

P-830	36.2	34.4
Pb-723	28.1	—
Feldspar	9.1	32.3
Flint	6.1	11.4
Clay	8.2	10.0
Barium Carb.	—	6.5
Zinc Oxide	12.3	5.4
	100.0	100.0

Cone 02-4

P-786	86 - 90
Clay	14 - 10
	100

Cone 02-2

P-991	45
Pb-349	45
Clay	10
	100

Cone 02-2

P-991	45
Pb-197	45
Clay	10
	100

Cone 02-4

Pb-716	43.7
Feldspar	28.2
Clay	8.0
Flint	11.6
Whiting	8.5
	100.0

Cone 02-2

P-25	26.8	26.4
Pb-316	11.0	—
Feldspar	39.2	41.3
Clay	8.0	8.1
Calcined Clay	5.0	5.2
Whiting	10.0	10.0
Zinc Oxide	—	9.0
	100.0	100.0

Cone 02-8

Pb-349	86 - 92
Clay	14 - 8
	100

Cone 01-2

P-54	27.9	29.6
Feldspar	33.9	30.5
Flint	9.6	16.6
Clay	11.7	11.7
Whiting	3.4	0.4
Zinc Oxide	—	8.5
Fluorspar	2.6	2.7
Talc	4.2	—
Barium Carb.	6.7	—
	100.0	100.0

Cone 01-2

P-926	30.1	32.7
Pb-545	12.3	13.4
Feldspar	31.8	34.5
Flint	8.7	9.5
Clay	8.1	8.8
Barium Carb.	1.0	1.1
Zinc Oxide	8.0	—
	100.0	100.0

Cone 01-4

P-609	58.1
Whiting	7.2
Nepheline Syenite or Feldspar	11.8
Clay	10.3
Flint	12.6
	100.0

Cone 1-6

P-311	51.5
Pb-316	25.1
Whiting	4.0
Clay	4.7
Flint	14.7
	100.0

Cone 01-2

P-54	25.8
Pb-316	16.9
Feldspar	24.2
Flint	13.2
Clay	11.0
Whiting	0.2
Zinc Oxide	8.7
	100.0

Cone 01-2

P-926	20.3
Flint	16.9
Feldspar	29.5
Clay	9.3
Whiting	12.9
Barium Carb.	1.1
Zinc Oxide	10.0
	100.0

Cone 1-6

P-54	31.5
Pb-316	24.6
Whiting	5.4
Clay	16.6
Flint	21.9
	100.0

Cone 1-6

Pb-704	88-86
Clay	12-14
	100

Cone 2-4

P-54	22.5
Feldspar	36.0
Flint	11.5
Clay	10.6
Whiting	8.4
Zinc Oxide	11.0
	100.0

Cone 3-5

P-830	34.7
Nepheline Syenite or Feldspar	26.4
Flint	16.9
Clay	10.5
Barium Carb.	6.3
Zinc Oxide	5.2
	100.0

Cone 3-5

P-1403	80.0
Clay	10.0
Flint	5.0
Zinc Oxide	5.0
	100.0

Cone 4-8

P-609	58.1
Whiting	7.2
Nepheline Syenite or Feldspar	11.8
Clay	10.3
Flint	12.6
	100.0

Cone 2-5

P-586	57.7
Pb-316	9.5
Feldspar	12.6
Flint	9.4
Clay	7.6
Zinc Oxide	3.2
	100.0

Cone 3-5

P-926	31.2
Feldspar	32.9
Flint	16.5
Clay	9.4
Barium Carb.	1.0
Zinc Oxide	9.0
	100.0

Cone 4-7

P-586	63.5
Potash Feldspar	12.6
Zinc Oxide	3.2
Clay	6.6
Flint	14.1
	100.0

Cone 7-10		Cone 9-12	
P-586	62.8	P-688	43.7
Potash Feldspar	12.6	Potash Feldspar	19.9
Zinc Oxide	3.1	Clay	8.1
Clay	11.2	Flint	12.0
Flint	10.3	Zirconium Silicate	
	100.0	Opacifier	16.3
			100.0

High-Temperature Glazes
for Sanitary Ware

Cone 9-12	
P-827	78.0
Clay	11.0
Flint	11.0
	100.0

Cone 9-12

P-675	54.0
Potash Feldspar	19.5
Clay	6.8
Flint	16.8
Tin Oxide	2.9
	100.0

Low-Expansion Glazes
for Vitreous Bodies

Cone 01-6

Pb-658	88.4
Clay	11.6
	100.0

Cone 9-12

P-404	21.3
Potash Feldspar	22.5
Flint	34.6
Wollastonite	10.8
Clay	10.8
	100.0

Cone 1-4

Pb-821	88.3
Clay	11.0
Bentonite	0.7
	100.0

Batches for Fast-Fire Glazes on Talc or Wollastonite Bodies

P-786	99.0	—	—
P-1B25	—	99.0	—
P-1P93	—	—	99.0
Bentonite	1.0	1.0	1.0
	100.0	100.0	100.0

Pb-1710	89.0	89.0	94.0	95.0	99.0
Pb-674	10.0	—	—	—	—
Feldspar	—	10.0	—	—	—
Silica	—	—	5.0	—	—
Setit "A"	—	—	—	5.0	—
Bentonite	1.0	1.0	1.0	—	1.0
	100.0	100.0	100.0	100.0	100.0

These glazes are self-opacified and are recommended for white only

P-1421	99.0	60.0	—	—	—
Pb-545	—	30.0	30.0	—	49.0
Pb-1492	—	—	69.0	99.0	—
Pb-74	—	—	—	—	50
Bentonite	1.0	—	1.0	1.0	1.0
Ball Clay	—	10.0	—	—	—
	100.0	100.0	100.0	100.0	100.0
* Opacifier	15.0	15.0	15.0	15.0	15.0

Fast-Fire Glazes on Prepared, Partially-Fritted Bodies**

Pb-1421	60.0	—	—	—	90.0
Pb-704	—	45.0	90.0	99.0	—
Pb-545	30.0	45.0	—	—	—
Ball Clay	10.0	10.0	10.0	—	10.0
Bentonite	—	—	—	1.0	—
	100.0	100.0	100.0	100.0	100.0
* Opacifier	15.0	15.0	15.0	15.0	15.0

* Zirconium Silicate Type
** Pemco Prepared Bodies DT-1 and DT-2

Pb-704	90.0	99.0		Pb-674	59.0	—
Ball Clay	10.0	—		Pb-461	—	39.0
Bentonite	—	1.0		P-941	30.0	40.0
	100.0	100.0		P-1855	10.0	20.0
				Bentonite	1.0	1.0
* Opacifier	15.0	15.0			100.0	100.0
P-1409		99.0		* Opacifier	15.0	15.0
Bentonite		1.0				
		100.0				
* Opacifier	15.0	15.0				

* Zirconium Silicate Type

Chapter IV

COSMETICS AND DRUGS

Antiperspirants
Formula No. 1

a	"Kessco" Wax A-45	18
	Mineral Oil 65/75	3
	Propylene Glycol	3
	Titanium Dioxide	1
	Water	25
b	Chlorhydrol	20
	Water	30

a is heated to 95°C and allowed to cool to 35°C while agitating. Chlorhydrol is dissolved in the water at room temperature. This solution b is then stirred into a in small portions and the whole mixture is stirred until a smooth cream is obtained.

No. 2

a	Stearic Acid	14
	Beeswax	2
	Mineral Oil	1
	"Myrj" 52	5
	Atlas "G-2162"	5
b	Water	51
	Preservative	q.s.
c	Chlorhydrol	22
	Perfume	q.s.

Melt a to 70°C; add b at 85°C; cool to 35°C and add all of c. Agitate until c is completely dissolved.

No. 3

a	"Tegacid"	18.00
	"Protegin" X	4.00
	Spermaceti	4.00
	Propylene Glycol	7.00
	Glycerol	5.00
	Water	41.55
b	Titanium Dioxide	0.25
	"Tegosept" M	0.20
	Chlorhydrol	20.00

98

Place all ingredients of division a in one container. Stir while heating to 90-100°C, and cool to about 35°C. Now add ingredients of division b, agitating until completely dissolved or dispersed and a smooth cream is formed. Milling optional.

No. 4

a	Cetyl Alcohol	1.20
	Atlas "G-2151"	3.00
	Stearic Acid	10.00
	Propylene Glycol	4.00
b	"Veegum"	2.50
	Water	59.30
c	Chlorhydrol	20.00

Heat a to 70°C. Prepare b by adding the Veegum to the water slowly, continually agitating until smooth. Add b to a and mix until smooth and cool. Add c and mix until thoroughly dispersed. Homogenize formula.

No. 5
(Water-in-Oil Emulsion)

"Amerchol" CAB	24.0
Spermaceti	1.5
Water	54.5
Chlorhydrol	20.0

Add water containing dissolved Chlorhydrol to waxes at 60°C. Mix rapidly until cool. Homogenize for best results.

No. 6

"Neocol" 5190	
or 5192	5.5
Water	54.5
Chlorhydrol (50%	
w/w sol.)	40.0
Perfume	q.s.

The "Neocol" is heated to 75°C. One part of water at 75°C is added to the Neocol with lively agitation. Thereafter, the rest of the water is added in a steady slow stream with continued agitation until the temperature drops to 37°C. Add perfume and then the Chlorhydrol in a slow steady stream until all is incorporated.

No. 7

"Kessco" X-211	2
Polyethylene Glycol	
6000 Monostearate	4
"Kessco" Wax A-21	5
Water	49
Aluminum Chlorhydroxide Complex	
(50% Solution)	40

Place all the ingredients, except the aluminum chlorhydroxide solution, in one container. Heat to 90-95°C. Allow the mixture to cool to 35°C, with stirring. Slowly add the aluminum chlorhydroxide solution, with stirring.

No. 8

a Glycerol Monostearate Pure   3.5
   Polyethylene Glycol
   1540 Monostearate   3.7
   Cetyl Alcohol   1.0
   Propylene Glycol
   Monolaurate
   D & D   1.0
   Water   30.8
b Aluminum Chlorhydroxide Complex   25.0
   Water   35.0

Heat ingredients of a to 95°C. Allow the mixture to cool to 40°C with stirring. Make a solution of b in cold water, add this solution to part a and stir well.

No. 9

a "Kessco" Wax A-33   8
   Cetyl Alcohol   4
   Polyethylene Glycol
   1540 Distearate   5
   White Mineral Oil
   (65-75 visc)   3

Propylene Glycol   3
Water   24
b Aluminum Chlorhydroxide Complex
   (50% Solution)   30
   Water   23

Heat part a to 85°C and then allow it to cool to 35°C while stirring. At room temperature stir part b in small portions into part a. Continue stirring until a smooth cream is obtained.

No. 10

a "Kessco" Wax A-21   4
   Stearic Acid   15
   White Mineral Oil
   (65-75 visc)   3
   Propylene Glycol   5
   Water   23
b Aluminum Sulphate   25
   Urea   13
   Water   12

Heat all the ingredients of a to 90-95°C. Allow the mixture to cool to 45°C with stirring. Place the ingredients of b in another container and agitate at room temperature until completely dissolved. Pour part b into part a and agitate until a smooth cream is obtained.

No. 11

"Amerchol" L-101	3
"Modulan" (or "Acetulan")	2
Spermaceti	4
Propylene Glycol	3
Atlas "G-2162"	2
"Myrj" 52	2
Glyceryl Monostearate (acid-stable type)	17
Water	28
Chlorhydrol (50% solution)	38
Titanium Dioxide	1
Preservative	q.s.

No. 12

"Amerchol" L-101	2.0
Stearic Acid	18.0
"Arlex"	2.0
Mineral Oil (70 visc)	2.0
Propylene Glycol	7.0
"Amerdex"	3.0
Water	64.5
Triethanolamine	1.5
Perfume	q.s.

Melt the oil-soluble materials in one container and heat to 85°C. Heat the water-soluble materials, except the titanium dioxide, chlorhydrol, and perfume, in another kettle to 85°C. Add the water solution to the oils while agitating, and when well mixed add the chlorhydrol where required. Continue mixing to 40°C; add the titanium dioxide, preservative, and perfume; mix well and pack.

No. 13

"Amerchol" CAB	24.0
Spermaceti	1.5
Water	59.5
* Aluminum Salt	15.0

* Aluminum sulfate, chlorhydrol, etc.

Add the water containing dissolved salts to the oil phase with the temperature of both phases 60°C. Mix rapidly while warm and at slower speeds as the batch cools. For best results, homogenize.

No. 14

Oil Phase:

"Solulan" 98	2.0
"Amerchol" L-101	4.0
Cetyl Alcohol	1.5
Glycerol	2.0
"Myrj" 52	4.0
Stearic Acid XXX	2.0

Water Phase:

"Veegum" HV	1.0
Water	47.5
Chlorhydrol (50% soln)	36.0

Perfume and
Preservative   q.s.

Combine all ingredients of oil phase and heat to 65-70°C. Add "Veegum," well dispersed in water, at 70°C to the oil phase with moderate mixing. Cool to 40°C and add the 50%-solution of Chlorhydrol. Continue cooling with slow agitation to room temperature. Remix the following day.

No. 15

a "Arlacel" 165 acid-
    stable g.m.s.   15
    Water   45
b Aluminum Chlorhy-
    droxide Complex
    (50% solution)   40

Heat a to 70°C, b to 72°C; add b to a with agitation; stir until cool.

No. 16

a "Arlacel" 165 acid-
    stable g.m.s.   5
    Cetyl Alcohol   5
    Water   50
b Aluminum Chlorhy-
    droxide Complex
    (50% solution)   40

Heat a to 70°C, b to 72°C; add b to a with agitation; stir until cream has set.

No. 17
(Gel)

1 volume of a 40% w/w aqueous solution of sodium lactate mixed with 3 volumes of 50% w/w aluminum chlorhydrate.

A clear, solid gel is obtained on standing for three days at room temperature; or in five minutes, by heating to 60°C.

No. 18

A mixture of 10 parts by weight of 50% w/w of aluminium chlorhydrate, 10 parts of glycerol, and 4 parts by weight of 55% w/w sodium lactate solutions, after heating for five minutes, sets to a clear gel, which remains stable in air without drying or crumbling.

No. 19

3 volumes of 50% w/w aluminium chlorhydrate mixed with 1 volume 50% w/w potassium acetate solution and 4 volumes of water sets to a clear, colorless, soft plastic gel suitable for packing in collaps-

ible tubes or plastic squeeze-bottles.

No. 20

3 volumes of 50% w/w aluminium chlorhydrate mixed with 1 volume 50% w/w potassium acetate, 2 volumes of water, 1 volume of 95% alcohol and 1 volume of glycerol sets on warming to 60°C to a clear, colorless plastic solid.

No. 21

Powdered Salicylic Acid	4
Magnesium Carbonate	10
Calcium Carbonate	10
Titanium Dioxide	26
Talc	50

No. 22
(For Feet)

Chinosol	10
Boric Acid	40
"Aerosil"	50

No. 23

Paraformaldehyde	3
Lactic Acid	5
Boric Acid	10
Magnesium Carbonate	10
"Aerosil"	20
Talc	50

No. 24

Magnesium Peroxide	20
"Aerosil"	40
Talc	30
Boric Acid	10

No. 25

Beeswax, Yellow	8
Tallow	25
Anhydrous Lanolin	65

Melt together and in this dissolve:

Salicylic Acid	2
Benzoic Acid	1

Stick
Formula No. 1

Stearic Acid	8.00
"Eutanol" G	28.75
"Comperlan" 100	3.00
"Dehydag" Wax 0	3.00
Propylene Glycol	2.00
Essential Oil	2.00
Hexachlorophene	0.25
Ethanol, 96%	50.00-55.00
Sodium Hydroxide, 38%	3.00
Coloring	As desired

Melt stearic acid, "Eutanol" G, "Comperlan" 100, "Dehydag" Wax 0, and hexachlorophene, together with the greater part of the ethanol, in a water-jacketed vessel at 140°F. Mix the propylene glycol and the remaining ethanol with the sodium hydroxide solution, likewise heated to 140°F, and then stir this into the alcoholic fatty components. When cooled to about 122°F add the essential oil and possible coloring, and pour the whole mixture into molds.

The mass sets within 5 minutes and can be removed from the molds before completely cool.

It is advisable to work with a greater quantity of ethanol than prescribed. This means about 10% for small batches and less for larger ones. The finished stick should be wrapped in tin-foil and packed in a case to prevent evaporation.

No. 2

a Stearic Acid	
(low I.V.)	3.00
Palmitic Acid	
(low I.V.)	2.00
Ethanol	34.35
"Sorbo" (70%	
Sorbitol	
Solution)	5.00

b Sodium Hydroxide	0.65	
Water	5.00	
c 40% w/w Sodium		
Aluminum Chlor-		
hydroxy Lactate		
in Water	50.00	

Heat a to 65°C, and b to 70°C. Add b to a with agitation. Heat c to 60°C, and add it to the a-b mixture when this mixture has cooled to 62°C. Pour at 60°C into heated molds. Remove from molds when cool, wrap in aluminum foil, and package in air-tight containers.

No. 3
(Cream)

Stearic Acid	150
Polyoxyethylene	
Sorbitan	
Monostearate	10
Diglycol Stearate	23
Sorbitan Sesquioleate	20
"Actamer"	10
Isopropyl Palmitate	15
Anhydrous Lanolin	20
Sorbitol Solution	15
Potassium Hydroxide	7
Distilled Water	730

No. 4

a Methyl "Tuads"	0.5
Ethanol	74.6

b Sodium Stearate   6.9
c "Carbowax" 4,000   8.0
  Water   10.0

Procedure:
Add the Methyl "Tuads" to the ethanol.
Add b to a.
Dissolve the Carbowax in the water with a little heat. Add c to a-b and heat to 70°C. Pour into molds.

This stick is very easily applied and has a cooling effect on the skin. It can be used on the feet as well as for an underarm deodorant.

No. 5

"Cetrimide"   1
Aluminum Lactate   15
Urea   5
Ethanol   25
Perfume   To suit

No. 6

"Cetrimide"   1.0
Water-Soluble
  Chlorophyll   0.2
Ethylene Glycol
  Monolaurate   4.0
White Petrolatum   5.0
Mineral Oil   13.0
Precipitated Chalk   18.0
Calcium Stearate   0.5
Distilled Water   17.0

Aerosol
Formula No. 1

Aluminum Chlor-
  hydrate   15.0
Water   17.0
SDA #40 Anhydrous
  Ethanol   60.0
"Tween" 80   2.0
"Arlacel" 80   1.3
Dichlorodifluoro-
  methane   4.7

No. 2

Aluminum Chlor-
  hydrate   12.5
Water   12.5
"Emcol" 14   2.5
Glycerol Mono-
  stearate   2.5
Isopropyl Myristate   1.0
Propellent Mixture*   69.0

---

* 15% dichlorodifluoromethane and 85% tetrafluorodichloroethane. The propellent is pressure-filled.

With formulations for both the oil-in-water and the water-in-oil types of emulsions it is necessary to use a "mechanical break-up" type of valve actuator.

No. 3

a Aluminum Chlor-
    hydroxide complex 5.0
  Distilled Water   5.0
b Isopropyl Myristate 1.0
  Dipropylene Glycol   4.5

Tartaric Acid 170 in
SDA Ethanol #40　0.5
Hexachlorophene
USP　0.3
Perfume Oil　0.5
Ethanol SDA #40　83.2

*Filling Charge in Glass
Aerosol:*
70% Concentrate Formula
(as above)
30% Propellent Mixture
60% "Freon" 114 or
"Genetron" 320
40% Propellent 12

No. 4

Glycerol　0.70
Perfume　0.70
Hexachlorophene　0.28
Isopropyl Myristate　0.70
Dipropylene Glycol　3.50
Anhydrous
Ethanol　64.12
Dichlorodifluoro-
methane　12.00
Dichlorotetra-
fluoroethane　18.00

No. 5

PVP K-30　0.5
Ethanol (85%)　57.4
"Actamer"　0.5
Propylene Glycol　1.0
"Emcol" E 607
(n-(acyl colamino

formylmethyl)-
pyridinium
chloride)　0.1
Perfume　0.5
Propellent 114　30.0
Propellent 12　10.0

Zirconium-Compound Deodorants

Formula No. 1 (Liquid)

U.S. Patent 2,498,514

Sodium Zirconium
Lactate　15
Glycerol　5
Ethanol　20
Water　To make　100
Perfume　To suit

No. 2 (Stick)

Sodium Zirconium
Lactate　5
Sodium Stearate　8
Propylene Glycol　5
Ethanol　To make　100
Perfume　To suit

No. 3 (Powder)

Sodium Zirconium
Lactate　5
Talc　50

Zinc Stearate	5
Precipitated Chalk	25
Colloidal Kaolin	15
Perfume	To suit

No. 4
(Stick)
U.S. Patent 2,857,315

Propylene Glycol	75.00
Sodium Stearate	7.50
"Carbitol"	1.50
Sodium Zirconium Lactate (50%)	9.90
Perfume	0.06
Hexachlorophene	0.25

Dissolve at 70°C and pour into molds.

Nonirritating Deodorant
U.S. Patent 2,498,514

Neutralized Zirconium Lactate	4 g
Glycerol	12 cc
Hard Hydrogenated Castor Oil	10 g
Soft Hydrogenated Castor Oil	70 g

Aqueous Cosmetic Deodorant

Aluminum Chloride	15 g
Urea	5 g
Water	79 cc
"Cetrimide"	1 g

Foot Deodorant

Aluminum Chloride	30.0 g
Urea	10.0 g
"Cetrimide"	1.0 g
Water To make	60.0 cc
Terpeneless Perfume Compound	1.5 cc
Benzyl Cresol	0.4 cc
p-Chloro-m-cresol	0.2 g
Menthol	0.5 g
β-Phenoxyethyl Alcohol	2.0 cc
Ethanol To make	140.0 cc

Thioglycolic Emulsion Depilatory
German Patent 723,020

Thioglycolic Acid	8
Calcium Oxide	8
Sodium Dodecyl Xanthate	15
Purified Sperm Oil	5
Perfume Oil	1
Water	64

Hand Cream (O/W)
Formula No. 1
(Liquid)

a "Cerawax"	3.00
Isopropyl Myristate	12.00
Oleic Acid	0.50
"Fluilan"	4.00
Triethanolamine	0.25
b Glycerol	2.50
Water	90.00

Heat a to 65-70°C; bring b to 60°C. Add a to b with high-speed agitation, stir until cold. Homogenize; add perfume at 25°C.

No. 2
(Liquid)

"Vigilan"	6
Glyceryl Monostearate	13
Glycerol	7
Sesame Oil	3
Isopropyl Myristate	1
Water	70

Mix glycerol and water to 165°F. In a separate container melt other components to 165°F. Add water-glycerol solution slowly to oils with continuous agitation. Continue mixing till cool.

No. 3
(Paste)

a "Arlacel" 165 Acid-stable g.n.s.	5
Stearyl Alcohol	5
b "Sorbo"	5
Water	85
Mild acid to adjust pH	q.s.

Heat a to 70°C, b to 72°C. Add b to a with agitation. Stir until cream has set.

No. 4

Stearic Acid	10
Igepal CO-850	9
Propylene Glycol	5
Water	76
Preservative and Perfume	q.s.

No. 5

"Tegin"	8.0
Stearic Acid	8.0
Cetyl Alcohol	4.0
Lanolin	4.0
"Robane"	8.0
Propyl Glycol	6.0
Triethanolamine	2.0
Water	160.0
Propylparaben	0.4
Perfume	q.s.

No. 6
(No Mineral Oil)

"Lantrol"	5.0
Myristyl Alcohol	1.0
Glyceryl Mono-stearate Pure	10.0
"Tegosept" P	0.1
"Deltyl" Extra	5.0
Water	72.7
"Arlex"	5.0
"Tegosept" M	0.2
"Duponol" WA Paste	1.0
Perfume	q.s.

Heat oil phase and water phase to about 75°C. Add water phase to oil phase with rapid agitation. Stir down to about 40°C, and add perfume.

No. 7

Stearic Acid	40
Isopropyl Myristate	140
Paraffin Oil	140
Glycerol	10
Triethanolamine	10
Distilled Water	660

No. 8
(Acid O/W)

"Amerchol" L-101	6.0
"Modulan"	2.0
Glyceryl Mono-stearate (Neut)	12.0
Petrolatum	4.0
Mineral Oil (70 visc)	5.0
Stearic Acid	2.0
Glycerol	5.0
Water	63.5
Sodium Lauryl Sulfate	0.5
Preservative and Perfume	q.s.

No. 9
(Aerosol)

a	"Acetulan"	3.0
	Stearic Acid	2.6
	Glyceryl Mono-stearate	3.0
b	Polyoxyethylene Lanolin Derivative	3.0
	Preservative	0.1
	Triethanolamine	1.0
	Propylene Glycol	3.0
	Water	84.0
c	Perfume, "Fleuralia" #73	0.3

92% Actives

8% Propellent
F-114/F-12
(40:60)

Heat a and b separately to 80°C. Add b slowly to a while mixing; when cool add c and mix.

Skin Cream

a "Arlacel" 165 Acid-
       Stable g.m.s.   12
    Lanolin   1
    Cetyl Alcohol   3
    Mineral Oil   4
b Propylene Glycol   1
    Water   79

Heat a to 65°C; b to 67°C.
Add b to a. Stir until cool.

Cold Cream

Beeswax, White   40.0
Spermaceti   10.0
"Robane"   100.0
Borax   2.0
Lanolin   2.0
Propylparaben   0.4
Perfume   q.s.

Aerosol Cold Cream

"Amerchol" L-101   4.00
Polyoxyalkylene
    Lanolin   1.60
"Tegin"   2.40
Stearic Acid   3.20
Mineral Oil
    (80-90 visc)   9.60
"Tegosept" P   0.08
Triethanolamine   0.40
Distilled Water   78.72

Concentrate, 92.0% by wt;
Genetrons 114A/12 (43:57)
8.0% by wt

170 gm net fill in 6 oz can
for shaving cream

O/W Emollient
All-Purpose Cream

Beeswax   13.0
Petrolatum   12.0
Mineral Oil
    (light visc)   20.0
"Lantox" 55   20.0
Preservative   0.5
Water   34.0
Borax   0.8
Perfume   q.s.

Heat waxes and oils to 70°C
separately. Heat water with
borax to 75°C. Add borax solution to waxes and oil phase.
The addition of water should
be slow but the agitation
should be continuous. Perfume to be added when the
mixture reaches 50°C.

Anionic Emollient
Cream O/W

"Polylan"   1.0
"Amerchol" L-101   6.0
"Modulan"   2.0

Glyceryl Mono-	
stearate C	12.0
Petrolatum	4.0
Mineral Oil	
(70 visc)	5.0
Stearic Acid XXX	2.0
Glycerol	5.0
Water	62.5
Sodium Lauryl	
Sulfate	0.5
Preservative and	
Perfume	q.s.

Emulsify by adding the water phase at 85°C to the oil phase at 85°C, and mix until 40°C. Add preservative and perfume. Continue mixing until cool.

Washable W/O Creams

	No. 1	No. 2
"Amerchol"		
L-101	2	4.0
Mineral Oil		
(80-90		
visc)	40	27.5
Petrolatum	35	38.0
Beeswax	—	2.0
"Emulphor"		
VN-430	—	3.0
Sorbitan		
Sesqui-		
oleate	3	—
Water	20	25.0

Borax —	0.5
Preservative q.s.	q.s.

Add the water containing the water-soluble ingredients to the oils with the temperature of both phases at 70°C. Mix rapidly until cool.

Day Cream

Cetyl Alcohol	7.0
Stearyl Alcohol	14.0
Sodium Lauryl	
Sulfate	4.0
"Robane"	17.0
Sweet Almond Oil	10.0
Glycerol	10.0
Propylparaben	0.4
Water	138.0
Perfume	q.s.

Moisture Cream

Water	46.7
Glycerol	3.0
"Tegosept" M	0.2
"Duponol" WA	
Paste	1.0
"Nimlesterol"	10.0
Mineral Oil	
(70 visc)	16.0
Cetyl Alcohol	2.0
Glyceryl Mono-	
stearate Pure	8.0
"Brij" 35	8.0

"Tegosept" P 0.1
Microcrystalline
Wax (155°) 5.0
Perfume q.s.

Add perfume at 50°C and agitate to 40°; pour.

Moisturizing Night Cream

"Lantrol" 10.0
Stearic Acid, Triple
Pressed 1.5
Cetyl Alcohol 0.2
Mineral Oil
(65-75 visc) 20.0
Beeswax 4.0
"Span" 60 5.0
"Tegosept" P 0.1
Borax 0.3
Glycerol 5.0
Triethanolamine 0.7
Water 52.0
"Tween" 60 1.0
"Tegosept" M 0.2
Perfume q.s.

Modern Emollient
Moisturized Cream

Beeswax 6.5
Stearic Acid, Triple
Pressed 2.0
"Span" 40 2.5
"Span" 60 2.5

Mineral Oil
(light visc) 20.0
"Lanoile" 12.0
Preservative 0.2
Water 50.0
Borax 0.8
Triethanolamine 0.5
Propylene Glycol 3.0
Perfume q.s.

Dissolve the ingredients in water, bring the temperature to 80°C. Separately melt waxes and oils to 80°C. Pour water solution into the waxes. Use stirring or homogenizing at 65-75°C.

Cleansing Cream (W/O)
Formula No. 1

a Lanolin 8
"Hartolan" Wool
Wax Alcohols 3
Spermaceti Wax 4
Cetyl Alcohol 5
White Beeswax 6
White Mineral Oil 25
b Water 49

Heat a together until just melted and allow to cool to about 50°C. Heat b to 50°C and add to a. Agitate with a paddle-type mixer until cold. Perfume at 25-30°C.

No. 2

"Vigilan"	6
Stearic Acid	3
Triethanolamine	
Lauryl Sulfate	2
Glycerol	1
Isopropyl Myristate	18
White Mineral Oil	3
Olive Oil	5
Water	62

Dissolve triethanolamine lauryl sulfate and glycerol in water and heat to 170°F. In a separate container heat remaining components to 170°F. Slowly add water solution to oils with continuous agitation. Continue agitation till emulsion reaches room temperature.

No. 3

	(lb)
Stearic Acid	25.00
White Mineral Oil	57.00
Anhydrous	
Lanolin	34.00
Terpineol	0.35
Triethanolamine	9.50
Propylene Glycol	75.00
Quince-Seed	
Mucilage	19.00
Water	315.00

Melt the stearic acid in the mineral oil, add the lanolin and terpineol, and heat to 70°C.

In a separate container, heat the solution of the triethanolamine in water to 70°C. Add the hot oil mixture to the heated amine solution, stirring vigorously until a good emulsion is formed.

Add the quince-seed mucilage. [This is made beforehand by adding 9.5 oz of quince seed to 20 lb of water at 80°C, soaking overnight, and then straining through a cloth. A suitable preservative should be added to the quince-seed mucilage to prevent its growing mold over a period of time. The mucilage can then be stored for use as needed.]

Mix the perfume in the propylene glycol and stir this solution into the cream when it has cooled to about 50°C. The stirring should be fast enough to keep the cream well mixed but not to aerate it. Stirring should be continued at low speed until the emulsion has cooled to room temperature. If the cream is allowed to cool without stirring, it will thicken on standing a few days.

Fluffy Cleansing Cream
Formula No. 1

Stearic Acid	24.00
Water	64.00
Glycerol	10.51
Potassium	
Hydroxide	0.99
Perfume	0.50

Melt the stearic acid. Dissolve the potassium hydroxide in the water, add the glycerol, and heat to 85°C. Then add the solution slowly to the melted stearic acid, with constant stirring. Perfume at 40°C.

No. 2

Stearic Acid	15.0
Cetyl Alcohol	0.5
Glycerol	5.0
Potassium	
Hydroxide	1.0
Water	78.5

No. 3

Stearic Acid	30.00
Triethanolamine	1.75
Potassium	
Hydroxide	1.30
Water	69.00
Glycerol	5.50
Perfume	0.50

Melt the stearic acid. Make a hot solution of the alkali, triethanolamine, and water, and add the glycerol. Add the alkaline solution to the stearic acid and mix until emulsified. Perfume at 40°C.

No. 4

Stearic Acid	8
"Robane"	56
Triethanolamine	2
Glycerol	2
Water	132
Perfume	q.s.

No. 5
(Antiseptic)

"G-11"	0.5
"Deltyl" Extra	10.0
Mineral Oil	
(visc 65-75)	40.0
Cetyl Alcohol N.F.	1.0
Beeswax	10.0
"Arlacel" #20	3.0
Borax	1.0
Water	34.0
Perfume Oil	0.5

Bleaching Cream
Formula No. 1

Diglycol Stearate	120
Paraffin Oil	10
Isopropyl Myristate	10
Stearyl Alcohol	40
Cetyl Alcohol	40

Sorbitol Solution	40
Triethanolamine Lauryl Sulfate	13
Distilled Water	667
Zinc Peroxide	60

No. 2

a Glyceryl Mono-stearate	110
White Petrolatum	20
Mineral Oil	30
Beeswax, White	10
b Glycerol	100
Distilled Water	625
c Bismuth Oxychloride	100
d Rose White 08272 (Perfume)	5

Heat a and b separately to about 70°C, and stir b into a. Take a small part of this mixture, pre-mix it with c, and then stir this pre-mixed batch into the rest of the mixture. Then add c and continue stirring until cool.

Cream Sachet

a "Arlacel" 165 Acid-Stable g.m.s.	16
Spermaceti	5
Cetyl Alcohol	5
Isopropyl Myristate	3

b Atlas "G-2162" (Polyoxyethylene oxypropylene stearate)	2
Water	59
c Perfume	10

Heat a to 75°C; heat b to 75°C; add b to a, stirring until the cream sets. When the product has cooled to 45°C, add the perfume and continue stirring until the product is homogeneous.

Anhydrous Lanolin Cosmetic

U.S. Patent 2,840,509

Lanolin	30
"Ucan" LB	25

Mix and heat to 49°C; add

Mineral Oil, Medicinal	45

Mix until temperature falls to 24-27°C.

Protective Skin Cream

"Cetrimide"	0.5
Cetostearyl Alcohol	10.0
Mineral Oil	10.0
Water and Perfume To make	100.0

Skin Cream

French Patent 1,142,927

Lanolin	6-12
Cetyl Alcohol	0.5-1.5
Shark Oil	15-20
Propylene Glycol	5-15
Glyceryl Mono-	
Oleate	20-30

and 50,000 I. U. of vitamin A; 25,000 I. U. of vitamin D; water, perfume, and preservative. [The shark oil must be deodorized by catalytic hydrogenation before incorporation in cosmetic specialties.]

Emollient Skin Cream (Aerosol)

Lanolin U.S.P.	
Anhydrous	19.1
Atlas "G-1425"	0.9
Castor Oil AAA	
Grade	70.5
Water	9.5

The water and G-1425 are mixed and heated to 150°F, then slowly added with agitation to the mixture of castor oil and lanolin which has been heated to 160°F. After emulsification is complete, the product is filled into cans. After the valve has been crimped in place, ten percent by weight of dichlorodifluoromethane is pressure-filled into the container. Although the foam stability of this product is poor, this does not interfere with its functioning.

Where it is necessary or desirable to have a water-in-oil foaming aerosol emulsion with a good stable foam, it can be accomplished in a circuitous manner. If the propellent is first emulsified in water using a suitable emulsifier and then pressure-filled into the water-in-oil emulsion, there results a system which can be designated as an oil-in-water/water-in-oil emulsion. The dispersed phase is droplets of water and droplets of propellent emulsified in water (oil-in-water phase). The continuous phase is the oil which now does not contain the propellent. Although there is only a small amount of water present, this type of emulsion gives a rich, stable foam.

Barrier Creams
Formula No. 1

Stearic Acid	150
Isopropyl Palmitate	20

Silicone Fluid		Glyceryl Mono-	
(1000 cps)	100	stearate	8
Potassium		Beeswax	3
Hydroxide	10	Petrolatum	10
Sorbitol Solution	183	Mineral Oil	5
Distilled Water	537	Water	60

No. 2

Glyceryl Mono-	
stearate	140
Beeswax	40
Stearyl Alcohol	20
Lanolin, Anhydrous	60
Isopropyl Myristate	20
Glycerol	40
Sodium Lauryl	
Sulfate	10
Zinc Stearate	130
Distilled Water	540

No. 3

Stearic Acid	10.000
Synthetic Japan	
Wax	2.000
"DC" 200	
(1000 cts)	20.000
Potassium	
Hydroxide	0.500
Methylparaben	0.025
Propylparaben	0.015
Distilled Water	67.500

No. 4

British Patent 797,992

PVP K-30	1

No. 5
(Aerosol)

a	Glycerol	15.0
	PVP K-30	1.7
	Water	878.3
b	Myristic Acid	10.0
	Stearic Acid N.F.	40.0
	Cetyl Alcohol N.F.	3.3
	Modified Lanolin	3.3
	Isopropyl Myristate	11.7
	Triethanolamine	11.7
	"Silicone Oil" DC	
	555 Fluid	20.0
c	Perfume Oil	5.0
d	Propellents 12/114	
	(40/60)	111.0

Heat parts a and b separately to 70°C to melt and dissolve. Add a to b while stirring; add c when cool. Pressure-fill d (10 parts propellent to 90 parts concentrate).

Diaper Cream

Glyceryl Mono-	
stearate	12.0
Spermaceti	6.0

Oleyl Alcohol	2.0
Glycerol	12.5
Perfume	q.s.
Diaphene	0.1
Purified Water q.s.	100.0

Dissolve the diaphene in the oil phase consisting of glyceryl monostearate, spermaceti and oleyl alcohol at 80°C.

Massage Cream
(Semi-liquid)

"Polawax"	4
Wool Wax Alcohols	0.4
Propylene Glycol	5.0
Olive or Arachis Oil	8.0
Preservative and Perfume	0.6
Witch Hazel Extract (Dist)	33.0
Water	49.0

Pore Cream

"Polawax"	10.0
Water	65.0
Arachis Oil	6.0
Ext. Witch Hazel	20.0
Glycerol	4.0

Powder Base Cream

"Polawax"	8.0
Water	75.0
Glycerol	9.5
Powder Base	10.0
Stearic Acid	8.0
Sorbitol	6.0

Cream Rouge

"Polawax"	12.0
Water	80.0
Dyestuff	1.0
Glycerol	7.0
Potassium Hydroxide	0.2
Pigment	7.0
Stearic Acid	2.0

Leg Make-up Base

"Polawax"	12.0
Water	75.0
Liquid Paraffin	15.0

Cosmetic Stick

Stearic Acid	9.3
38% Sodium Hydroxide Solution	3.5
"Comperlan" HS	12.0
"Dehydag" WAX O	3.0
Glycerol	10.0
"Eutanol" G	50.2
Essential Oil	2.0
Ethanol 96%	10.0

Heat stearic acid, "Comperlan" HS, "Dehydag" WAX O,

"Eutanol" G, and ethanol in a water-jacketed vessel to about 158°F. Mix the sodium hydroxide solution and glycerol, heat likewise to 158°F, and then stir this into the alcoholic fatty mixture. When this has cooled to about 140°F, add the perfume, mix, and pour the whole into molds.

Cream Make-Up
Formula No. 1
(Matte)

a	"Nytal" 400	18.5
	Kaolin	1.3
	TiO$_2$	3.7
	Iron Oxide	1.5
b	Propylene Glycol	5.0
	"Darvan" No. 1	0.3
	Water	18.5
c	"Veegum"	1.5
	Water	28.5
d	"CMC" (Low Visc.)	0.1
	Water	9.9
	"Arlacel" 80	0.5
	"Tween" 80	1.5
e	Stearyl Alcohol	2.0
	"Amerchol" L-101*	2.0
	Isopropyl Myristate	5.0
	Perfume	0.2
	Preservative	q.s.

Micropulverize a and add to b. Add the Veegum to the water slowly, agitating continually until smooth. Disperse the CMC in water; mix with c. Add a-b to preceding and mix until smooth; heat to 65-70°C. Heat e to 70°C; add this other mixture and stir until cool. Add perfume.

This product is designed to be a complete make-up with a matte finish.

No. 2

a	Polyethylene Glycol 400 Monostearate	1.0
	Glycerol Monostearate Pure	2.0
	Stearic Acid	12.0
	Isopropyl Palmitate	1.0
	Preservative	0.1
b	Glycerol	2.5
	Propylene Glycol	12.0
	Titanium Dioxide 1:5*	10.0
	Pigment	1.0
	Water	58.4

* A commercial dispersion of 20% Titanium Dioxide in Talc.

Heat parts a and b separately to 90°C, and add b to a slowly with constant agitation. If the mixture is not smooth and uniformly dispersed, pass it through a homogenizer or ball mill.

Liquid Make-Up Base

Oil Phase:

"Amerchol" L-101	6.0
"Acetulan"	4.0
Glyceryl Mono-	
stearate, Neut.	2.5
Stearic Acid, XXX	5.0

Water Phase:

"Solulan" 98	2.0
Propylene Glycol	5.0
Triethanolamine	1.0
Water	74.5
Perfume and	
Preservative	q.s.

Add the water phase at 85°C to the oil phase at 85°C with mechanical agitation and cool with stirring to 40°C. Add perfume and continue cooling with stirring to room temperature.

Finished Make-up is prepared by adding 10 or more parts of solids (color, titanium dioxide and talc) to 90 parts of formula.

Foundation Cream

Polawax	13.0
Water	80.0
Glycerol	8.5
Liquid Paraffin	3.5

Cream Talc

a	"Neocol" 57-E	6.00
	Cetyl Alcohol N.F.	5.50
	Perfume	.50
b	Water	45.00
	Methyl *p*-hydroxy-	
	benzoate	.10
c	Talc Mix #4	25.00
	Water	17.90

Heat *a* to 60°C in a water bath. Heat *b* to 80-85°C until solution is complete. Then add *c* and stir gently until talc is wet. Adjust temperature to 60°C. With *a* and *b-c* at 60°C, add *b-c* slowly to *a* with gentle mixing. After mixing with a motor-driven propeller for about 5 minutes, add the water, which has been previously warmed to 45°C. Mix while cooling. Mixing is then discontinued at 40-42°C, when the cream starts to set.

Talc Mix #4 is Trinity Superfine Talc, 96.0%; Mistron Micropaque, 4.0

Face Powder Compact

"Acetulan"	3
Zinc Stearate, USP	6
Kaolin #2457	2
Talc, Alabama #140	88
Magnesium Carbon-	
ate #690	1

Pigments & Tita-
nium Dioxide   q.s.

Add the "Acetulan" to the
combined solids blended with
equal portion of talc. Mix until uniform. Add to the remaining talc and blend thoroughly. Micronize to ensure
uniformity and press.

Nongreasy Cream Bases

Nongreasy cream bases can
be made very simply from self-
emulsifying glyceryl monostearate which contains soap as
the emulsifier. As some medicaments are not compatible
with soap, it is advisable to
include in the formula either
bentonite or cetyl alcohol to
stabilize the emulsion.

Glyceryl Mono-	
stearate S	10
Glycerol	25
Bentonite	2
Water	63

Hand-Protective Cream
No. 1

a	"Dow-Corning"	
	XEF-220	25
	"Aldo" 33	7

b	Water	67¾
	Methyl "Parasept"	¼

Heat *a* to 80°C and, while
stirring, add *b*. Stir until cool.

No. 2

Gelatin		4
Glycerol		30
Water	To make	100

No. 3

Sodium Polyacrylate		10
Glycerol		30
Water	To make	100

No. 4

Cellulose Ether		
("Polyfibron")		4
Glycerol		30
Water	To make	100

Hand-Lotion Cream
No. 1

Glyceryl Mono-	
stearate	10.0
Mineral Oil	20.0
Olive Oil	10.0
Cetyl Alcohol	6.0
Butyl *p*-Hydroxy-	
benzoate	0.2
Glycerol	2.5
Stearic Acid	3.0
Water	600.0

| Blue Color | To suit |
| Perfume | 0.2 |

No. 2

a Glyceryl Mono-
stearate .... 17 lb
Stearic Acid .... 30 lb
Absorption Base .... 7 lb
Lanolin .... 10 lb
Cetyl Alcohol ...... 8.5 lb
Mineral Oil .... 32.25 gal
Olive Oil .... 4.25 gal
Cottonseed Oil ..... 0.75 gal
Propyl Gallate ..... 5 oz

b Triethanol-
amine .... 11.25 lb
Butyl
p-Hydroxy-
benzoate .... 20 oz
Water .... 177.5 gal

c Isopropanol .... 2.5 gal
Water .... 2.5 gal
Perfume .... 71 oz

Heat a to 90°C. Run the water into a separate container, add the preservative, and heat it to the same temperature. Add the triethanolamine and pour the hot waxes into b. Stir the cream until it cools to 45°C. Add c and perfume, and continue stirring until the perfume is evenly distributed throughout the emulsion.

Hand Cream

Stearic Acid	5.0
Oleic Acid	1.5
Cetyl Alcohol	0.5
Lanolin	1.0
Almond Oil	5.0
Isopropyl Myristate	10.0
Triethanolamine	1.2
Glycerol or Glycol	5.0
Water	70.8
Perfume and Preservative	q.s.

Day Cream

Stearic Acid	17.00
Diglycol Stearate	3.00
Cetyl/Stearyl Alcohol	1.00
Lanolin	2.00
Isopropyl Myristate	5.00
Triethanolamine	1.00
Sodium Hydroxide	0.25
Glycol or Glycerol	8.00
Water	62.75
Preservative	Sufficient

Skin-Conditioning Cream

Self-Emulsifying Monostearin	10.0
Glycerol	6.0
Triethanolamine Ricinoleate	1.0
Lanolin Alcohols	6.0

Chlorocresol	0.2
Oleyl Alcohol	3.0
Distilled Water	
To make	100.0

Isopropyl	
Myristate	4.0 cc
"Cetrimide"	0.1 g
Water	11.9 cc

After-Shave Cream

"Lan-Aqua-Sol"	
(W/S Lanolin)	5.3
Ethanol	19.0
Hexachlorophene	0.2
Menthol	0.3
Benzocaine	0.2
Water	75.0

Dissolve hexachlorophene, menthol, and benzocaine in ethanol; mix "Lan-Aqua-Sol" 50% with water. Add aqueous solution to ethanol solution, mix thoroughly, age, and filter.

Antiseptic Cream
Formula No. 1

Extra-pale Lanolin	4 g
Cetyl Alcohol	4 g
Isopropyl	
Myristate	4 cc
"Cetrimide"	1 g
Water	11 cc

No. 2

| Lanolin | 4.0 g |
| Cetyl Alcohol | 4.0 g |

Medicated Cream Base

Cetyl Alcohol	6.4
Stearyl Alcohol	6.4
Sodium Lauryl	
Sulfate	1.5
White Petrolatum	14.3
Mineral Oil	21.4
Water	50.0

Warm the fatty alcohols on a water bath to 65°C, and stir in the sodium lauryl sulfate. Add the mineral oil and petrolatum. Cool to room temperature and add the water slowly, with constant stirring.

Washable Ointment Base

a	Polyethylene Glycol	
	200 Distearate	3.5
	Triethanolamine	0.7
	Glycerol	1.7
	Water	77.7
b	Stearic Acid XXX	13.4
	Mineral Oil (65-75)	2.0
	Butyl Stearate	1.0
	Preservative	0.1

Heat separately the ingredients of *a* and of *b* to 70°C. Add *a* slowly to *b*, stirring continuously. Stir to below 35°C, and add perfume to suit. Avoid rapid stirring. Pack the cream at 30 to 35°C.

Vanishing Creams

	Formula No. 1	No. 2	No. 3	No. 4	No. 5
Oil Phase:					
"Amerchol" L-101....	0.5%	— %	— %	4.0%	2.5%
"Amerchol" CAB....	—	5.0	4.5	—	—
"Modulan"..........	—	2.0	—	2.0	—
"Acetulan"..........	3.0	—	0.5	—	0.5
Cetyl Alcohol........	0.5	—	—	—	—
Spermaceti..........	—	—	1.5	—	—
Stearic Acid, XXX...	18.0	22.8	18.0	18.0	—
Glyceryl Monostearate	—	—	—	—	10.0
Mineral Oil, (70 vis)..	—	—	—	2.0	—
Silicone Fluid, (350 cstk).............	—	—	—	—	10.0
Water Phase:					
Glycerol.............	—	4.0	4.5	5.0	—
Propylene Glycol.....	14.0	—	—	—	—
Water..............	63.2	65.0	70.0	65.0	77.0
Triethanolamine.....	—	1.2	—	2.0	—
Potassium Hydroxide.	—	—	1.0	—	—
Sodium Hydroxide...	0.8	—	—	—	—
Borax, USP..........	—	—	—	2.0	—
Perfume and Preservative..............	q.s.	q.s.	q.s.	q.s.	q.s.

Add the water phase at 85°C to the oil phase at 85°C while mixing. Cool with mixing to 40°C. Add perfume and continue mixing. Cool to room temperature. Remix the next day.

No. 6

a Stearic Acid (T.P.) 10
"Polawax" 1
"Isocreme" Absorp-
tion Base 3
Diglycol Laurate 1
Glyceryl Mono-
stearate (S.E.) 2
Mineral Oil 1
Glycerol 2
b Potassium Hydroxide 1
Water 79

Heat a together to a tem-
perature of 75°C. Heat b to
the same temperature. Add b
to a with constant agitation,
stir until cold. Perfume at
40°C.

No. 7

"Arlacel" 165 Acid-
Stable g.m.s. 18
Lanolin 2
Cetyl Alcohol 1
Mineral Oil 1
"Sorbo" 2
Water 76

Heat the ingredients to
90°C. Stir thoroughly to form
an emulsion.

No. 8

Glyceryl Mono-
stearate 14.0
Lanolin 2.0
Glycerol 4.0
Perfume 0.5
Methylparaben 0.1
Water 79.4

Heat all the ingredients, except the perfume, together to
80°C; stir until cold, adding
the perfume at 50°C.

No. 9
(Liquid)

Paraffin Wax 60 g
White Beeswax 120 g
Liquid
Petrolatum 540 g
Stearic Acid 10 g
Sodium Borate 10 g
Perfume Bouquet 2 cc
Distilled Water 260 cc

Melt the paraffin, white
wax, mineral oil, and stearic
acid together and heat to
70°C. Dissolve the borax in
the water and heat the solution to 70°C. Slowly add the
lipoid phase to the aqueous
phase with constant stirring.
Stir until congealed, then add

the perfume, mix thoroughly, and pass through a mill.

Light-Bodied Moisturizing Cream Base (Vanishing Type)

		% by weight
a	Stearic Acid (T.P.)	1.5
	Glyceryl Mono-stearate (Pure)	2.0
	Stearyl Alcohol	2.0
	White Petrolatum	1.0
	Mineral Oil	3.6
	Lanolin	0.4
	Allantoin	0.05
	Propylparaben	0.1
b	Propylene Glycol	4.0
	Methylparaben	0.2
	Potassium Hydroxide	0.15
	Water	85.0

Combine the ingredients in a, and heat to 175°F. Heat water-phase components of b to 180°F, while stirring with a Lightnin' Mixer. Add oil phase, a, slowly while continuing to mix. At about 100°F, the cream begins to set.

Recommended for dry skin, particularly for psoriasis and/or 'itching' skin.

Antiflash Cream

This is a smooth, creamy white paste for the protection of hands and face from flash burns.

Titanium Dioxide	600.00
PVP K-90 (20% soln)	400.00
"Methocel" 400 (10% soln)	400.00
Diethylene Glycol	240.00
Trichlorophenyl Acetate	0.06
Stabilizer #1	0.06

The components are added in the order indicated with mixing after each addition followed by milling for two hours in a Werner-Pfleiderer mill. Higher glycols or plasticizers may be substituted for diethylene glycol.

Cream for Rashes and Insect Bites

Glyceryl Mono-stearate	20.00
Propylene Glycol	5.00
Spermaceti	5.00
Water	67.90
Camphor	0.50
Menthol	0.50

Phenol	0.50
Clove Oil	0.25
Eucalyptol	0.25

Heat the first four ingredients together to boiling, remove from the heat, and stir until cool. Mix the last five ingredients together and stir or grind in a mortar until a clear mixture results. When the emulsion is cool, add the second mixture to it slowly, with stirring, and mix thoroughly.

Antiseptic Gel
Formula No. 1

Gelatin	5 g
"Cetrimide"	2 g
Water	193 cc

No. 2

Gelatin	5 g
"Cetrimide"	2 g
β-Phenoxyethyl Alcohol	4 cc
Urea	10 g
Water	179 cc

The melting point of these gels can be raised by the addition of compatible synthetic thickening agents or they can be made into creams by the addition of nonionic waxes, such as cetyl alcohol and polyethylene glycol 600 monostearate. A useful gel can also be made by dissolving a high enough percentage of "Cetrimide" in hot water, filtering if desired, and allowing to cool.

Antiseptic Facial Mask

a	"Veegum"	6.00
	Water	82.25
b	Ethanol	4.00
c	Glycerol	4.00
	Sulfonated Castor Oil	3.00
d	"Vancide" BL	0.25
e	"Truodor" Floral Note #16 (Perfume)	0.50
	Color	q.s.

Add the "Veegum" to the water slowly, agitating continually until smooth. Add b to a; then add c. Add a small portion of b to d; paste out and add to the remainder of b. Add e to b-d.

This product is designed to be packaged in a tube. The bacteriostatic activity of "Vancide" BL in this formula is confirmed when tested against *Staphylococcus aureus* (antibiotic-resistant) and *Bacillus subtilis*.

Decontaminant for Vesicant (Military) Gas

U.S. Patent 2,518,345

Bleaching Powder	1250.00
Sugar	6.30
Asbestos Powder	3.10
Sodium Dodecyl Sulfate	0.31
Water	1870.00

Powdered Ointment Base

Polyethylene Glycol 4000	236.00
Sodium Stearate	236.00
Stearic Acid	468.30
Cholesterol	11.80
Sodium Lauryl Sulfate	47.20
Methylparaben	0.45
Propylparaben	0.25

Mix 6 parts of this with 4 parts of hot water; stir until uniform.

Calamine Lotion
Formula No. 1

Calamine		150.0 g
Zinc Oxide		50.0 g
Bentonite		30.0 g
Sodium Citrate		7.5 g
Glycerol		50.0 ml
Liquefied Phenol		5.0 ml
Distilled Water	To make	1000.0 ml

Triturate the calamine, zinc oxide, and bentonite with a solution of the sodium citrate in 700 ml water; add the liquefied phenol, then the glycerol, and finally the remainder of the distilled water.

No. 2

Finely Powdered Calamine	150.0 g
Zinc Oxide	50.0 g

Sodium Carboxymethylcellulose		10.0 g
Diisopropyl Naphthalene Sulfonate		0.5 g
Glycerol		50.0 ml
Liquefied Phenol		5.0 ml
Distilled Water	To make	1000.0 ml

Disperse the cellulose in about 60 ml of distilled water at about 70°C, with constant stirring. The preparation is otherwise standard, with the wetting agent incorporated with the powders and the cellulose solution used as the vehicle.

No. 3

Calamine		8.0 g
Zinc Oxide		8.0 g
Polyvinyl Alcohol ("Elvanol" 50-42)		2.5 g
Glycerol		3.0 cc
Water	To make	100.0 cc

Triturate the powders in a mortar, add the glycerol, and bring the preparation to volume by adding water in divided portions.

Emollient (Absorption) Bases

Formula No.	Hydrous (Water-in-oil)				Anhydrous	
	1	2	3	4	1	2
"Amerchol" L-101	12.5	—	—	15.0	36.0	—
"Amerchol" CAB	—	—	20.0	—	—	40.0
"Amerchol" H-9	—	43.0	—	—	—	—
"Modulan"	—	—	10.0	—	—	20.0
Mineral Oil (70 vis)	16.7	—	8.0	20.0	16.0	16.0
Microcrystalline Wax	12.5	—	—	15.0	24.0	—
Spermaceti	—	1.8	—	—	—	—

Formula No.	1	2	3	4	1	2
Beeswax	—	—	7.0	—	—	14.0
Cetyl Alcohol	—	—	2.0	—	—	4.0
Lanolin	8.3	—	—	10.0	24.0	—
Petrolatum	—	—	3.0	—	—	6.0
Vegetable Oil	—	7.0	—	—	—	—
"Carbowax" 1500	5.0	9.0	—	—	—	—
Water	45.0	39.2	50.0	40.0	—	—
Preservative	q.s.	q.s.	q.s.	q.s.	q.s.	q.s.

Add the water to the melted fats with the temperature of both phases at 65°C. Mix rapidly until cool and homogenize for best results. The anhydrous bases should be heated to dissolve the ingredients and mixed slowly until well mixed.

Waterless Hand Cleaner
Formula No. 1

a Polyethylene Glycol
   600 Monostearate   6.2
Polyethylene Glycol
   300 Distearate   6.2
Deodorized
   Kerosine   52.9
Stearic Acid   4.0
Preservative   0.1
b Water   28.4
Triethanolamine   2.2

Heat a and b to 50°C in separate containers. Add b to a with stirring, and continue to stir the mixture until it cools to 35°C.

No. 2

a Polyethylene Glycol
   600 Monostearate 12.5
Polyethylene Glycol
   300 Distearate   2.5
Deodorized
   Kerosine   67.5
Preservative   0.1
b Water   17.4

Heat a and b to 50°C in separate containers. Add b to a with stirring and continue to stir the mixture until it cools to 35°C.

No. 3

a Deodorized
   Kerosine   30-45
   Lanolin   3
b "Arlacel" 40   2.5
   "Tween" 40   7.5
c Water   42-57
   Preservative   q.s.

Mix *a* and *b* and heat to 60°C. Heat *c* to 62°C and add to the first mixture with moderate agitation, continued until emulsion is cool.

Slightly more or less lanolin may be used in such a formula.

Although emolliency is often a desirable feature, most users of waterless hand cleaners object to any greasy or slippery "after-feel." The formula as shown above should give a pleasant non-greasy feeling.

(Aerosol)

	No. 4	No. 5
"Amerchol" L-101	1.00	0.6
Deodorized Kerosine	42.50	43.0
Oleic Acid	6.00	6.1
Cetyl Alcohol	0.50	—
Triethanolamine	3.00	3.0
Propylene Glycol	2.35	3.2
"Tergitol" 4	1.40	1.8
Perfume	0.25	0.3
Distilled Water	43.00	42.0

Concentrate, 85.0% by wt; "Genetron" 12 15.0% by wt. 150 gm net fill in 6-oz can.

No. 6

a "Veegum"	2.5
Water	30.0
b "Tween" 60	8.0
"Span" 60	2.0
Deodorized Kerosine	35.0
c "Methocel" 4000 cps	0.5
Water	22.0
Preservative	q.s.
d Propellent: 50%	

Nitrous Oxide/ 50% Nitrogen	1.10

Add the "Veegum" to the water slowly, agitating continually until smooth; heat to 62°C. Heat *b* to 60°C, cautiously, add *a* and mix until cool. Prepare solution *c* by dispersing the "Methocel" in one-third of the water at 90°C. Add the rest of the water cold and mix thoroughly.

————————

Liquid Eye Liner

a "Veegum"	2.5
Water	75.5
b Polyvinyl-	
pyrrolidone	2.0
Water	10.0
c Pigment	10.0
Preservative	q.s.

Add the "Veegum" to the water slowly, agitating continually until smooth. Dissolve the polyvinylpyrrolidone in water using a little heat. Add b to a; then add c. Mix well.

For a product with a soft cream-like viscosity that can be applied readily with a brush, increase the amount of "Veegum" to 3.5%. This product forms a good film, washes off readily and is easily applied in either the liquid or cream form.

————————

Eye-Shadow Cream

Glycerol Monostearate, (not emuls.)	200
Cetyl Alcohol	40
Beeswax, White	90

Isopropyl Myristate	40
White Petrolatum	400
"Iso-lan"	150
Color Lakes	80

————————

Mascara
Formula No. 1
(Block)

a Isopropyl Myristate	120
Stearic Acid	200
Beeswax, White	110
Emulsifier	300
b Triethanolamine	20
Distilled Water	200
"Nipagin M"	1
c Color Pigment	120

Heat a and b separately to 85°C. Stir first b into a and then add c. The batch is milled and poured into the mold while still warm.

No. 2
(Cream)

Diglycol Stearate	80
Anhydrous Lanolin	30
Polyethylene Glycol-	
400 Distearate	100
Stearyl Alcohol	130
Isopropyl Palmitate	20
Triethanolamine	
Lauryl Sulfate	15
Distilled Water	545
Color	80

No. 3
(Waterproof Cream)

a	Beeswax	10.0
	Light Mineral Oil	5.0
	Carbon Black	3.0
b	Stearic Acid	1.5
	Carnauba Wax	7.5
c	"Veegum"	2.0
	Water	38.2
d	"CMC" (Low Visc)	0.1
	Water	9.9
e	Propylene Glycol	1.5
	Water	20.5
	"Darvan" No. 1	0.2
f	Morpholine	0.6
	Perfume	q.s.
	Preservative	q.s.

Heat *a*; mill. Add *a* to *b*. Heat to 70°C, and mix thoroughly. Add the "Veegum" to the water slowly, agitating continually until smooth. Disperse the CMC in water. Add it to preceding; then add *e*, and heat to 65-70°C. Add morpholine to preceding, then immediately emulsify, adding *a-b* with constant stirring until cool. Finally, add perfume.

Lotion Vehicle
Formula No. 1

Methylparaben	0.25
Propylparaben	0.15
Sodium Lauryl Sulfate	10.00
Propylene Glycol	120.00
Stearyl Alcohol	25.00
Light Liquid Petrolatum	250.00
Distilled Water	595.00

Melt the stearyl alcohol on a water bath, add the petrolatum, and heat to 70°C. Dissolve the two parabens in hot water; then add the sodium lauryl sulfate and the propylene glycol. Adjust the aqueous solution to 70°C. Pour the aqueous mixture into the oil phase in a fine stream, with continuous stirring. Continue to stir until the temperature is below 45°C.

No. 2

Methylparaben	0.25
Propylparaben	0.15
Sorbitan Monolaurate	55.00
Polyoxyethylene Sorbitan Monolaurate	25.00
Sodium Lauryl Sulfate	10.00
Stearyl Alcohol	21.00
Light Liquid Petrolatum	400.00
Distilled Water	500.00

Melt the stearyl alcohol on a water bath, add the sorbitan monolaurate, and light liquid petrolatum and heat to 70°C. Dissolve the two parabens in hot water; then add the polyoxyethylene sorbitan monolaurate and the sodium lauryl sulfate. Adjust the aqueous solution to 70°C and continue as for No. 1.

Both these lotion vehicles are compatible with zinc oxide, sulfur, salicylic acid, oil of cade, ammoniated mercury, phenol, resorcinol, coal-tar solution, ichthammol, ethyl aminobenzoate, and sodium borate. Practically no sedimentation or creaming occurs in the bottles that stand undisturbed, or in those that are shaken daily.

Lotions
Formula No. 1
(Nonionic)

Polyethylene Glycol 400 Monostearate	4.0
Propylene Glycol Monostearate Pure	1.0
Lanolin	1.0
Propylene Glycol	2.0
Water	91.9
Preservative	0.1

Heat all of the ingredients to 70-75°C in one container. Allow the mixture to cool to 35°C with continual stirring.

No. 2
(Pearlescent)

a	Cetyl Alcohol	1.0
	Butyl Stearate	1.25
	Stearic Acid	4.0
	Preservative	0.1
b	Water	93.0
	Potassium Hydroxide	0.15
	Glycerol	0.5

Heat a and b separately to 70°C; add b to a with agitation; stir the mixture until it cools to 35°C.

No. 3

a	Stearic Acid	6.4
	Preservative	0.1
	Isopropyl Myristate	1.5
	Polyethylene Glycol 600 Monostearate	0.3
b	Potassium Hydroxide	0.2
	Water	50.3
c	Sodium Carboxymethylcellulose High Viscosity Grade (0.8% Solution)	41.2

Heat *a*, *b*, and *c* to 65-70°C in separate containers. Add *b* to *a* and mix well; add *c* with stirring. Allow the mixture to cool to 35°C with continual agitation.

No. 4
(Silicone)

a	Glycerol Mono-	
	stearate S.E.	3.4
	Stearic Acid	2.4
	Lanolin	1.0
	Propylene Glycol	4.7
	Preservative	0.1
b	Triethanolamine	1.0
	Water	77.4
c	Isopropanol	5.0
	Silicone Oil*	5.0

* Dimethyl Polysiloxane — Viscosity @ 25°C = 1000 cs.

Heat *a* and *b* separately to 95°C; add *b* to *a* and stir well for several minutes. Allow the mixture to cool to 35°C. Add *c*, mix well and then homogenize. This formulation is suitable for packaging in squeeze-type containers.

No. 5

a	Diethylene Glycol	
	Monostearate	
	Pure	2.0

	Cetyl Alcohol	3.0
	Isopropyl Palmitate	3.0
	"Kessco" Wax A-21	1.0
	Preservative	0.1
b	Water	87.9
	Polyethylene Glycol	
	600 Dilaurate	3.0

Heat *a* and *b* separately to 95°C. Add *b* to *a* with agitation and allow the mixture to cool to 35°C with continual stirring. This formula is suitable for packaging in squeeze-type containers.

No. 6

a	Glycerol	2.0
	Stearic Acid	2.0
	Glycerol Mono-	
	stearate Pure	3.5
	Propylene Glycol	
	Monolaurate	8.5
b	Triethanolamine	1.0
	Water	83.0

Heat *a* and *b* to 85°C in separate containers. Add *b* to *a* with stirring until the mixture cools to 35°C. This formulation is suitable for packaging in squeeze-type containers.

No. 7

a Stearic Acid	6.0
Butyl Stearate	1.0
Polyethylene Glycol	
600 Monostearate	3.0
Preservative	0.1
b Potassium	
Hydroxide	0.2
Water	89.7

Heat a and b to 65-70°C in separate containers. Add b to a and stir until the mixture cools to 35°C. This formulation is suitable for packaging in squeeze-type containers.

No. 8

a Glycerol Mono-	
stearate, S.E.	3.5
Stearic Acid	2.5
Lanolin Anhydrous	1.0
Propylene Glycol	5.0
Preservative	0.1
b Triethanolamine	1.0
Water	81.9
c Isopropanol	5.0

Heat a and b to 90-95°C in separate containers. Add b to a while stirring until the mixture cools to 35°C. Add the alcohol and stir well. This formulation is suitable for packaging in squeeze-type containers.

No. 9

a White Mineral Oil	
(65-75 visc)	35
Polyethylene Glycol	
1540 Distearate	8
Glycerol Mono-	
stearate, Pure	5
b Water	45
Propylene Glycol	7

Heat a and b to 80°C in separate containers. Add b to a with continuous stirring until the mixture cools to 35°C.

No. 10

a Stearic Acid	2.0
Mineral Oil	
(65-75 visc)	15.0
Preservative	0.1
Isopropyl Myristate	2.0
Polyethylene Glycol	
400 Monostearate	10.0
Lanolin	4.0
Beeswax	4.0
Propylene Glycol	4.0
b Triethanolamine	1.0
Water	57.9

Heat a and b to 70-75°C in separate containers. Add b to a and stir until the mixture cools to 35°C. This formulation is suitable for packaging in squeeze-type containers.

No. 11

a	Polyethylene Glycol 300 Monostearate	3.8
	Glycerol Monostearate Pure	1.3
	Stearic Acid	2.9
	White Mineral Oil (visc, 65-75)	8.4
	Butyl Stearate Cosmetic Grade	1.3
	Paraffin Wax M.P. 135°F	2.1
	Preservative	0.1
b	Triethanolamine	0.3
	Water	79.8

Heat a and b to 75°C in separate containers. Add b to a and stir until the mixture cools to 35°C. This formulation is suitable for packaging in squeeze-type containers.

No. 12

Propylene Glycol Monolaurate D & D	46
Lanolin USP	25
White Mineral Oil (65-75 visc)	25
"Thixcin"	4

Heat all the ingredients except "Thixcin" to 75°C in one container, then add the "Thixcin" with stirring. Allow the mixture to cool to 35°C with continual stirring. Homogenize to develop maximum viscosity.

No.	13	14	15	16	17	18	19	20
"Amerchol" L-101	8.0	9.0	8.0	5.0	8.0	7.00	10.0	8.0
"Modulan"	—	2.0	—	—	12.0	—	—	—
"Acetulan"	0.5	—	1.0	—	—	1.00	—	8.0
Stearic Acid	5.0	2.5	2.5	4.0	2.0	3.50	5.0	2.0
Glycerol Monostearate (neut)	—	1.5	2.0	—	0.5	0.75	—	0.5
Paraffin Wax	1.5	—	—	0.3	—	1.00	—	—
Microcrystalline Wax	—	—	—	—	—	—	—	4.0

(Continued)

	No. 13	14	15	16	17	18	19	20
Mineral Oil (70 visc)	—	—	4.5	—	17.0	—	—	17.0
Beeswax	—	—	—	—	—	—	2.0	—
Ceresin	—	—	—	1.0	—	—	—	—
Petrolatum	3.5	—	—	—	—	1.00	—	—
Glycerol	—	4.5	—	—	4.0	—	5.0	4.0
Water	80.5	75.0	76.5	88.7	56.0	85.00	77.0	56.1
Propylene Glycol	—	—	4.5	—	—	—	—	—
Triethanol-amine	1.0	1.0	1.0	0.5	0.5	0.75	1.0	0.4
Ethanol	—	4.5	—	—	—	—	—	—
"Veegum"	—	—	—	0.5	—	—	—	—
Preservative and Perfume	q.s.	q.s.	q.s.	q.s.	q.s.	q.s.	q.s.	q.s.
Consistency	thin	thin	med	med	heavy	heavy	very heavy	med

Melt together all fat-soluble materials in one kettle and heat to 90°C.

Where Veegum is called for, add this to the water while agitating until thoroughly dispersed. Add remaining water-soluble materials (except the ethanol, perfume, and preservative) and heat to 90°C.

Add the aqueous solution to the oil solution while stirring at a rate that will not cause the inclusion of air. Continue mixing to 40°C, add the ethanol where called for, and the preservative and perfume. Continue mixing until cool; remix the next day and pack.

No. 21

"Lanoile"	5.0	Isopropyl Myristate	3.5
Propylene Glycol Stearate	4.0	Preservative	0.5
		Water	80.0
Stearic Acid (Triple Pressed)	3.0	Triethanolamine	1.0
		Glycerol	3.0
		Perfume	q.s.

Melt oils and waxes separately at 80°C; heat aqueous solution to 85°C. Add aqueous solution to the oil phase and keep the temperature for about 1 hour in order to saponify. During this time mix slowly and then faster in order to obtain a smooth emulsion. Perfume at 50°C.

No. 22
(No Mineral Oil)

"Lantrol"	3.0
"Tegin" P	5.0
"Tegosept" M	0.2
"Tegosept" P	0.1
Cetyl Alcohol	1.0
Oleic Acid	2.5
Water	82.2
Propylene Glycol	5.0
Triethanolamine	1.0
Perfume	q.s.

Heat water phase and oil phase separately to about 75°C. Add water phase to oil phase with rapid agitation. Stir down to about 40°C; add perfume.

No. 23

Glyceryl Mono-stearate	10.0
Stearic Acid	5.0

"Robane"	12.5
Lanolin	2.5
Sodium Lauryl Sulfate	1.0
Propylene Glycol	12.5
Propylparaben	0.5
Water	195.0
Perfume	q.s.

No. 24

a	"Arlacel" 165 Acid-Stable g.m.s.	5
	Cetyl Alcohol	5
b	"Sorbo"	5
	Water	85
	Preservative	q.s.
	Mild Acid to adjust pH	q.s.

Heat a to 70°C, b to 72°C; add b to a with agitation and stir until cool.

No. 25

a	Mineral Oil	10
	"Arlacel" 165 Acid-Stable g.m.s.	10
b	Water	80
	Preservative	q.s.
	Mild Acid to adjust pH	q.s.

Heat a to 60°C, b to 62°C; add b to a with agitation and stir until cool.

No. 26
(Acid O/W)

"Amerchol" L-101	6.0
"Modulan"	2.0
Stearic Acid	2.0
Glycerol Mono-stearate (Neut.)	2.0
Propylene Glycol	5.0
Sodium Lauryl Sulfate	1.0
"Veegum"	0.5
Water	81.5
Preservative and Perfume	q.s.

Add the water containing the water-soluble ingredients to the melted oils, with the temperature of both phases at 85°C. Mix until cool and remix the following day.

(Aerosol)

	No. 27	No. 28
"Modulan"	1.5	0.2
"Acetulan"	1.0	—
Stearic Acid, TP	5.0	5.5
Myristic Acid	—	1.5
Cetyl Alcohol	1.0	0.5
"Deltyl" Extra	0.5	1.0
Triethanol-amine	2.0	3.5
Propylene Glycol	4.0	5.0
PVP, Type NP-K30	0.3	0.3
Perfume	0.5	0.5
Distilled Water	84.2	82.0

Concentrates, 92.0% by wt; Genetrons 114A/12 (43:57), 8.0% by wt.

170 gm net fill in 6-oz shaving cream can.

No. 29
(Aerosol)

a	Stearic Acid (Triple pressed)	5.0
	Cetyl Alcohol	0.5
	Stearyl Alcohol	0.5
	"Acetulan" (Isopropyl Myristate)	1.5
	Isopropyl Linoleate	0.5
b	Triethanolamine	2.0
	Propylene Glycol	4.0
	PVP K-30	0.3
	Perfume	0.5
	Water	83.7

Add the oil phase to water phase at 80°C with stirring. Charge 92% of the concentrate with "Genetron" propellents 12/114a (57/43), 8.0% by wt.

No. 30
(Silicone Water-Repellent)

a Mineral Oil | 35.0
Lanolin Anhydrous | 1.0
"Lanette" Wax SX | 4.0
Dimethylpolysiloxane (350 cps) | 5.0
b Water | 54.9
Methyl-p-Hydroxy Benzoate | 0.1

Heat the two phases separately to 65°C, add b slowly to a with stirring; stir to room temperature, homogenize.

No. 31
(Silicone Barrier)

a "Dow" Silicone Fluid 555 | 5.00
Stearic Acid | 5.00
"Tween" 20 | 4.50
"Span" 20 | 3.50
b "Veegum" | 1.75
Water | 80.00
c "Vancide" BN | 0.25

Heat to 75°C. Add the "Veegum" to the water slowly, agitating continually until smooth and heat to 70°C. Add a to b and mix until cool. Add a small portion of a-b to c, triturating well. Add the rest of a-b and mix well.

No. 32
(Cationic, Antiseptic)

Oil Phase:

"Amerchol L-101 | 6.0
"Solulan" 16 | 1.0
"Modulan" | 1.0
Cetyl Alcohol | 1.5
"Hyamine" 10X | 0.1
"Emcol" E-607S | 0.6
"Propylparaben" | 0.125

Water Phase:

Propylene Glycol | 6.0
Water | 72.425

Salt Solution:

Sodium Chloride | 0.125
Sodium Benzoate | 0.125
Water | 11.0
Perfume | q.s.

Add the water phase at 70°C to the oil phase at 70°C while mixing. Add the salt solution, mix while cooling to 40°C, add the perfume. Continue mixing and cool to room temperature.

Antiseptic Massage Lotion

"G-11"	0.5
Diethylene Glycol	
Monostearate "C"	2.0
Stearic Acid T.P.	2.0
Cetyl Alcohol N. F.	0.5
Deltyl Extra	10.0
Lanolin U.S.P.	1.0
"Trisamine"	1.0
Water	82.5
Perfume Oil	0.5

The results of tests show that the cream and lotion containing G-11 are highly effective against the antibiotic-resistant staphylococci.

Astringent Lotion
No. 1

Sage Tincture	5.0
Hamamelis Extract	25.0
Borax	5.0
Glycerol	80.0
96% Alcohol	347.5
Perfume and	
Water	537.5

No. 2

Boric Acid	4
Glycerol	1
Eau de Cologne	15
Dilute Alcohol	80

No. 3

Glacial Acetic Acid	40
Rose Water	100
96% Alcohol	300
Perfume	5

No. 4

Alcohol	425
Distilled Water	415
Borax	5
Glycerol	150
Perfume	5

No. 5

Gum Arabic	37.00
Dextrin	2.00
Boric Acid	0.50
Alcohol	9.00
Glycerol	1.00
Cetyl Alcohol	2.00
p-Hydroxymethyl	
Benzoate	0.15
Camphor	0.20
Lavender Oil	0.30
Menthol	0.10-0.20
Distilled Water	
To make	100.00

The presence of dextrin and starch will appreciably increase the contracting action. Face packs and face masks generally include these substances. The whole mode of action should be absorptive as well as astringent.

Alcohol, cold water, and other substances are apparent or artificial astringents. These differ from the true astringents, because they do not react chemically with the proteins in the skin. It is well known that the application of cold water promotes the contracting of the pores and enhances the resistance of the skin. Alcohol is much more effective than water because it increases the rate of evaporation of aqueous solutions and the subsequent cooling has a special astringent effect.

Skin Astringent

Tincture of Benzoin	2.5 oz
Denatured Alcohol No. 40	1.0 pt
Orange Blossom Perfume	0.5 oz
Water	To make 1 gal

Beauty Milk

"Polawax"	8.0
Water	90.0
Arachis (Peanut) Oil	5.0

Cholesterol	0.5
Liquid Paraffin	10.0

Hazy, Colloidal Lanolin Lotion

"Acetulan"	24.5
"Modulan"	45.0
Mineral Oil (200 visc)	30.5

Heat all the materials together until clear. Mix thoroughly and allow to cool to room temperature.

This is a stable, golden, hazy preparation of lanolin product in mineral oil.

Soluble Lanolin

U.S. Patent 2,900,307

Lanolin	0.5
Coconut Diethylamide	5.0

This dissolves in:

Water	90.0

to give a clear product.

Soothing Lotion

Olive Oil	1.00
Lanolin	0.50
Light Mineral Oil	1.50
Menthol	0.30

Glycerol Mono-stearate	10.00
"Tween" 80	1.50
"Tween" 20	1.50
Methyl p-Hydroxy-benzoate	0.10
Propyl p-Hydroxy-benzoate	0.12
Lilac Perfume	To suit
Water To make	100.00

On a water bath, melt and mix the ingredients except the menthol, water, propyl p-hydroxybenzoate, and perfume. Then add the menthol, which dissolves readily. Now add slowly, with constant agitation, a small portion at a time, hot water with the methyl p-hydroxybenzoate dissolved in it, until the first-formed water-in-oil emulsion inverts to oil-in-water. Then add the remaining water, hot, rapidly to volume, with agitation. When cooled down somewhat, stir in the perfume.

The ingredients can be varied slightly to individual likes and dislikes. The finished product can be modified by reheating on a water bath, more of melted oil-phase ingredients or more hot water

added, stirred, and then allowed to cool.

Lilac perfume blends well with the predominant menthol odor, but other scenting agents may also be used.

Antiseptics can be incorporated by dissolving them in either the oil or the water phase, depending on solubility. For example, 1% of hexachlorophene ("G 11," Sindar) can be dissolved in the oil phase before emulsification.

Emulsification occurs very easily so that manual stirring is sufficient except in manufacture on commercial-scale.

Temperature control is not strict. The oil-phase ingredients are kept on the water bath to the point of melting and dissolving and the water is heated to approximately the same temperature.

Baby Oil
Formula No. 1

Mineral Oil USP	90
XXX Lanolin	3
Ethyl Stearate	4
Isopropyl Myristate	3

Warm the mixture until the

lanolin is completely dissolved. Mix well and package.

No. 2

Mineral Oil	
(65-75 visc)	79
Olive Oil	20
Lanolin	1
Antioxidant	To suit

No. 3

Olive Oil	40
Sesame Oil	30
Avocado Oil	20
Isopropyl Palmitate	10

No. 4
(Antiseptic)

Mineral Oil	75.0
Isopropyl Palmitate	21.9
Lanolin	3.0
Hexachlorophene	0.1

Warm the isopropyl palmitate and lanolin to complete solution. Add the hexachlorophene and stir until it is dissolved. Pour this mixture into the mineral oil and agitate until uniform.

Baby-Skin Lotion

Sodium Alginate	3.5
Butyl p-Hydroxy-	
benzoate	0.1
Methyl p-Hydroxy-	
benzoate	0.9
Distilled Water	670.0
Triethanolamine	5.0
Stearic Acid	9.0
Stearyl Alcohol	9.0
Cetyl Alcohol	5.0
Anhydrous Lanolin	10.0
Light Liquid	
Petrolatum	320.0

Dissolve the preservatives in hot distilled water. Transfer this solution into a continuous mixer and dust on the sodium alginate while mixing. Add the triethanolamine and stir until the liquid cools to 70°C. Heat the stearic acid, stearyl alcohol, cetyl alcohol, lanolin, and petrolatum together until melted, and cool to 70°C. With the temperature of both liquids at 70°C, slowly pour the oil phase into the water phase while agitating continuously with a mixer. Stir until cool.

Superfatted Baby Powder

Talc	67
Glyceryol Mono-	
stearate	2

Cetyl Alcohol	2
Magnesium Stearate	6
Kaolin	18
Boric Acid	5

Sunscreen Preparations
Formula No. 1
(Clear Type)

Isopropanol	56.0
Water	36.5
Isopropyl Myristate	5.0
Isobutyl p-Amino-benzoate	2.0
"Carbopol" 934	0.5
Diisopropanol-amine	To pH 6

Dissolve the "Carbopol" in the water at room temperature. Add the remaining ingredients, except diisopropanolamine, and mix well. Add diisopropanolamine with agitation to pH 6.

No. 2
(Cream Type)

a Cetyl Alcohol	1.0
Butyl Stearate	1.25
Stearic Acid	4.0
Preservative	0.1
Isobutyl p-Amino-benzoate	2.0

| b Potassium Hydroxide | 0.15 |
| Water | 91.0 |

Heat parts a and b, in separate containers, to 70°C. Add b to a with agitation. Stir until the mixture cools to 35°C. This formulation is suitable for packaging in squeeze-type containers.

No. 3

"Acetulan" (or "Modulan")	5.0
Sunscreen	2.0
Isopropyl Myristate	43.0
Mineral Oil	50.0

Where the "Modulan" is used, heat slightly in the mineral oil until clear. Remove from heat, add the remaining ingredients, and mix cold.

This formula is a typical anhydrous solution which remains brilliantly clear.

Sun-Tan Cream
Formula No. 1

a Mineral Oil	43.5
Beeswax	10.0
Anhydrous Lanolin	8.0
"Filtrosol" A	5.0
Cetyl Alcohol	4.0

Preservative	To suit
b Borax	1.0
Distilled Water	28.0
c Perfume	0.5

Heat *a* to 60°C and slowly add solution *b*, heated to the same temperature, to *a* with constant stirring. Cool to 50°C, add *c*, and homogenize the cream.

No. 2

Ethyl Amino-	
benzoate	30.0
Stearic Acid	63.0
Triethanolamine	8.0
Stearyl Alcohol	45.0
"Carbowax" 1540	30.0
Titanium Dioxide	10.0
Neocalamine	10.0
Brown Ferric Oxide	1.5
Methylparaben	2.0
Coumarin	0.5
Glycerol	100.0
Distilled Water	
To make	1000.0

No. 3

Paraffin Oil	230
Beeswax, White	10
Sorbitan Mono-	
stearate	20
Sorbitan Sesquioleate	17
POE-Sorbitan	
Monostearate	17

Oil-Soluble	
"Prosolal"	30
"Extrapone" VC	5
Water-Soluble	
"Prosolal" WL	70
p-Hydroxybenzoic	
Methyl Ester	2
Distilled Water	594
Perfume Oil	5

No. 4

Stearic Acid,	
Triple Pressed	70
Anhydrous Lanolin	5
Sorbitan Monooleate	5
Polyoxyethylene	
Sorbitan	
Monostearate	25
Oil-Soluble	
"Prosolal"	50
Water-Soluble	
"Prosolal" WL	50
"Nipagin" M	2
Distilled Water	788
Perfume Oil	5

No. 5

Paraffin Oil	400.0
Isopropyl	
Myristate	100.0
Beeswax, White	70.0
Polyoxyethylene	
Sorbitol Beeswax	
Compound 1726	80.0
POE-Sorbitan-	
Monopalmitate	20.0

Oil-Soluble
"Prosolal"	50.0
"Nipagin" M	1.8
Distilled Water	273.2
Perfume Oil	5.0

These emulsions are prepared by separately heating the fatty phase and the aqueous phase to about 75°C. The aqueous phase is then slowly stirred into the fatty phase, and stirring is continued until room temperature is reached. Perfume is added at 40°C.

No. 6

Propylene Glycol	50
"Extrapone" 1, Special	5
"Extrapone" Tormentilla, Special	5
"Extrapone" Camomile, Special	5
Water-Soluble "Prosolal" WL	150
"Nipagin" M	2
Distilled Water	773
Perfume Oil	10

No. 7

Alcohol (95%)	300.0
"Extrapone" Hamamelis	

Distilled Colorless, Special 20.0

"Extrapone" Sage, Special	10.0
"Extrapone" Tormentilla, Special	10.0
Water-Soluble "Prosolal" WL	150.0
p-Hydroxybenzoic Methyl Ester	1.5
Distilled Water	503.5
Perfume Oil	5.0

No. 8

Glycerol Monostearate	130
Isopropyl Palmitate	20
Anhydrous Lanolin	50
Oil-Soluble "Prosolal"	50
Silicone Oil	50
Propylene Glycol	50
"Nipagin" M	2
Distilled Water	645
Perfume Oil	3

No. 9
(Gel)

Isopropanol	52.0
Water	34.8
Polyethylene Glycol 600 Distearate	4.0
Isopropyl Myristate	4.0

Isobutyl *p*-Amino-
  benzoate   2.0
"Carbopol" 934   3.2
Diisopropanol-
  amine   To *p*H 6

Dissolve the "Carbopol" in
the water at room temperature. Add the remaining ingredients except diisopropanolamine, warm to 45°C and
mix well. Add diisopropanolamine with agitation to *p*H
4.5. Avoid any excess of amine.
Allow the mixture to cool to
35°C with gentle stirring.

No. 10
(Jelly)

Methylcellulose
  M 450   15
Distilled Water   825
Water-Soluble
  "Prosolal" WL   100
Glycerol   50
"Nipagin" M   2
Perfume Oil   8

Sun-Screening Oils
Formula No. 1

Paraffin Oil   500
Isopropyl Myristate 440
Oil-Soluble
  "Prosolal"   50
"Extrapone" VC   5
Perfume Oil   5

No. 2

Olive Oil, Preserved 400
Peanut Oil,
  Preserved   200
Walnut Oil,
  Preserved   200
Isopropyl Myristate 145
Oil-Soluble
  "Prosolal"   50
Perfume Oil   5

No. 3

Paraffin Oil   300
Olive Oil Preserved 200
Isopropyl Myristate 440
Oil-Soluble
  "Prosolal"   50
"Extrapone" VC   5
Perfume Oil   5

No. 4

Isopropyl Myristate 500
Paraffin Oil   435
Oil-Soluble
  "Prosolal"   50
"Extrapone" VC   10
Perfume Oil   5

No. 5

Propylene Glycol
  Monolaurate   160
Dipropylene Glycol
  Salicylate   40
"Robane"   100

Heavy Mineral Oil   700
Perfume and Color

No. 6
(Aerosol)

"Prosolal,"
Oil-Soluble   3.0
Paraffin Oil   17.0
Isopropyl Myristate 9.7
Perfume Oil   0.3
Frigene 11-12 (1:1) 70.0

Skin Tan, Quick

U.S. Patent 2,949,403

Alcohol   50
Acetone   1
Water   45
Didyhroxyacetone   4
Perfume   To suit

Skin Darkening Lotion

Dihydroxyacetone   3.0
"Polawax" G.P 200
  (Nonionic
  Emulsifier)   8.0
Menthyl Salicylate   10.0
Glycerol   2.5
Alcohol (660 P)   33.0
Water, Distilled   100.0

Lipsticks
Formula No. 1

The pouring, cooling, and finishing of lipsticks involves many tricks, expedients and manual skills, and they are not easily described. The temperature at which the excipient should be poured into molds varies, as it depends on the melting point. The molds must be absolutely clean; this is of far greater importance than the old question of whether the mold should be lubricated with paraffin oil or silicone oil, or whether it is better to sprinkle them lightly with talc (a practice which has not been successful at all).

"Flaming" of lipsticks in order to efface spots and fingerprints is done mechanically in the bigger plants: placed on sockets, the lipsticks travel along moving belts, slowly turning on their own axis, through an electric arc whose heat radiation is adjusted to the melting point of the lipsticks. In all positions the distance from the heat sources inside the electric arc at which they are placed is about the same. By this means a uniform surface is obtained.

Modern lipstick molds make it possible to place the cases upon the lipsticks without touching them with the hands.

Ozocerite (70-72°C)	130
Petrolatum First grade, slightly stringy	40
Glycerol Monomyristate	50
Cetyl Alcohol	50
Anhydrous Lanolin	60
Spermaceti	30
Hydrogenated, Hardened peanut oil	30
White Beeswax	40
Carnauba Wax	30
Glycerol Monostearate, (Not self-emuls.)	30
Ceresin (54-56°C)	50
Cacao Butter	20
"Loramin" Wax, OM 101	150
Castor Oil	100
Isopropyl Myristate	60
Tetrabromofluorescein	20
Color Lakes	95
Rose 06506 (Perfume)	20
p-Oxybenzoic Methyl Ester	1

Dissolve the bromo acids in the melted fat mixture of "Loramin" wax, castor oil, and isopropyl myristate, and add to the melted fats. When the mass has solidified, homogenize on the roller mill. The perfume oil, which is to aromatize the lipstick well, is added shortly before the melted excipient is poured. The aroma should amount to at least 1.5% in lipsticks. Raspberry, violet, and special rose notes are excellent for the purpose.

No. 2

Ozocerite (70-72°C)	190
Petrolatum, 1° grade, slightly stringy	80
Glycerol Monomyristate	110
Anhydrous Lanolin	70
Spermaceti	60
Ceresin (54-56°C)	60
Propylene Glycol Monomyristate	30
Carnauba Wax, Light	30
Stearone	20
Beeswax	20
Silicone Oil, AK 50	10
Stabilized Castor Oil	132
Isopropyl Myristate	40
Color Lakes	68
Tetrabromofluorescein	20
Tetrahydrofurfuryl Alcohol	40
Peach 06755 (Perfume)	20

A few drops of azulene does not impair the color of the excipient and it can help to prevent irritation of the lips.

No. 3

Castor Oil	36
Glycerol Mono-stearate (not self-emuls.)	45
Mineral Oil	6
Petrolatum USP	4
Carnauba Wax	4
Anhydrous Lanolin	3
Bromo Acids	1
Color Lakes and Pigments	9
"Fantasy" GV 58 Perfume Oil	2

No. 4
(Creamy)

Castor Oil	65.0
Lanolin	10.0
Isopropyl Myristate	5.0
Beeswax	7.0
Candelilla Wax	7.0
Ozocerite	3.0
p-Hydroxybenzoic Acid Propyl Ester	0.2
Halogenated Fluoresceins	3.0
Color Lakes and Pigments	12.0
Perfume	To suit

No. 5
(High Stain)

Castor Oil	60.0
Propylene Glycol Monoricinoleate	10.0
Lanolin	5.0
Polyethylene Glycol 400	5.0
Beeswax	7.0
Candelilla Wax	7.0
Ozocerite	3.0
p-Hydroxybenzoic Acid Propyl Ester	0.2
Halogenated Fluoresceins	3.0
Color Lakes and Pigments	12.0

No. 6

Carnauba Wax	70
Beeswax, White	160
Isopropyl Palmitate	150
Ceresin, White	250
Petrolatum White	200
Paraffin Oil	40
Eosin Color Acid	30
Color Lakes	100

No. 7

Carnauba Wax	40
Castor Oil	350
Isopropyl Myristate	90
White Petrolatum	40
Glyceryl Mono-	

stearate (not
emuls.)   410
Eosin Color Acid   20
Color Lakes   50

No. 8

Beeswax	360
Paraffin Oil	60
Anhydrous Lanolin	40
Cetyl Alcohol	20
Palmoil, Hardened	230
Isopropyl Myristate	100
Diethyl Sebacate	100
Eosin Acid	20
Color Lakes	140
Titanium Dioxide	20
Antioxidants	1

No. 9

Stearyl Alcohol	7.00
White Beeswax	7.00
Stearic Acid	1.75
Paraffin Wax (162°F)	12.25
Wool Fat, Anhydrous	2.80
Carnauba Wax	2.80
Liquid Paraffin	1.40
"Comperlan" HS	20.00
"Eutanol" G	45.00
Eosinic Acid	1.50
Pigment	6.00
Perfume	As desired

The eosinic acid is dissolved in the "Comperlan" HS in a water-jacketed vessel at about 203°F. The pigment coloring is dissolved in the "Eutanol" G. The remaining fatty substances and waxes are melted in a water-jacketed vessel and then combined with the other preparation. The finished mass is passed through a roller mill and then poured into molds. The finished sticks should be passed quickly through a small flame.

No. 10

"Lantox" 110	7.0
Beeswax	8.5
Lanolin	3.0
Carnauba Wax	2.0
Candelilla Wax	5.0
Paraffin Wax	2.7
Castor Oil	57.0
Isopropyl Myristate	5.5
Bromo Acid	2.5
Lipstick Lake Colors	6.8
Perfume	q.s.

Heat the waxes to 80-90°C separately. Grind bromo acid in oils. Next day add the waxes to the oil in melted form and bring the temperature to 80°C. Add lipstick color and perfume. Grind finally, cool, and mold as required.

No. 11

U.S. Patent 2,548,970

Isopropanol	60-90
Gum Benzoin	1-12
Cetyl Alcohol	1-12
Sulfonated Castor Oil	1-12
Lecithin	3-20
Dye	0.05-3
Perfume	0.03-2

No. 12

Ethanol	93.0
Ethyl Cellulose	3.5
Oleic Acid Mono- or Diglyceride	3.5

Add 5% of a dye to this solution.

Lipstick Base

Castor Oil or Bromo Acid Solvent	30
Mineral Oil	15
Beeswax	15
Paraffin Wax	10
Carnauba Wax	10
Ceresin Wax	10
General Electric Silicone Fluid SF-96I (1000 centistokes)	10
Perfume-Flavor Compound	To suit

Chapped-Lip Stick

a	Beeswax	35.0
	Mineral Oil	50.4
	Petrolatum	10.0
	Silicone Oil	2.0
b	"Vancide" BN	0.1
	Water	1.0
c	"Emulphor" EL-719	1.0
d	"Truodor" Jasmin #6 (Perfume)	0.5

Blend *a*. Add *c* to *b*; add to *a*. Heat mixture to 90°C and hold this temperature for about five minutes with continued mixing. Cool to 70°C and add *d* with mixing. Continue cooling. Pour into molds and allow to harden.

Liquid Lip-Rouge

U.S. Patent 2,230,063

Formula No. 1

Ethyl Cellulose	3.5
Ethanol	90.7
Castor Oil	5.0
Eosine	0.8

No. 2

Ethyl Cellulose	3.0
Bleached, Wax-Free Shellac	2.5

Hydrogenated		
Methyl Abietate	7.5	
Petroleum Ether	11.0	
Ethanol	77.0	
Oil-Red O	1.0	

No. 3
(Water-Base)

a	Lanolin	6.0
	Petrolatum	6.0
	Isopropyl Myristate	3.0
	D&C Red #35	1.0
	Titanium Dioxide	1.0
	"Nytal" 400	7.5
b	Oleic Acid	2.0
	Carnauba Wax	2.0
	Beeswax	4.0
c	"Veegum"	0.5
	Water	9.5
d	"CMC" (low visc)	0.2
	Water	19.9
e	Propylene Glycol	5.0
	Water	31.0
	"Darvan" #1	0.3
f	Morpholine	0.8
	"Truodor" Jasmin #6	
	(Perfume)	0.3
	Preservative	q.s.

Mill a, add to b; heat to 70°C. Add the "Veegum" to the water slowly, agitating continually until smooth. Disperse the CMC in water; add to "Veegum." Add e; heat to 60-65°C. Add f to preceding, then immediately emulsify, adding a-b with constant stirring. Homogenize. Cool and add perfume.

This is a O/W emulsion of medium viscosity with the pigments and other solids ground into the inner phase. The "Veegum" plays an important part in stabilizing the emulsion. The formula is designed to be somewhat waterproof with a matte finish.

Nail-Polish Remover
Formula No. 1

Castor Oil	3
Industrial Methylated	
Spirit	25
Ethyl Acetate	20
Butyl Acetate	15
Butyl Alcohol	7
Amyl Acetate	15
Acetone	15

No. 2

Butyl Stearate	4
Isopropyl Myristate	2
Diethylene Glycol	
Monoethyl Ether	18
Acetone	79

No. 3

British Patent 683,149

Triethanolamine Stearate	6.5
Methyl Ethyl Ketone	25.0
Water	4.0

Warm together and mix until uniform.

No. 4

"Ceroxin"	15
"Emulgade" F	5
Acetone	50
Amyl Acetate	10
Water	20
Preservative and Perfume	As desired

"Ceroxin" and "Emulgade" F are melted in a water-jacketed vessel at approximately 158-176°F. The water is likewise heated to 158-176°F, and then stirred into the fatty components. When this has cooled to about 86°F, the mixture of acetone and amyl acetate is gradually added.

No. 5

German Patent 1,026,919

Cellulose Acetate	10
Diethylene Glycol	10
Phenyl Salicylate	10
Acetamide	2
Methanol	23
Ethylene Dichloride	45

This is coated on nails; dried well, and stripped off. This removes old polish.

Perfume Oil Solubilizer Formula No. 1

Monoethanolamine Laurate	2
80% Potassium Tallate Soap	2
Isopropanol	1
Water	10

This will solubilize 5 parts of certain aromatic substances, e.g., anisic aldehyde or bromostyrol.

No. 2

Petroleum Sulfonate	2
80% Potassium Tall-Oil Soap	2
Isopropanol	1
Water	10

This will solubilize 5 parts of amyl cinnamic aldehyde, amyl salicylate, benzaldehyde, citronellal, ionone, methyl salicylate, safrol, terpineol, etc.

No. 3

Light Mineral Oil	88
Triethanolamine	3.8-4.0
White Oleic Acid	8-10
Perfume Oil	To Suit

Oil-in-Water
Cream Sachet

a	"Tegacid" Special	15.0
	Spermaceti	5.0
	Titanium Dioxide*	4.0
	Talc*	15.0
	Preservative	0.1
	Water	48.9
b	"Atlas" Emulsifier	
	G2160	2.0
	Perfume Oil	10.0

* Add last

Heat *a* to 76°C. Warm *b* and add to *a*; stir until cold. Mill and pack cold.

As sachet powders and creams contain much more perfume than ordinary cosmetic powders and creams, they are more likely to produce allergic reactions in sensitive individuals. It is important, therefore, to avoid the use of perfume oils containing ingredients that are known to have irritating or sensitizing properties.

Perfumes for Soap
Amber Royal

Ambropur Dragoco	6
Ambrofix	50
Amber synth. liq.	
1049	170
Ambracret 8232	64
Mousse de Perse	164
Chypre 8267	120
Rose de Mai	
Dragoco 2000	30
Bulgaryol III	60
Sandalwood Oil, E. I.	40
Santalol	10
Carnatin 1045	20
Iromuskon DP	10
Extractol Castoreum	24
Jasmonon (decol.)	38
Methyl Iridon 100%	10
Vanillin	80
Coumarin	10
Heliotropine	10
Vetivert Oil	34
Labdanum Extract,	
decol. "Standard"	30
Tolu-Extract	
"Standard"	20

Azalea

Citronellol	100
Benzyl Phenyl-acetate	100
Sandalwood Oil, E.I.	100
Terpineol	100
Hydroxycitronellal Dimethyl Acetal	100
Orange Oil	50
Phenylethyl Alcohol	50
Phantolide	50
Linalool *ex* Bois de Rose	50
Acetanisole	50
Indoflor (H & R)	50
Moskène	30
Cananga Oil, Java	30
Trichlormethyl Carbinyl Acetate	30
Guaiacwood Oil	30
Benzyl Isoeugenol	20
Diphenyl Methane	20
Anise Alcohol	20
Ethyl Laurate	10
Methyl Amyl Ketone	10

Burmatia

Menthanyl Acetate	200
Lavender Oil	200
Cedarwood Oil	100
Coumarin	80
Phenylacetaldehyde Dimethyl Acetal	60
Lilial	60

Phantolide	50
Petitgrain Oil (Paraguay)	40
Heliopal	40
Peppermint Oil	30
Vetiver Oil (Bourbon)	30
Moskène	30
Clove Oil	20
Sauge Sclarée (Robertet)	20
Cassia Aldehyde (H & R)	20
Patchouly Oil	20

Cavenia

Diheptyl Acetate	250
Geranium Oil (Syn.)	200
Orange Oil, California	140
Linaloe Oil, Brazil	140
Heliocrete	85
Bromelia	50
Phenylacetaldehyde Dimethyl Acetal	45
Phantolide	30
Moskène	30
Vetiver Oil, (Bourbon)	25
Patchouly Oil	5

Coptical

Diheptyl Acetate	200
Citronellol	100

Linalool *ex* Bois de
   Rose 100
Sandalwood Oil,
   Australia 80
Heliopal-S 80
Lilial 80
Cananga Oil, Java 70
Cedarwood Oil
   (Florida) 40
Cinammic Alcohol 40
Phantolide 40
Musk Xylene 40
Patchouly Oil 40
Petitgrain Oil
   (Paraguay) 40
Vetiver Oil, Bourbon 20
Phenylacetaldehyde
   Dimethyl Acetal 10
Coumarin 10
Cedrol 10

Fleurs de Chine

Patchouly Oil 150
Geraniol 100
α Isopropyl
   Muguetton 100
Cananga Oil, Java 80
Geranium Oil
   Bourbon 80
Musk Ketone 60
Parmiron 50
Vetiver Oil, Bourbon 50
Sandalwood Oil,
   Australia (Plaimar) 50

Mousse de Chêne,
   decol. (EMA) 40
Cinnamyl Cinnamate 40
Cassia Oil 40
Labdanum Resinoid 40
Phantolide 40
Coumarin 20
Styrax Resinoid 20
Cypress Oil 15
Santalyl Phenyl-
   acetate 15
Resinoid Castoreum 10

Ivoralia

Geranium Oil
   (Synth.) 250
Citroviol 130
Linaloe Oil (Brazil) 120
Cassia Oil 60
Phenylacetaldehyde
   Dimethyl Acetal 50
Benzyl Isoeugenol 50
Petitgrain Oil
   (Paraguay) 50
Sandalwood Oil
   (W.I.) 40
Vetiver Oil, Bourbon 40
Butylphenyl Acetate 40
Moskène 40
Trichloromethyl
   Carbinyl Acetate 30
Almond Fixative 30
Phantolide 20
Patchouly Oil 10

Sauge Sclarée	10
Santalyl Phenyl-acetate	10

Majunal

Geraniol, Pure	200
Linaloe Oil (Brazil)	150
Phenylethyl Alcohol	100
Jonon-S	90
Iso-Bergamot	80
Indoflor	50
Acetisoeugenol	50
Lactoscaton	50
Sandalwood Oil (E.I.)	35
Geranium Oil, Bourbon	35
Jacinthol-S	35
Vetiver Oil, Bourbon	25
Musk Xylene	25
Cinnamic Alcohol	25
Patchouly Oil	20
Aurantiol	20
Coumarin	10

Pengari

Sandalwood Oil (E.I.)	200
Clove Oil	100
Geraniol, Pure	100
Cedarwood Oil, Florida	100
Iso-Bergamot	80
Patchouly Oil	80

Vetiver Oil, Bourbon	50
Rosottone-Savon	50
Geranium Oil (Synth.)	50
Musk Xylene	50
Lavandozon	40
Cedrol	40
Mousse-S, (decol.)	30
Coumarin	20
Cassia Aldehyde, (Spez.)	10

Sumatral

Patchouly Oil	215
Citroviol	125
α-Isopropyl Muguetton	100
Sandalwood Oil, Australia, (Plaimar)	100
Coumarin	100
Isobutyl Phenyl-acetate	65
Phenylacetaldehyde Dimethyl Acetal	50
Vetiver Oil, Bourbon	45
Parmiron	45
Musk Ambrette	30
Dimethyl Hydroquinone	25
Mace Oil	25
Bay Oil	25
Methyl Phenyl-acetate	20
Bromelia	15

Eugenol Methyl Ether	10	Citronellol	75
Mousse de Chêne (decol.)	5	Citronal-S	75
		Petitgrain Oil, (Paraguay)	60
Tarrazal		Orange Oil Exchange	50
		Coumarin	50
Citroviol	200	Mousse de Chêne abs. (Yugosl.)	35
Opoponax Oil	150	Patchouly Oil	30
Clove Oil	80	Vetyrisia	30
Diphenylmethane	60	Labdanum Resinoid	30
Lavandozon	50	Benzoin Siam Resinoid	30
Orange Oil, California	50	Cinnamic Aldehyde	30
Cinnamic Alcohol	50	Terpinyl Acetate	30
Trichloromethyl Carbinyl Acetate	50	Guaiac-wood Oil	30
Petitgrain Oil (Paraguay)	40	Versalide	30
Patchouly Oil	40	Cinnamic Alcohol	25
Aurantiol	40	Diheptyl Acetate	25
Geranium Oil (African)	40	Geranium Oil, Bourbon	25
Moskène	30	Bornyl Acetate, cryst.	20
Musk Xylene	30	Citral Acetal	19
Musk Ambrette	25	Spike Oil (Spanish)	15
Resinoid Opoponax	20	Mace Oil	15
Sauge Sclarée	20	Linaloe Oil (Cayenne)	10
Mousse-S, decol.	20	Ethyl Laurate	9
Resinoid Tonka	5	Ethyl Vanillin	8
		Benzyl Acetate	5
Mirana		Trichlorrosenkorper (Roseacetol)	2.5
Geraniol	160	Methyl Amyl Ketone	1.5
Methylionone (100%)	75		

Spike Lavender

Lavender synth., 9863	300
Lavandin	100
Spike Oil, Spanish	100
Gurjun Balsam Oil	50
Peppermint Terpenes,1815	30
Oil of Camomile	20
Oil of Juniper Berry	20
Oil of Spearmint	20
Oil of Rosemary	20
Oil of Sage	20
Oil of Thyme	20
Melonia	20
Linalyl Acetate	50
Coumarin	30
Oil of Patchouly, Singapore	10
Oil of Origanum, French	10
Musk Xylene	50
Mousse Ambrée	30
Benzyl Benzoate	100

Sandalwood

Cedarwood Oil, Florida IA	250
Oil of Sandalwood (W. I.)	100
Oil of Sandalwood (E. I.)	100
Noval	100
Carnatin S	50

Linalool	50
Centifiol S	50
Oil of Patchouly, Singapore	50
Phenylethyl Alcohol	50
Amyl Salicylate	50
Extract Storax "Standard"	40
Opoponax 3046	30
Vanillin	20
Musk Xylene	40
Cinnamic Alcohol	20

Esmeralia-S

Lavender Oil,MB1 40-42%	135
Opoponax Resinoid	100
Linaloe Oil, Brazil	80
Synalyl	75
Cetonia-S	75
Vetiver Oil,Bourbon	50
Dihydrocarveol	50
Mousse de Chêne-S (decol.)	40
Toncarine	50
Benzyl Acetate	40
Parmiron	40
Diheptyl Acetate	40
Mousse de Chêne absol. (Yugosl.)	35
Cumaryl	35
Labdanum Resinoid	30
Heliopal-S	25
Moskène	20

Sandalwood Oil	
(E.I.)	20
Versalide	15
Isobutylquinolin	15
Patchouly Oil,	
Singapore	15
Geranium Oil,	
Bourbon	15

Floralia

Hydroxycitronellal	
Dimethyl Acetal	100
Geraniol	100
Diheptyl Acetate	100
Parmiron	100
Geranium Oil,	
Bourbon	90
Benzyl Acetate	50
Ylangoil	50
Isobutyl Phenyl-	
acetate	50
Cumaryl	50
α-Amyl Cinnamic	
Aldehyde	50
Citronellal	
Dimethyl Acetal	40
Lactoscatone	40
Acetanisol	30
Versalide	30
Mace Oil	
(Nardenized	30
Benzoin Siam	
Resinoid	30
Coriander Oil,	
(synth.)	20

Ethyl Vanillin	20
Vetyrisia	15
Almond Fixative	5

Fougèral

Citronellol	120
Toncarine	100
Lavender Oil,MB1	
40-42%	100
Lavandin Acetyle	100
Parmiron	100
Vetyrisia	60
Cetonia	50
Petitgrain Oil	
(Paraguay)	50
Cedrenen	50
Diheptyl Acetate	50
Coumarin	40
Orange Oil Exchange	40
Labdanum Resinoid	30
Mousse-S (decol.)	30
Versalide	30
Patchouly Oil	30
Ethyl Vanillin	20

Fougère American

2-Cyclohexyl	
Cyclohexanone	220
Dorisyl	220
Rosettone	200
Nonylacetate	110
Lavender Oil	
(synth.)	50
Musk Xylene	40

4-tert-Butyl	
Cyclohexanone	40
Coumarin	40
Rosaryl	30
Rogepel	30
4-Cyclohexyl	
Cyclohexanone	10
Velvetine	10

La Paloma

Geranium Oil,	
Bourbon	100
Citronellol	100
Diheptyl Acetate	100
Spike Oil, Spanish	80
Parmiron	60
Toncarine	50
Trichlorrosenkorper	50
Labdanum Resinoid	50
Linaloe Oil, Cayenne	50
Moskène	40
Bay Oil,	
(Nardenized)	40
Orange Oil Exchange	30
Heliocrete	30
Citronella Oil, Java	30
Lactoscaton	30
Coriander Oil	30
Patchouly Oil,	
Singapore	30
Petitgrain Oil,	
Paraguay	20
Galbanum Resinoid	20
Cumaryl	20

Ethyl Vanillin	20
Vetyrisia	20

Merinal

Linaloe Oil,	
Cayenne	100
Lavender Oil, MB1.	
40-42%	100
Diheptyl Acetate	80
α-Isopropyl	
Muguetton	70
Phenylethyl Alcohol	50
Drago-Jasimia	50
Parmiron	50
Moskène	40
Cumaryl	40
Sandalwood Oil	
(E.I.)	40
Mousse-S (decol.)	40
Patchouly Oil,	
Singapore	40
Orange Oil Exchange	40
Labdanum Resinoid	40
Heliopal-S	40
Toncarome	35
Geranium Oil,	
Bourbon	30
Vetyrisia	20
Lactoscaton	20
Vetiver Oil, Bourbon	20
Mace Oil	
(Nardenized.)	20
Fixative 404	20
Ethyl Vanillin	15

Muguet

Convial	250
Hydroxycitronellal	180
Jasmonon (decol.)	20
Nerol	20
Heliotropine	20
Linalool	100
Reuniol	50
Phenylethyl Alcohol	30
Ionone	10
Geraniol	50
Citronellol	80
Terpineol	50
Cyclamol	20
Iromuskon DP	50
Jasminet 6019	70

Muguet des Bois

Hydroxycitronellal	40
Rhodinol	10
Linalool	10
Linalyl Acetate	15
Heliotropin	8
Benzyl Acetate	5
α-Amyl Cinnamic Aldehyde	2

Tea-Rose

Algerian Rose Geranium	7.00
Pelargol	7.00
Geraniol	5.00

Phenyl Ethyl Alcohol	15.00
Cinnamic Alcohol	7.50
Diphenyl Ether	2.75
Geranyl Formate	2.50
Guaiac-Wood	10.00
Guaiac-Wood Acetate	1.00
Fixative	3.00

For 100 lb of soap, 14 oz perfume will be found satisfactory.

Jasmin

α-Isopropyl Muguetton	150
Benzyl Acetate	250
Colorless Terpineol	205
Java Cananga Oil	60
Terpenyl Acetate	100
Musk Xylene	50
Linalool	25
Cinnamic Alcohol	50
α-Amyl Cinnamic Aldehyde	50
Paraguay Petitgrain Oil	60

Amber Fixative

Resinodor of Labdanum	150
Resinodor of Tolu	100

Resinodor of Tabacine	100
Superior Fraction of Cypress Oil	100
Natural Civet (Abs.)	25
Resinodor of Ambrette	125
Tongking Musk (Abs.)	25
Coumarin	40
Heliotropin	40
Essence of Sauge Sclarée	90
Ethyl Vanillin	35
Thibetolide	20
Ambrettolide	30
Muscone BRB	20
Jasmin Abs.	50
Rose Abs.	50

Jasmine Perfume Base (Concentrate)

Formula No. 1

Benzyl Acetate	350
Terpineol Extra	200
α-Amyl Cinnamic Aldehyde	200
Hydroxycitronellal	100
Phenyl Ethyl Alcohol	100
Linalol	50
Amyl Salicylate	50
Isoeugenol	50
Olibanum Resinoid	25

Terpinyl Anthranilate	25
Orange Flower Abs.	15
Aldehyde C11	5
γ-Undecalactone	5
Indolflor (or indole)	15
Benzyl Alcohol	0-110
	1190-1300

This is a relatively inexpensive composition of great strength. Rounded off with a little isojasmone and natural jasmin absolute, together with some ylang-ylang absolute, it gives very acceptable results.

No. 2
(Swiss Type)

Hydroxycitronellal	230
Phenyl Ethyl Alcohol	200
Benzyl Acetate	200
Ylang-ylang Oil, III	60
Linalool	70
Aurantiol	40
Linalyl Acetate	40
Terpineol	20
Rhodinol	15
α-Ionone	6
Benzyl Formate	6
α-Amyl Cinnamic Aldehyde	6
Paracresol	3
Benzoin Resinoid	4

γ-Undecalactone	0.5
Ethyl Methylphenyl	
Glycidate	0.7
Aldehyde C8	0.2
Benzyl Alcohol	100
	1001.4

Synthetic Neroli Oil
Formula No. 1

Methyl Anthranilate	200
Petitgrain Oil	200
Phenylethyl	
Alcohol	100
Shiu Oil (or Other	
Source of	
Linalool)	100
Geranium Oil	
(synth.)	100
Terpineol	75
Linalyl Acetate	50
Citronellol	50
Hydroxycitronellal	50
Pinene	20
Orange Oil	25
Phenylacetic Acid	20
Cananga Oil	5
Aldehydes: C9, C10,	
C11, C12 (at 50%	
strength)	5

It is absolutely necessary to leave this sort of compound for at least a month after blending—and preferably for two months.

No. 2

Linalool (from	
bergamot or	
petitgrain)	355.0
Petitgrain	
Bigarade	
(rectif.)	200.0
Phenylethyl	
Alcohol	90.0
Geraniol Extra	50.0
Geranyl Acetate	35.0
Citronellol	35.0
Hydroxycitronellal	25.0
Bergamot Oil,	
Natural	50.0
Linalyl Acetate	55.0
Terpinyl	
Anthranilate	25.0
Linalyl	
Anthranilate	20.0
Methyl	
Anthranilate	12.0
Orangeflower	
Water, Abs.	20.0
Nerol	15.0
Citral	2.2
Pinene, rectified	10.0
Aldehydes	0.8

The aldehydes may, for example, consist of the following blend: Peach pseudo-aldehyde

0.1, C8 0.2, C10 0.1, C11 0.1, and phenylacetaldehyde 0.3.

drop to 20°C and mix for half hour.

NOTE: An excellent solid cream may be made by reducing the mineral oil content to 5 parts by weight.

Hair Groom
Formula No. 1

a "Alcolan" 40	14.25
b Mineral Oil (60 vis.)	31.45
c Water	54.00
d Borax	0.30

Melt a and b together at 70°C. Dissolve d in c at 70°C. Add c-d to a-b gradually, mixing until completely emulsified. Permit temperature to

No. 2

"Ethylan"	0-5
PVP	0-10
Ethanol	To make 100

The optimum amount for an aerosol hair lacquer is within the range of 0.2% to 0.5% based on the total weight of the formulation.

No. 3
(For Collapsible Tubes)

"Nimo" Base (45-26 Modified)	10.0	I	Heat to 80°C
Isopropyl Palmitate	5.0		
"Tegosept" M	.2		
"Tegosept" P	.1		
Cetyl Alcohol	10.0		
Petrolatum	28.0		
Water	45.7	II	Heat to 80°C
Magnesium Sulfate	1.0		
Perfume	q.s.		

With moderate agitation, slowly add II to I. Perfume at 50°C and stir to room temperature.

No. 4

"Cetrimide"	0.5
Cetostearyl Alcohol	3.0
Glyceryl Monostearate	5.0
Mineral Oil	20.0

Beeswax	3.0
Water, Perfume	
To make	100.0

No. 5
(Nonionic)

"Amerchol" L-101	6.0
"Ricilan" B or C	1.0
Mineral Oil	
(125 visc)	25.0
Microcrystalline	
Wax (170° m.p.)	5.0
White Petrolatum	5.0
Aluminum	
Monostearate	1.0
"Arlacel" 83	0.2
Glycerol	5.0
Water	51.8
Preservative and	
Perfume	q.s.

Disperse the aluminum stearate in mineral oil at room temperature. Heat to at least 85°C, as required for solution and gelling. Add the remainder of the oil phase ingredients and continue heating until clear. Adjust to 70°C and emulsify by adding the aqueous phase (water and glycerol) to the oil phase, both at 70°C. Mix at moderate speed and cool to 30°C. Allow emulsion to stand over night and then remix.

No. 6

a	"Polawax"	35.0
	Olive Oil	25.0
	Lanolin	25.0
	White Petroleum	
	Jelly	11.0
	Liquid Paraffin	35.0
b	Sodium Lauryl	
	Sulfate	0.5
	Water	45.0

Heat a to 70°C. Bring b to 70°C. Add a to b slowly with moderate, but thorough agitation, stir until cold. Perfume at 40°C.

No. 7

Stearic Acid	10
Beeswax	15
Diglycol Mono-	
stearate	30
Isopropyl Palmitate	80
Petrolatum	30
Sorbitol Solution	50
Distilled Water	785

No. 8

Isopropyl Palmitate	100
Castor Oil	210
Olive Oil	130
Mineral Oil	560

No. 9

Heavy Mineral Oil	15.0
"Igepal" CO-530	7.5

"Igepal" CO-430	7.5
"Alipal" CO-436	1.5
Water	68.5
Perfume	q.s.

Add the surfactants to the mineral oil and mix well. Then add the water slowly with stirring to give a rich white stable emulsion, which is useful as a hair groom.

No. 10
(Stick)

"Lanwax" #120	11.5
Paraffin Wax	4.0
Isopropyl Palmitate	9.5
Petrolatum	70.0
Cetyl Alcohol	5.0

Melt cetyl alcohol, paraffin wax, petrolatum, and "Lanwax," and add isopropyl palmitate, mixing thoroughly.

	No. 11	No. 12
"Amerchol" L-101	9.0	4.0
"Modulan" (or "Acetulan")	3.5	—
Glyceryl Monostearate (self-emulsifying)	13.5	12.0
Spermaceti	1.5	—
Mineral Oil (70 visc)	8.5	2.0
Lanolin	—	2.0
Stearic Acid	—	6.0
Glycerol	4.5	—
Water	59.5	73.0
Triethanolamine	—	1.0
Preservative	q.s.	q.s.

Add the water containing the water-soluble materials to the oils with the temperature of both phases 85°C. Mix at moderate speed until cool and remix the following day.

No. 13

"Amerchol" L-101	6.00	Mineral Oil (125 visc.)	25.36
Stearic Acid	0.64	Petrolatum	5.00

Microcrystalline
Wax (170 m.p.)   5.00
Lime Water-
Glycerol Soln.*   58.00
Preservative   q.s.

To oils at 68°C, add the slightly warmed lime water-glycerol solution. Mix until cool and remix thoroughly the following day.

The addition of 0.2% sorbitan sesquioleate increases the stability with very little whitening effect.

If preferred, calcium stearate may be used in place of the stearic acid and lime water-glycerol solution.

---

* To make lime water-glycerol soln.—
To each 10.6 g cold water add 1.0 g glycerol. Add USP calcium hydroxide to make a supersaturated solution. Stopper and let stand at least one hour, stirring occasionally. Filter: use immediately.

No. 14

PVP K-30   1.0
"Emulphor"
EL-719 (GAF)   1.5
"Emulphor"
ON-870 (GAF)   0.5
Ethanol (SDA-40)   15.0
Water   82.0

Dissolve PVP in ethanol-water mixture. Add the other ingredients and shake mixture to yield a clear solution. If desired, a bacteriostat, such as hexachlorophene, and a modifier, such as a water-soluble lanolin derivative, may be added to enhance the performance of the product.

No. 15

"Amerchol" L-101   5.00
Petrolatum   5.00
Mineral Oil
(70 visc)   33.50
Lanolin   3.00
Sorbitan
Sesquioleate   3.00
Beeswax   2.00
"Veegum"   0.25
Borax   0.50
Water   47.75

Add the water containing the water-soluble ingredients to the melted oils, with the temperature of both phases 70°C. Mix rapidly until cool, and homogenize for best results.

No. 16

"Amerchol" L-101   7.00
"Modulan"   10.50
Isopropyl
Palmitate   15.00

Stearic Acid	1.65
Glyceryl Mono- stearate	0.50
Glycerol	3.50
Triethanolamine	0.35
Water	61.50
Preservative and Perfume	q.s.

Add the water containing the water-soluble materials to the oils with the temperature of both phases 85°C. Mix slowly until cool and remix the following day.

No. 17
(Diphase)

This preparation contains an oil phase and an aqueous phase from which an emulsion is formed on agitation but it breaks readily:

Oil Phase:

Olive Oil	7.64
Oil-Soluble Emulsifier	0.33
Concentrate of Natural Mixed Tocopherols (34% concentra- tion)	0.08
Perfume	0.21

Aqueous Phase:

Ethyl Alcohol	40.50
Water	49.80
"Cetrimide"	0.34
Sodium Chloride	1.07
Sodium Bitartrate	0.03

Aerosol Hair Groom
Formula No. 1

Anhydrous Lanolin	3.0
Isopropyl Myristate	9.0
Liquid Paraffin	3.0
Dichlorodifluoro- methane	42.5
Trichloromono- fluoromethane	42.5

This formulation is suitable for packing into metal containers. The pressures may be reduced to a level suitable for glass bottles by the use of a 1:3 mixture of dichlorodifluoromethane and dichlorotetrafluoroethane as propellent.

No. 2

Isopropyl Myristate	18.0
Liquid Paraffin	24.5
Perfume	0.5
"Freon" 11/12 (1:1)	57.0

Avoid using nylon dispensers for hair oil aerosols, as they are liable to leak. Only those dispensers that have been specially designed for the purpose will prove to be trouble-free.

Dissolve PVP in alcohol. Add all other ingredients but propellent. Heat slightly, allow to stand overnight, filter. Add propellent.

No. 3

PVP	5.0
Liquid Lanolin Fraction	0.5
Polyethyleneglycol #400 Monolaurate	0.5
Perfume	1.0
Anhydrous Alcohol	93.0
	100.0

Charge: Concentrate 40; propellent 60

No. 4

Dimethylhydantoin-Formaldehyde Resin	33
Polyvinyl Acetate	165
Monoethanolamine	22
Absolute Ethanol, USP	220

Refluxing of the mixture is necessary to bring the polyvinyl acetate into solution.

	Formula No. 5	No. 6
"Acetulan"	0.75	0.75
PVP, Type NR-K30	3.00	—
Poly (VP/VA) (60:40) 50% Ethanol Sol.	—	6.00
Dimethylphthalate	0.15	0.15
Diisopropylthiourea	0.10	0.10
Perfume	0.25	0.25
SDA 40, Anhydrous	25.75	22.75
"Genetrons" 11/12 (60:40)	70.00	70.00
Approx. psi at 70°F	30	30

170 gm net fill to 6 oz container.

	No. 7	No. 8	No. 9	No. 10
"Amerchol" L-101	—	—	—	0.16
"Acetulan"	0.75	0.75	0.2	—
PVP, Type NR-K 30	3.00	—	—	—
Poly (VP/VA) (60:40), 50% Ethanol Soln	—	6.00	—	—
Dimethylhydantoin-Formaldehyde Resin	—	—	2.0	—
Ciba Resin 325, (50% in Ethanol)	—	—	—	3.23
Dimethylphthalate	0.15	0.15	—	—
Diisopropylthiourea	0.10	0.10	—	—
Polyethyleneglycol 400 Monolaurate	—	—	0.2	—
Methylene Chloride	—	—	—	14.25
Perfume	0.25	0.25	0.2	0.16
SDA 40, anhydrous	25.75	22.75	27.4	—
"Genetrons" 12/11 (40-60) .	70.00	70.00	—	—
"Genetrons" 12/11 (35:65) .	—	—	70.0	—
"Genetrons" 12/11 (30:70) .	—	—	—	82.20

Propellent for Cosmetics

No. 1
Hydrocarbon

	Wt/Vol
"Polyox" Resin WSR-35 or 205	1.0 g
Methylene Dichloride	20.0 ml
"Santicizer" M-17	2.0 g
Isopropyl Myristate	2.0 ml
Ethanol	5.0 ml
Perfume	0.1 ml
Methyl Chloride	70.0 ml

Disperse "Polyox" resin in the alcohol and then dissolve slurry in methylene dichloride, being careful to avoid any agglomeration. Warm, if necessary to aid solution. Add "Santicizer" M-17, isopropyl myristate, and perfume. The methyl chloride can then be added either by pressure-filling or by refrigeration.

resin in nonsolvent before solubilizing.

No. 2
Fluorocarbon

	Wt/Vol
"Polyox" Resin WSR-35 or 205	1.5 g
Methylene Dichloride	12.5 ml
Ethanol	14.3 ml
Isopropyl Myristate	1.0 ml
Water	0.7 ml
Fluorocarbon 12	70.0 ml

As for No. 1; disperse the

Hair Cream
Formula No. 1

Glyceryl Mono-stearate	12.0
Cholesterol	1.5
Lanolin	1.5

No. 2

Cholesterol	0.4
Vitamin F	1.0
Lanolin	2.0
"Polawax"	9.0

Water-in-Oil Hair Lotion

	Viscosity	
	Medium	Heavy
"Amerchol" L-101	11.00	15.00
Lanolin	3.00	2.50
Beeswax	3.00	7.50
Mineral Oil (70 visc)	31.50	25.00
Petrolatum	5.50	—
Borax	0.75	0.50
Water	46.50	49.00
Perfume and Preservative	To suit	To suit

Dissolve the "Amerchol," lanolin, beeswax, mineral oil, and petrolatum in one kettle and heat to 70°C. Dissolve the borax in the water, heat to 70°C, and add the water solution to the oils, with rapid agitation. The emulsion forms immediately but it should be mixed until cool and homogenized for best results. The perfume may be added when the emulsion is cool.

Cream Oil Hair Lotion
No. 1
(W/O)

Oil Phase:

"Amerchol" L-101	5.0
Petrolatum, USP White	5.0
Mineral Oil (70 Vis)	33.5
Lanolin	3.0
Sorbitan Sesquioleate	3.0
Beeswax, USP	2.0

Aqueous Phase:

"Veegum"	0.25
Borax, USP	0.50
Water	47.75
Perfume and Preservative	q.s.

Disperse the "Veegum" thoroughly in the water at room temperature. Add the aqueous phase at 70°C to the oil phase at 70°C while stirring. Mix while cooling to 40°C. Add the perfume and continue mixing and cooling to room temperature. Homogenize.

No. 2
(O/W)

Oil Phase:

"Amerchol" L-101	5.0
"Acetulan"	1.0
Stearic Acid, XXX	2.5
Glyceryl Monostearate, Neut.	2.0
Mineral Oil (70 Vis)	29.5

Aqueous Phase:

Triethanolamine	1.0
Propylene Glycol	4.5
Water	54.5
Perfume and Preservative	q.s.

Add the water phase at 85°C to the oil phase at 85°C while mixing. Mix while cooling to 40°C; add perfume. Continue mixing and cool to 30°C.

Liquid Hair Dressing
(Hydroalcoholic)

Ethanol, SDA-40	68.0
Water	16.0
"Ucon" LB-1145	16.0
	100.0

Mix ingredients. Color and perfume as desired. Other LB type "Ucons" (Methyl ethers of polypropylene glycols) may be substituted for "Ucon" LB-1145.

Gel Hair Dressing
No. 1
(Waterless) %

"Ucon" LB-1145	45
Dipropyleneglycol	42
2-Ethylhexanediol-1,3	7
"Cab-O-Sil," M-5	6

Combine the liquids in a closed system, such as an Abbe or a Dopp mixer. Slowly add "Cab-O-Sil" with a minimum of dusting. Stir until homogeneous. The product will be a clear, colorless gel. Perfume and color as desired.

No. 2
(Hydroalcoholic)

"Carbopol" 934	0.5
"Methocel" 65HG 4000	1.0
"Solulan" 98	2.0
Dipropyleneglycol	4.0
Alcohol SDA-40	60.0
Triethanolamine	0.4
Water	32.1

Wet the gums well, adding glycol and lanolin derivative ("Solulan" 98) to them. Add water, agitate well with an Eppenbach-type mixer. Then add alcohol (together with any perfume and color, if desired) effecting complete dispersion of the "Methocel" and "Carbopol." Allow mixture to deaerate. Add triethanolamine with slow, but thorough stirring.

The dipropylene glycol may be substituted by sorbitol, glycerol, etc.

Hair Pomade
Formula No. 1

"Modulan"	10
"Amerlate" P	5
Polyethylene Glycol 400 Monostearate	5
Mineral Oil (180 visc)	30
Petrolatum, Snow White	50
Perfume	q.s.

Warm with stirring until clear. Stir while cooling. Fill just above the setting point.

No. 2

"Amerchol" CAB	20
Microcrystalline Wax	
(m.p. 170)	20
Mineral Oil (70 visc)	60
Perfume	q.s.

Place all ingredients in one container and heat to dissolve. Mix well and pour while hot.

Hair Straightener

U.S. Patent 3,017,328

Caustic Soda	4
Cetyl Alcohol	16
Petrolatum	4
Mineral Oil	8
Sodium Lauryl	
Sulfate	4
Water	To make 100

Hairdressers' Lemon Rinse

Benzoate of Soda	3.5 g
Citric Acid	5 oz
Terpeneless	
Lemon Oil	5 cc
FDC Yellow	
No. 5	To suit
Water	To make 1 gal

Lactic Acid Hair Lotion

Cologne Essence	5 g
Menthol	2 g
95% Ethanol	500 cc
Lactic Acid	2 g
Water	480 cc

Dissolve the cologne and menthol in the ethanol and add to the lactic acid; finally, add the water.

For dry hair, 3 g glycerol may be added.

Hair Tonic
Formula No. 1

Cholesterol	0.2
Glycerol	3.4
Water	20.0
95% Ethanol	70.0
Isopropanol	5.5
Lecithin	0.2

No. 2

Cholesterol	1.0
Oxyquinoline	
Sulfate	0.5
Salicylic Acid	0.2
95% Ethanol	275.0
Lecithin	0.3
Castor Oil	3.0

No. 3

Cholesterol	1
Glycerol	5
Water	12

95% Ethanol	75
Carbon Tetrachloride	5
Lecithin	1

No. 4

Cholesterol	0.25
Glycerol	2.65
Water	13.00
95% Ethanol	75.00
Isopropanol	5.00
Carbon	
Tetrachloride	3.00
Lecithin	0.10

No. 5

Cholesterol	0.3
Glycerol	2.0
Water	13.6
95% Ethanol	80.0
Carbon	
Tetrachloride	3.0
Lecithin	0.1

Bleaching Lotion

a	"Neocol" 5192	5.00
	Cetyl Alcohol, NF	0.63
	Stearic Acid, TP	0.20
b	Water, Distilled	3.70
c	Water	30.00
	Phenacetin	0.05
d	Water	40.32
	"Sequestrene" NA2	0.10
e	"Neocol" F-45	2.85
f	35% Hydrogen	
	Peroxide soln.	17.15

Heat *a* to 90-95°C; heat *b* to 90-95°C. While mixing *a*, slowly add *b*. While stirring, slowly add *c*, which has been heated to 85°. (The phenacetin is previously dissolved in the water at 90-95°C with stirring.) Dissolve *d* at room temperature and add this cold to the warm emulsion. Continue stirring while rapidly cooling. Around 35-40°C add *e*. At 25°C slowly add *f*. After addition of the hydrogen peroxide, continue stirring for about 5 minutes.

Cold Wave Lotions

Most of the cold wave kits now on the market for home waving contain the equivalent of 6% thioglycolic acid, although a few contain as little as 4.6%, and one contains as much as 7%. All are at a pH of about 9.3.

Beauty shop waving lotions for bleached and dyed hair usually contain the equivalent of 4.6% thioglycolic acid, while lotions for waving regular or normal hair contain the equivalent of anywhere from 6.5% to 8.3%, depending on the method of use and how fast they are designed to work. Generally, lotions containing the higher concentrations of thioglycolic acid are used on resistant or hard-to-wave hair. The pH is usually between 9.2 and 9.3 for all types of lotions.

Ammonia is used as the alkali to adjust the pH of the thioglycolic acid to 9.2-9.3 in most of the various brands. Monoethanolamine (MEA), a non-volatile alkali, is used instead of ammonia in some lotions to adjust the pH. MEA is used both by itself and in combination with ammonia. Usually, in the latter case, the MEA is used to neutralize the acid, and then the ammonia is added to adjust the pH. The principal reason for using MEA is probably because its slight odor can be masked more effectively than can the odor of ammonia.

Sodium hydroxide, a strong alkali, is also being used in some cold wave lotions, but usually in combination with ammonia or some other weak alkali.

Most of the *no*-neutralizer kits on the market for home waving contain exactly the same lotion as the regular kits. The only difference seems to be in the method of use. For a while certain *self*-neutralizing lotions containing manganese salts were marketed. The manganese salts were added to accelerate oxidation in air, which is normally slow. In recent years, manufacturers have found that the addition of such metallic salts was not necessary and thus this type of lotion is no longer being marketed.

Clouding Agents

Most cold wave lotions contain clouding agents either for their aesthetic value or for the wetting and conditioning properties that they can impart to the lotion. If no clouding agent is used, then a suitable wetting agent, such as sodium lauryl sulfate paste (50 g/liter of lotion), should be added.

Two basic types of clouding agents are used: resin clouding agents and oil emulsifiers. Resin clouding agents are economical and extremely stable in cold waving solutions. But they generally do not contain conditioning agents that will enhance the waving properties.

Oil emulsifiers, on the other hand, produce a white, creamy effect and contain ingredients that condition the hair. They are preferred by many, and today are finding wide acceptance in the field.

There are now available clouding agents of both resin and oil types which will obtain the effect desired.

Neutralizers

Neutralizers are packaged both in powder and liquid form. There are two main powder types available; the "instant" type of neutralizer based on sodium bromate and the "Perdox" or *high speed* neutralizer containing sodium perborate monohydrate. Either type is available in plain or printed foil envelopes or as bulk powder.

For "instant" neutralizers, there are boosters to which sodium bromate is added to make a complete neutralizer. Originally, potassium bromate was used to a great extent, but in recent years it has been supplanted almost completely by the more water-soluble sodium bromate in the beauty shop field and to some extent in home kits. The neutralizers most widely used in the home-kit field are the hydrates of sodium perborate. In beauty shops, diluted hydrogen peroxide is still used by many as a neutralizer.

It is recommended that manufacturers and distributors of neutralizers realize the harmful effects that can result from the oral ingestion of such mixtures, particularly those of the bromate type. In the case of home permanent waving kits, a warning is generally prescribed in the directions for use to the effect that the product should not be kept where children can get at it, as it is harmful if taken internally.

Directions for Use

Each bottle of lotion must be accompanied by a folder containing adequate instructions as well as the necessary precautions to be taken when giving a cold permanent wave so that hair damage and skin irritation may be avoided.

Bottle Caps

An ordinary cork-backed Vinylite liner in a plastic cap is a satisfactory closure both for cold wave lotions and for neutralizers other than hydrogen peroxide neutralizers. If available, caps with white rubber liners or polyethylene plugs are excellent. Hydrogen peroxide neutralizers require a special closure. The suppliers of hydrogen peroxide can recommend the most suitable closure for their product.

Perfuming Cold Wave Lotions

The simplest basic perfume materials are invariably the best. Complex compounds are rarely satisfactory. Keep in mind that a cold wave lotion is alkaline and is a strong reducing agent which will react with many perfume compounds.

Any waving lotion containing a new perfume should be shelf-tested to make sure that it will not discolor and that it will retain its original scent.

Equipment

Although the formula may appear to be simple, the manufacture of cold wave lotion requires careful control. The equipment commonly used in a cosmetic plant is far from satisfactory because it is necessary to avoid contact with metals which cause discoloration and catalyze deterioration of cold wave lotions. It takes only three ppm of iron or copper to discolor a lotion.

Ceramic or glass-lined mixing tanks are preferred for the manufacture of cold wave lotions. Stainless steels 304, 316, and 317 are suitable for filling-heads and tanks as long as the surface is free from active metal particles which would contaminate the lotion. Such tanks should be prepared for use by wetting the steel surfaces with a 10% nitric acid solution containing 0.1% sodium chromate. After flushing out this solution, the steel surface is tested with ammonium thioglycolate for discoloration. Thioglycolic acid, either concentrated or dilute, should not be allowed to stand in contact with stainless steel; it should be neutralized with an alkali immediately after it is placed in the tank.

Stirrers for the tanks can be

of wood, "Haveg," or aluminum. Aluminum blades should be painted frequently with "Tygon." The following materials are suitable for pumps, valves and piping for the cold wave lotion or its raw materials: glass-lined steel, "Pyrex" glass, polyethylene, saran, and "Teflon."

Metallic pumps must not be used for the lotion or any of its component materials, such as thioglycolic acid, distilled water, and ammonia.

The raw materials used in the manufacture of cold wave lotion can also be a major source of contamination of the product by heavy metals. All raw materials should be carefully selected and checked for heavy metals before use in a lotion. If it is found that they will be a source of such contamination they should be further purified or not used at all.

Vacuum-distilled thioglycolic acid is best to assure purity and freedom from all heavy metals.

Cold Wave Lotions

The following schedules may be used to make 100-liter batches within the pH range of 9.2-9.3 at the various concentrations noted:

Conc.,% Desired	Thiovanic Acid (liters)			Ammonia (liters)	
	98%	80%	70%	26°Bé	24°Bé
4.6	3.55	4.56	5.35	6.63	7.53
5.5	4.24	5.46	6.45	7.67	8.71
6.0	4.62	5.95	6.98	8.19	9.28
6.5	5.00	6.45	7.54	8.74	9.92
7.0	5.48	6.94	8.13	9.35	10.06
7.4	5.70	7.34	8.59	10.00	11.35
8.3	6.39	8.24	9.63	11.02	12.51

Conc.,% Desired	Ammonium Thioglycolate 50% (liters)	Ammonia (liters)	
		26° Bé	24° Bé
4.6	7.65	3.10	3.54
5.5	9.18	3.30	3.75
6.0	10.00	3.58	4.08
6.5	10.83	3.98	4.54
7.0	11.67	4.30	4.90
7.4	12.34	4.56	5.18
8.3	13.88	4.95	5.64

1. Place 75 liters of distilled water in a suitable tank.
2. With constant stirring, slowly add the Thiovanic Acid or ammonium thioglycolate to the water.
3. Slowly add the ammonia with stirring.
4. Add a suitable clouding agent or wetting agent.
5. Add water-soluble perfume mixture, if desired.
6. Add sufficient distilled or deionized water to make 100 liters of finished cold wave lotion.
7. Before bottling, accurately check the finished cold wave lotion for pH and equivalent thioglycolic acid content to be certain the lotion meets the desired specifications.

Home Cold-Wave Lotion

Ammonium Thioglycolate (57.3% Active	10.50 lb	Anhydrous Sodium Sulfite	1.80 lb
		"Duponol" C	0.05 lb
Ammonium Hydroxide (29° Bé)	6.75 pt	"Dow 529K" Styrene-Butadiene Emulsion	0.20 lb
Ethyl Alcohol	4.00 pt	Water	78.00 lb

Cold-Waving Cream

U.S. Patent 2,479,382

Mineral Oil		3 lb 8 oz
Lanolin		1 lb 12 oz
Chlorinated Paraffin Wax		8 lb 12 oz
Sorbitan Monooleate		8 lb 12 oz
Sorbitan Monooleate Polyoxyalkylene Derivative		17 lb 8 oz
Gelatin		14 oz
Borax		14 oz
Ammonium Thioglycolate		70 lb
Ammonia (as NH$_3$)		14 lb
Water	To make	100 gal
Perfume		To suit

Mix together the mineral oil, lanolin, and chlorinated paraffin, warm to approximately 50°C, and add the emulsifying agents. Dissolve the gelatin and borax in 25 gal of water and heat to 45°C. Add the oil mixture and stir until the emulsion has cooled to room temperature. Dissolve the ammonium thioglycolate in the remainder of the water, adjust the pH to 9.0-9.5 with ammonium hydroxide, and add to the emulsion, with thorough mixing.

Permanent Waving Fixing Cream

Technical mixture of stearyl and cetyl alcohols	50 g
Mixture of glyceryl monostearate and phosphate of diethylamino-ethyl-stearyl amide	41 g
Wool Wax Alcohols	20 g
40% aqueous solution of the formate of triethanolamine monostearyl ester	10 g

These are melted together at 70°C and emulsified with water starting around 70°C and continuing agitation as the temperature falls. 10 ml of lactic acid (80%) is added with the water and sufficient water is used to make 500 g of cream.

For use as a fixative, 25 g of this cream are mixed with 25 ml of 6% hydrogen peroxide and 25 ml of water.

Wave Set

PVP K-30	5.0
Sorbitol	0.5
Ethanol (SDA-40)	44.3
Water	50.0
Perfume	0.2

Flexibility of the film from this formulation may be controlled by addition of a plasticizer in amount equivalent to 10-25% of the PVP content. Suitable modifiers are: dimethyl phthalate, "Santicizer" E-15, glyceryl monoricinoleate, "Corol," "Dytol" L-79, diethylene glycol, and dipropylene glycol. The addition of water-insoluble plasticizers will aid in reducing tack.

Aerosol Hair Spray

PVP K-30	2.0
Alcohol-Soluble Lanolin	0.5
Perfume Oil	0.3
Ethanol (SDA-40 anhydr)	22.2
Propellants 11/12 (70/30)	75.0

Dissolve PVP in ethanol. Add lanolin and heat slightly, if necessary, to effect solution. Add perfume oil. Allow to stand. Filter and charge with propellents.

To improve flexibility and reduce hygroscopicity of the PVP film, a compatible plasticizer, such as "Citroflex" A-2, dimethyl phthalate, glyceryl monoricinoleate, or oleyl alcohol may be added (10-25% based on resin solids). Suitable lanolin derivatives include "Acetulan" and "Amerchol" L-101, "Ethoxylan" 100, "Ethylan" and "Lanogel" 41, and "Lanethyl."

Hair Dyes
Organic: Two Bottles

Complex modern hair dyes are packaged in two bottles, one of which contains the coloring liquid, the other, the developer. The neck of the second bottle must be wide enough to take 0.5-g tablets of urea peroxide, now commonly used as the developer.

For each formula given here the following general directions should appear on the package:

Empty the tablets into a saucer. Fill the bottle that contained the tablets with water and pour this on the tablets, crushing and dissolving them in the water. Then mix this peroxide solution with the contents of the second bottle containing the dye and apply the mixture.

Formula No. 1
(Black)

p-Toluylenediamine 5 g
Sulfo-p-amino-
  diphenylamine 5 g
Sodium Sulfite 10 cc
50% Alcohol 100 cc

Bottle 1:

10 cc of this coloring solution for small size: 20 cc for large size

Bottle 2:

4 tablets of urea peroxide (0.5 g each)
Contact: 1 hour

No. 2
(Chestnut)

p-Toluylenediamine 2 g
Sulfo-p-
  aminophenol 2 g
Sulfo-p-amino-
  diphenylamine 1 g
Sodium Sulfite 5 cc
50% Ethanol 100 cc

Bottle 1:

10 cc of this solution

Bottle 2:

3 tablets urea peroxide
Contact: 40 minutes

No. 3
(Brown)

p-Toluylenediamine 3 g
Sulfo-p-amino-
  diphenylamine 3 g
p-Aminophenol 1 g
Sodium Sulfite 6 cc
50% Ethanol 100 cc

Bottle 1:

10 cc of this solution

Bottle 2:

3 tablets urea peroxide
Contact: 45 minutes

No. 4
(Blond)

Sulfo-*o*-aminophenol	3.0 g
p-Toluylene-diamine	1.0 g
Pyrogallol	0.5 g
Sodium Sulfite	4.0 g
Water	80.0 cc
Ethanol	20.0 cc

Bottle 1:

10 cc of this solution

Bottle 2:

2 tablets urea peroxide
Contact: 30 minutes

Hair Dye Base
Clear Type

Propylene Glycol	10.0
Oleic Acid	35.0
Polyoxyethlene Sorbitan Monooleate	10.0
Isopropanol	5.0
Water	29.0
Ammonia (26° Bé)	11.0

Metallic: Two Bottles
(Black)

Formula No. 1

Bottle 1:

Silver Nitrate	120 g
Copper Sulfate	10 g

10% Ammonia	To dissolve ppt
Water	To make 2.5 liters

Bottle 2:

Pyrogallol	50.0 g
Water	1.0 liter

No. 2

Bottle 1:

Copper Sulfate	6 g
Iron Sesqui-chloride	15 g
Nickel Sulfate	2 g
Water	1 liter

Bottle 2:

Pyrogallol	40 g
Water	1 liter

Medium Brown

Formula No. 1

Bottle 1:

Silver Nitrate	0.7 g
Copper Sulfate	1.0 g
Cobalt Nitrate	6.0 g
Nickel Nitrate	1.0 g
Ammonia (10%)	q.s. To dissolve the precipitate
Water	To suit

No. 2

Bottle 1:

Silver Nitrate	0.2 g
Copper Sulfate	2.0 g
Cobalt Nitrate	4.5 g
Nickel Nitrate	1.5 g
Ammonia (10%)	To dissolve the precipitate
Water	To suit

No. 3

Bottle 1:

Silver Nitrate	1 g
Copper Sulfate	1 g
Cobalt Nitrate	5 g
Ammonia (10%)	To dissolve the precipitate
Water	To suit

No. 4

Bottle 1:

Copper Sulfate	0.5 g
Cobalt Nitrate	5.0 g
Nickel Nitrate	1.0 g
Ammonia (10%)	To dissolve the precipitate
Water	To suit

Bottle 2 (for Formulas 1 to 4):

Pyrogallol	30 g
Water	1 liter

Or:

Potassium Sulfide	50 g
Alcohol	200 cc
Water	800 cc

Bismuth Dyes
(Standard Solutions)

No. 1

Glycerol	100 cc
Neutral Bismuth Nitrate	100 g
Silver Nitrate	50 cc
Glycerinated Water 1:9	To make 1 liter

No. 2

Glycerol	100 cc
Neutral Bismuth Nitrate	100 g
Glycerinated Water 1:9	To make 1 liter

General Instructions:

Introduce the crystals in a graduated cylinder and pour in 100 cc of glycerol. Crush the crystals in the glycerol with the aid of a strong glass rod, flattened at the end. Prepare in the meantime the glycerinated water 1:9 in a sufficient quantity and add, dissolving the ingredients slowly, making sure that this does not take place suddenly, in order to prevent any precipitation of the crystals. Stir briskly to dissolve the crystals. Finally complete the volume with glycerinated water to make 1000 cc, avoiding always any too sudden addition of the solvent.

Formula No. 1
(Blond)

Bottle 1:

Solution No. 2

Bottle 2:

Sodium	
Hyposulfite	200 g
Water	800 cc

No. 2
(Chestnut)

Bottle 1:

Solution No. 1

Bottle 2:

Potassium	
Sulfide	50 g
Water	800 cc
Alcohol	200 cc

OR

| Pyrogallol | 30 g |
| Water | 1 liter |

One-Bottle Dyes
Formula No. 1
(Blond)

Pyrogallol	40
Sodium Sulfite	80
Water	1200
10% Ammonia	400

No. 2
(Ash Blond)

Pyrogallol	40 g
Cobalt Nitrate	20 g
Sodium Sulfite	80 g
Water	To make 1 liter

No. 3
(Dark Brown)

Pyrogallol	40 g
Sodium Sulfite	80 g
Cobalt Nitrate	150 g
Copper Sulfate	50 g
Water	To make 2 liters

No. 4
(Chestnut)

Cobalt Nitrate	100
Sodium Sulfite	80
Pyrogallol	40
Water	2000

No. 5
(Black)

| Pyrogallol | 50 |
| Iron Sesquichloride | 100 |

Nickel Nitrate	200
Sodium Sulfite	100
Water	2000
Perfume	To suit

General Directions:

Dissolve the sulfite in the water; then the pyrogallol; and finally the metal salts.

These dyes should be packed in bottles filled to the neck and tightly stoppered.

The shade develops only very slowly in the air.

Blue-Gray Hair Dye

British Patent 889,813

Coconut Fatty Acids Diethanolamide	5
Stearic Acid Diethanolamide	5
Sodium Lauryl Sulfate (35%)	5

5,8-Dihydroxy-1,4- naphthoquinone	1
Water	To make 100

Hair Straightener

U.S. Patent 3,017,328

Caustic Soda	4
Cetyl Alcohol	16
Petrolatum	4
Mineral Oil	8
Sodium Lauryl Sulfate	4
Water	To make 100

Brilliantine

"Modulan" (or "Acetulan")	5
Mineral Oil	95

Place both ingredients in one container, heat slightly until dissolved, and mix.

Hair Rinse

	Formula No. 1	No. 2
a "Neocol" 5192	5.00	7.00
Cetyl Alcohol NF	0.30	0.30
b "Hyamine" 1622	0.30	0.30
Polyvinylpyrrolidone K30	0.50	0.50
Distilled Water	93.90	91.90

Warm *a* to 80°C. Warm *b* to 80°C. Add *b* to *a*. Stir rapidly. Cool with stirring to 40°C.

Depilatories

Thioglycolic acid has been widely used, not only in America, but throughout the world, for the formulation of depilatories. It is commonly accepted at the present time as "the mercaptan" to be used in depilatories. Either the straight thioglycolic acid or its calcium salt may be used.

Because thioglycolates are the principal ingredients of depilatories, it is necessary (as in the case of cold waving lotions) to keep to the minimum the metallic impurities in the other ingredients used in manufacturing the product. Properly formulated depilatory creams are stable and have a shelf life of over one year. The container must be tightly sealed or the cream will rapidly lose its strength through oxidation.

Although tin is the most satisfactory material for tubes, it is possible today to buy wax-lined lead tubes that give satisfactory protection.

Most depilatories on the market contain about 2.5%-3% thioglycolic acid and are adjusted to a pH around 11.

All four of the following formulas give approximately the same depilating effect. To manufacture 100-kilo batches use the following quantities:

	Formula No. 1	No. 2	No. 3	No. 4
Thioglycolic Acid (70%)	3.2 liters	3.2 liters	—	—
Calcium Thioglycolate	—	—	5.4 kg	5.4 kg
Calcium Hydroxide	8.8 kg	12.6 kg	6.6 kg	10.4 kg
Strontium Hydroxide·8H$_2$O	3.7 kg	—	3.7 kg	—
Cetyl Alcohol	6.0 kg	6.0 kg	6.0 kg	6.0 kg
"Brij" 35	1.0 kg	1.0 kg	1.0 kg	1.0 kg
Distilled Water	76.6 liters	76.6 liters	77.3 liters	77.2 liters
Perfume	As desired	As desired	As desired	As desired

Depilatory is prepared in two parts and then mixed together, as follows:

Part I: (For all four formulas)

Heat 60 kg of distilled water to 70°C.
Add the cetyl alcohol and the "Brij" 35 with constant stirring until completely dissolved. Discontinue stirring and allow to stand and cool to room temperature.

Part II: (For formulas 1 and 2)

To 14 liters of distilled water add the thioglycolic acid (70%) and stir. Slowly add the calcium hydroxide while stirring. Add the strontium hydroxide at this point (only to formula I). Add sufficient distilled water to make Part B weigh 33 kg.

Part II: (For formulas 3 and 4)

To 14 liters of distilled water add the calcium thioglycolate and mix well. While stirring slowly add the calcium hydroxide. Add the 3.7 kg of strontium hydroxide at this point, only if you are using formula 3. Add sufficient distilled water to make Part II weigh 33 kg.

Combine Part II with Part I and stir well. Add perfume at this point as desired.

The final product should be assayed for content of thioglycolic acid, pH, and depilation time.

In the manufacture of depilatories it is extremely important that the equipment used should not be a source of contamination of the product. In general, the same materials that are suitable for the handling of cold waves are also suitable for depilatories. The equipment should be of either stainless steel or glass. If stainless steel is used, it should be checked to ensure that the particular type used does not contaminate the product with metallic impurities.

It is necessary to avoid contact with heavy metals and also

exposure to air as much as possible. Thioglycolic acid should be vacuum-distilled in order to assure purity and freedom from heavy metals.

After-Shave Lotion, Aerosol

PVP K-30	0.5
Ethanol (85%)	60-70
Perfume	0.5
Propellent 12	30.0
Propellent 114	10.0

Brushless Shaving Cream

No. 1

Stearic Acid	17.5
Mineral Oil	3.0
Lanolin	2.0
Glycerol	4.0
Caustic Potash	1.0
Water	To make 100.0

Heat the oils and wax to 80°C. Add the water at the same temperature to the hot oil phase under constant and vigorous agitation. Cool down under agitation. Add perfume around 60°C and treat in a colloid mill or homogenize at 50 to 55°C.

No. 2

a Triple-Pressed	
Stearic Acid	17.5
"Antarox" B-390	2.5
"Antarox" B-290	1.5
Heavy Mineral Oil	5.0
b Propylene Glycol	5.0
Distilled Water	64.5
c Triethanolamine	2.0
Borax	2.0
Perfume	To suit

Melt a and heat to 80°C; heat b to 82°C; add c to b and quickly add this mixture to a, with moderate agitation. Stir until the batch cools to 55°C. Add perfume and stir slowly until the temperature is 45°C. Store for 2-3 days at room temperature. Remix and pack in suitable containers.

No. 3

Polyethylene Glycol 1500	15.0
Propylene Glycol	10.0
Triethanolamine	1.0
Stearic Acid	12.0
85% Potassium Hydroxide	0.5
Sodium Alginate (2% Mucilage)	1.3

Perfume	0.2
Water	60.0

No. 4

Triple-Pressed Stearic Acid	25 lb
Anhydrous Lanolin	3 lb
Ethanol	2 lb
Triethanolamine	0.75 lb
Borax	0.5 lb
Glycerol	3 lb
Distilled Water	70 lb
Perfume	4 oz

Aerosol Shaving Cream

Formula No. 1

Stearic Acid	8.4
Stearyl Alcohol	1.0
"Sipon" LS	4.1
Mineral Oil	6.0
Glycerol	3.0
"Avitex" C	30.0
Perfume	0.2
Water	47.4

Put 90 parts of this mixture in a container with 10 parts of a 1:1 mixture of "Freon" 12 and "Freon" 11.

No. 2

PVP K-30	10.0 lb
Stearic Acid	262.0 lb
Coconut Fatty Acid	75.0 lb
Cetyl Alcohol	61.0 lb
Petrolatum	94.0 lb
Perfume	5.5 lb
"Sorbo"	160.0 lb
Triethanolamine	169.0 lb
Potassium Hydroxide (Pellets)	7.5 lb
Water (Deionized)	q.s. 500 gal

The triethanolamine, potassium hydroxide, and "Sorbo" are dissolved in water at 160°F and mixed well. The PVP is dissolved in water (25 gal) and added. The oils are added to the aqueous phase and stirred for twenty minutes. The emulsion is allowed to cool with agitation. At 105-110°F add perfume and cool to room temperature. The concentrate is charged with 10% propellent 12 or a mixture of 12/114.

Acid Cream Shampoo (Liquid)

Polyethylene Glycol 400 Distearate	2.6
Ethylene Glycol Monostearate	1.4
Propylene Glycol	2.0
Ammonium Lauryl Sulfate	49.0
Water	45.0

In some formulations of an acid shampoo, the pH is adjusted and maintained through the addition of citric acid, as illustrated by the following formula.

Paste-Type Acid Shampoo

Sodium Fatty Alcohol Sulfate Paste	73.00
Citric Acid	2.50
Sodium Lauryl Sulfate Powder	6.50
Sodium Citrate	2.80
Stearyl Alcohol	2.25
Beeswax	2.30
Glyceryl Monostearate	4.00
Lanolin	1.25
Sodium Alginate (1% Solution)	2.50
Water	2.00

Heat the first four ingredients at 80-85°C, add the alginate, and stir thoroughly. Heat the stearyl alcohol, beeswax, and monostearate to 80°C and add to the first mixture. Stir until cool.

Paste Cream Shampoo

Formula No. 1

Sodium Lauryl Sulfate Paste	50.00
Stearic Acid	5.00
Sodium Hydroxide	0.75
Lanolin	1.00
Water	43.25

No. 2

Sodium Lauryl Sulfate Paste	60.0
Stearic Acid	6.0
Sodium Hydroxide	0.9
Lanolin	1.0
Water	32.1

This gives an excellent lather in hard or soft water and it is stable.

No. 3

"Triton" AS35	50
Sodium Stearate	8
Water	41
Lanolin	1

Heat the ingredients together to 85°C, with constant stirring. When the mixture is homogeneous, cool slowly with stirring. Perfume may be added to the molten mix.

Egg Shampoo

No. 1
(Milky-Lotion)

Sodium Fatty Alcohol Sulfate	45
Magnesium Stearate	2
Polyvinyl Alcohol (10% Solution)	6
Lanolin	1
Cetyl Alcohol	2
Glyceryl Mono-stearate	2
Water and Dehydrated Egg	To make 100

No. 2

1) "Maprofix" WAC	11.00
2) "Super-Amide" L-9	4.00
3) "Tegin"	3.00
4) Cetyl Alcohol	2.00
5) Water	76.35
6) Powdered Egg	1.50
7) Water	2.00
8) Formalin (40%)	0.15

Dissolve 3 in 5 at 85°C. Add 2, 4, and then dissolve. Then add 1. At 35 to 40°C add 6 slurried in 7; then add 8. Mild agitation throughout.

Coconut Shampoo

Formula No. 1
(35% Real Soap)

Distilled Coconut Fatty Acids (Sap. No. 266)	29.6
100% Potassium Hydroxide	7.9
Water	To make 100.0
Perfume	As desired

Dissolve the potassium hydroxide in the necessary amount of water, heating at 120-130°F. Then add the fatty acids in a slow, steady stream. Agitate during the mixing of the fatty acid and alkali and continue until saponification is complete. Heat the reaction mixture at 150-160°F during the final stages of saponification. Check the neutrality of the soap and adjust as necessary. Allow the soap to stand and settle for several days at temperatures near freezing, if possible, and decant or filter the soap.

This soap is characterized by its clarity and profuse sudsing. If it clouds on standing, a trace of sequestering agent (0.1 to 0.5% EDTA) will restore brilliance.

No. 2
(35% Real Soap)

Distilled Coconut Fatty Acids (Sap. No. 266)	15.1
"Emersol" 220 White Elaine (Sap. No. 200)	15.1
100% Potassium Hydroxide	7.1
Water	To make 100.0
Perfume	As desired

Directions as for No. 1.

No. 3
(40% Real Soap)

Distilled Coconut Fatty Acids (Saponification No. 263)	12.7
"Emersol" 211 Low-Titer Elaine (Saponification No. 199)	12.7
Triethanolamine	14.7
"Carbitol"	7.0
Water	To make 100.0

Blend the oleic and coconut fatty acids. Dissolve the triethanolamine in the required amount of water and proceed as for Formula No. 1. Heating is not so essential as in No. 1. Longer standing is required without heat, but the possibility of making soaps without heating may be valuable in particular situations.

Alcohol may be substituted for a portion of the water in any of these formulas. Alcohol will improve clarity and increase the resistance of soap to low temperatures (precipitation and formation of gel), but it will also reduce foaming.

"Carbitol," glycerol, and various glycols may be used to increase the viscosity of the soap. Perfumes are generally used in small amounts to impart a distinctive and pleasing odor.

No. 4

Coconut-Oil Fatty Acids	21.0
Oleic Acid	14.0
Triple-Pressed Stearic Acid	13.5
Monoethanolamine	7.5
Triethanolamine	14.2
40% Formalin	1.2

Titanium Dioxide 0.2
"Tergitol" Wetting
    Agent 7 3.0
"Cellosize" WSLM 35.0
Distilled Water 125.0

Melt the fatty acids together and adjust the temperature at 50-55°C. Add the "Tergitol" and the formalin. Heat the water to 60°C, add the amines and "Cellosize," and adjust the temperature of the solution to about 50°C.

Add the water solution to the melted fatty acids, stir constantly until a clear viscous mixture is obtained, and then at intervals until the temperature is about 35°C. Add the perfume and mix thoroughly.

Disperse the titanium dioxide in half its weight of "Tergitol," grind it to a smooth paste, and then blend it into the cream. It may be necessary to mill the final product in order to obtain a complete dispersion of the titanium dioxide in the cream.

Prepare a concentrated solution of a suitable dye. Add a small amount of this solution at a time, with thorough mixing until the cream is of the desired tint.

Part or all of the oleic acid may be substituted by an equivalent amount of coconut-oil fatty acids to produce more copious lathering. A combination of soaps, however, is less irritating than coconut-oil soap alone.

Mixed isopropanolamine may be used instead of triethanolamine in these formulas to produce more stable color.

Paste Cream Shampoo

Formula No. 1

"Ultrawet" K 18-22
Sodium Stearate 4-6
Magnesium Stea-
    rate (Optional) 2-3
Sodium Chloride 3-5
Sodium Citrate
    (Optional) 2-3
Water 71-61

Magnesium stearate may be used instead of one-half of the sodium stearate, to make solid creams more opaque and short. Larger doses tend to make the solid creams too grainy.

No. 2

Fatty Alcohol Sulfate
    ("Duponol" ST) 35

Magnesium Stearate 4
Polyvinyl Alcohol
(10% Solution) 6
Lanolin 1
Glyceryl Monolaurate 2
Cetyl Alcohol 2
Water 50

The magnesium stearate is pasted with the fatty-alcohol sulfate, adding the polyvinyl alcohol and the water. The mixture is heated to 77°C. The cetyl alcohol, lanolin, and glyceryl monolaurate are heated together to the same temperature. The aqueous phase is then added with mixing.

No. 3

"Tergitol" Wetting
Agent 7 33.3
"Nacconol" Deter-
gent NRSF 16.6
Light-Colored
Bentonite 7.4
"Cellosize" WSLH 5.7
Water 37.0

1. Add the bentonite to the water, stir well, and let it stand for several hours. Mix with a mechanical stirrer until a smooth paste is obtained.
2. Mix the "Tergitol" and "Nacconol." Stir at intervals until the "Nacconol" is completely dispersed and a clear gel is obtained.
3. Gradually stir 2 into 1. Avoid aeration caused by rapid stirring.
4. Add the "Cellosize" and stir without aeration until a smooth cream is obtained.
5. Add perfume and tint as desired.

No. 4

"Duponol" WA 30.0
"Ninol" 2012A 7.0
"Ninol" CB60 3.0
Sequestering Agent* 0.6
Water 59.4

No. 5

"Neo-Fat" 12 14.00
Triethanolamine
(10% Excess) 10.40
Sequestering
Agent* 0.13
Water To make 100.00

* Sodium Ethylenediaminetetraace-
tate

No. 6

"Neo-Fat" 12 16.8
Potassium Hydrox-
ide (39° Bé) 11.5

Thickener 1-2
Water To make 100

Clear Liquid Shampoo
Formula No. 1

"Triton" AS30	14.4
"Ninol" 1281	2.8
Salt	2.8
Water	80.0

No. 2

"Triton" AS30	35
"Ninol" 201	6
Polyethylene Glycol 600 Distearate	2
Salt	1
Water	56

No. 3

"Triton" AS30	50
"Ninol" 201	9
Polyethylene Glycol 600 Distearate	1
Salt	1
Water	39

No. 4

"Triton" AS30	50
Polyethylene Glycol 400 Distearate	5
"Ninol" 201	5
3% Tallow Sulfate Solution	38
Salt	2

No. 5
(Minimum Irritation)

A sparkling clear liquid of rich yellow color. The viscosity is about 200 cps; the cloud point 15°C.

"Sipon" LS	20
"Miranol" C2M	20
Hexylene Glycol	2
Polyoxyethylene Sorbitan Monolaurate	1
Lauric Diethanol-amide (90%)	1
Water	56
	100

Mix all ingredients together except perfume and heat to 70°C with occasional stirring. Stir and cool to 40°C before adding perfume.

No. 6
(For Dry Hair)

A low-viscosity, light amber shampoo with a cloud point of 0-5°C.

"Sipon" LT6	49.0
TEA Oleate (50%)	9.8
Propylene Glycol	2.0
"Siponol" OC (oleyl alcohol)	1.0
Water, distilled	38.2
	100.0

To prepare 100 pounds:

Prepare the 50% TEA Oleate. Mix 4.9 pounds of water and 1.7 pounds of triethanolamine. Heat to 70°C, add to 3.2 pounds of oleic acid at 70°C. Mix until uniform and cool, and add to "Sipon" LT6 which is at 80°C. Mixture will be lumpy, but when water is added the lumps will break. Add water at 80°C, mix, and maintain temperature for 10 minutes. Cool to 50°C separately, mix oleyl alcohol, propylene glycol, perfume, and add this to the mix.

Lotion Cream Shampoo

Formula No. 1

"Triton" AS30	
(9% Active)	32
"Triton" X200	
(10% Active)	36
"OPE" 1	2
Water	30

Add "OPE" 1 to "Triton" X200, with slow stirring, then add the "Triton" AS30 and water in order indicated. Mix thoroughly to prevent separation of the shampoo on prolonged standing.

This formulation is particularly effective for oily hair. For dry hair, reduce the amount of "Triton" X200 and add tallow sulfate. This will also reduce the cost of the formulation somewhat.

No. 2

"Triton" AS30	85
Polyethylene Glycol	
400 Distearate	5
Magnesium Stearate	2
Water	8

Thoroughly mix the magnesium stearate with part of the "Triton" AS30. Add the other ingredients, with stirring, and heat at 160 to 170°F until the distearate melts. Continue stirring until cool.

Dandruff Shampoo

1)	"Super-Amide" L-9	5.0
2)	"Lanogel" 41	4.0
3)	"Neutronyx" 333	15.0
4)	"Neutronyx" 640	2.0
5)	"Isothan" Q-15	1.0
6)	Water (distilled)	73.0

Should be filtered before bottling. Dissolve 1 and 3 in 6 at 70°C. Add 2 and 4, then add 5. Cool with agitation.

Bubble Bath

Formula No. 1
(Concentrated)

"Solulan" C-24	5
Hexyleneglycol	10
"Monamid" 150 AD	20
"Monapon" T	65
Perfume and	
Preservative	q.s.

Warm first four ingredients enough to melt "Solulan" C-24. Mix thoroughly; add perfume and preservative.

No. 2

"Solulan" 25	5
"Maprofix" TLS 500	15
Onyx-ol 336	10
Water	30
Perfume and	
Preservative	q.s.
Water	40

Warm and mix first three ingredients until clear. Cool and add perfume, stirring to dissolve. Add water.

Vegetable Foam Bath

Use 40 to 50 g of concentrated soapwort extracts (20 to 30° Bé) for one foam bath. This corresponds to about 25 g of pure saponin (100 liters of foam).

Bath Powder

Formula No. 1

Sodium Lauryl	
Sulfonate	40.0
Sodium Bicarbonate	55.0
Powdered Gum	
Arabic	0.5
Pure Saponin	5.0
Perfume	To suit

No. 2

Sodium Lauryl	
Sulfonate	60.0
Sodium Meta-	
phosphate	10.0
Powdered Traga-	
canth Gum	0.3
Borax	20.0
Soluble Starch	8.0
Perfume	To suit

No. 3

Sodium Lauryl	
Sulfonate	30
Powdered Coconut	
Oil Soap	40
Borax	10
Powdered Gelatin	2
Soluble Starch	8
Sodium Meta-	
phosphate	10
Perfume	To suit

For one bath, 40-50 g is needed.

Isopropanol or Ethanol ... 10

Pine Needle Bath Oil

Formula No. 1

"Texapon" KM 14
Special ... 70
Pine Needle Oil ... 20

No. 2

"Texapon" K 14 S
Special ... 40
Pine Needle Oil ... 40
Ethanol ... 20

Tooth Paste

	Formula No. 1	No. 2
Cellulose Gum, CMC-12HP	0.86	1.8
Glycerol C.P.	10.00	17.0
Sorbitol 70%	—	11.0
Propylene Glycol	20.00	—
Water	13.04	23.6
"Tegosept" M	0.10	0.1
Saccharin Solution, 50%	0.10	0.1
Mineral Oil, Heavy	0.90	0.9
Oil of Peppermint, Heavy	0.50	0.5
Soap (Coconut-Tallow)	—	2.0
Soap (Sodium Stearate)	—	1.0
Sodium Lauryl Sulfoacetate	2.50	—
Sodium Tripolyphosphate	—	1.0
Calcium Carbonate	—	41.0
Dicalcium Phosphate	52.00	—

Mix Cellulose Gum with one-fifth of the combined humectants. Preheat water to 70-80°C, then add it to the wetted gum with rapid stirring. Add remainder of the humectants.

a When soap is used as a foaming agent, it is added to the gum solution, heating if necessary, for complete solution. Cool the entire mixture to room temperature and transfer to a

Werner-Pfleiderer mixer. Add the saccharin solution, oil of peppermint, and mineral oil while mixing the gum solution.

b When synthetic detergent is used rather than soap, it is added at this point while mixing the gum solution. The abrasive is added slowly, allowing it to disperse well. After the abrasive is added, mix until the paste appears homogeneous, usually from fifteen to sixty minutes.

	No. 3	No. 4	No. 5	No. 6	No. 7
"Polyox" Resin WSR-35	0.5-1.0	0.3-1.0	1.0	0.3-1.0	0.3-1.0
Propylene Glycol	17.0	19.0	17.0	19.0	17.0
Glycerol	4.5	9.3	8.5	9.6	8.4
p-Hydroxy Methyl Benzoate	0.2	0.27	0.27	0.3	0.27
10% Saccharin Solution	0.5	0.63	0.63	0.7	0.63
Mineral Oil	1.0	1.2	1.1	1.2	1.1
Sodium Lauryl Sulfate	3.5	—	2.9	—	3.0
Sodium Lauroyl Sarcosinate	—	2.0	—	—	—
Tricalcium Phosphate	43.0	42.2	—	—	—
Dicalcium Phosphate	—	—	51.0	—	—
Calcium Carbonate	—	—	—	42.0	40.0
Sodium Soap of Tallow Fatty Acids	—	—	—	2.8	—
Sodium Soap of Coconut Oil of Fatty Acids	—	—	—	0.7	—
Flavor	1.0	1.2	1.1	1.2	1.1
Water Approx.	28%	23%	16%	21%	27%

The "Polyox" resin should be blended with all dry ingredients first and then the dry ingredients should be mixed into the non-solvents for the resin. With constant stirring, slowly add water to the mix and as the "Polyox" resin dissolves, the blend will become uniform. After resin is completely solubilized, a stable, smooth paste of light density results. It foams well and feels good in the mouth.

No. 8

a "Veegum" F	0.8
Sodium Carboxy-methylcellulose (med. visc.)	0.7
Saccharin	0.15
b Water	25.9
c Sorbitol (70% Conc.)	25.0
d Dicalcium Phosphate (Dentrifice Grade)	45.0
e Flavor*	1.0
f Sodium Lauryl Sulfate	1.5

* Suggestions for Flavoring:

Oil Peppermint, double-distilled	70
Eucalyptol	17
Anethol	8
Methyl Salicylate	3
Oil Cassia (or Cinnamic Aldehyde)	2

Make a uniform dry blend of a and add a to b at room temperature, continually agitating until smooth. Add c and d alternately to a-b. Add e to preceding and finally add f, mixing until smooth.

NOTES: Avoid incorporation of air after addition of wetting agent or vacuum deairing may be necessary for a smooth paste.

This formula should have a thin mixing viscosity but, because of the thixotropic nature of "Veegum," it will set in the tube after agitation has ceased.

Another common procedure is to prepare the organic gum mucilage with the available water. The dry "Veegum" is dispersed in this mucilage after which a humectant and abrasives are added.

Non-Staining Chlorophyll Dental Cream

U.S. Patent 2,783,182

Polyvinylpyrrolidone (av.m.w. 40,000)	0.50
Sodium Copper Chlorophyllin	0.10
Dicalcium Phosphate Dihydrate	48.25
Glycerol	28.20
Detergent	3.75
Tetrasodium Pyrophosphate	1.00

Enzyme-Stimulating Dentifrice

U.S. Patent 2,574,659

Sodium Fumarate	10
Calcium Lactate	18
Calcium Carbonate	10
Glycerol C. P.	7
Sodium Carboxy-methylcellulose	2

Water	52
Essential Oil	1
Saccharin	Trace

This cleans by stimulating enzymatic action in the mouth rather than by the action of soaps, detergents, and abrasives. The plaque or film that forms on teeth, which is the forerunner of tartar and dental caries, consists mainly of mucin, a precipitated component of saliva. Ingredients in the new dentifrices are said to react with enzymes in mucin and saliva in such a way that this plaque becomes readily soluble and is easily rinsed away. One of the advantages claimed for this dentifrice is that it reaches the mechanically less accessible places in and between the teeth where cleansing is most important.

Anti-Enzyme Dental Preparations

South African Patents 16272 and 17041

Dental Cream

Calcium Carbonate	12.1
Dicalcium Phosphate Dihydrate	36.2

Sodium *m*-Lauroyl Sarcoside	2.0
Glycerol	30.6
Water	15.3
Irish Moss	1.0
Sweetener agent, flavor, preservative	

The glycerol, Irish Moss, water, and minor amounts of preservative and sweetener are mixed and heated to about 170°F to yield a gel-like mass. The sarcoside is added with agitation to form a homogeneous mass, after which the polishing agents are incorporated in powder form. After cooling and adding the flavors the mass is milled, deaerated, and strained to produce a smooth, homogeneous cream paste, which is filled into collapsible aluminum or lead tubes. This dental cream is practically neutral.

Ammoniated Dental Paste

U.S. Patent 2,588,992

Tricalcium Phosphate	26.67
Glycerol	45.40
Water	15.40

Peppermint
Flavoring   0.58
Gum Tragacanth   0.96
Saccharin   0.10
"Lathanol" LAL   2.89
Dibasic Ammonium
Phosphate   5.00
Urea (100 Mesh)   3.00

Mix together and add flavoring to suit.

No. 2

Dicalcium
Phosphate   86.0
Saccharin   0.1
Sodium Glyceryl
Monolaurate
Sulfate   5.0
Sodium Methyl
Salicylate   2.0
Methyl Salicylate   1.0
Adsorbed on:
"Magnesol"   6.0

Tyrothricin Toothpaste

U.S. Patent 2,723,217

Tyrothricin   0.05
Insoluble Sodium
Polymeta-
phosphate   26.89
Dicalcium
Phosphate   26.89
Gum Tragacanth   1.33
Saccharin   0.20
Flavor   0.90
Sodium Lauryl
Sulfate   1.14
Glycerol   18.90
Propylene Glycol   1.00
Distilled Water   22.70

No. 3

Table Salt   4 tsp
Baking Soda   8 tsp
Peppermint Oil   4 drops

Mix well, especially to distribute the oil.

No. 4

Sodium N-Lauroyl
Sarcoside   2.8
Soluble Saccharin   0.2
Flavor   2.0
Calcium Carbonate   25.0
Dicalcium Phos-
phate Dihydrate   70.0

Tooth Powder

Formula No. 1

Sodium Perborate   15
Magnesium Benzoate   5
Precipitated Chalk
USP   80

Liquid Dentifrice

Sodium N-Lauroyl
Sarcoside   2.0

Sodium Carboxy-
methylcellulose 4.0
Flavor 0.5
Water To make 100.0

Antienzyme Chewing
Gum

Gum Base, e.g.
Chicle 20.0
Sucrose 60.0
Corn Syrup 18.5
Flavor 1.0
Sodium N-Lauroyl
Sarcoside 0.5

Mouthwash

Formula No. 1

Sodium Alkylaryl
Sulfonate 40.0 g
Urea 80.0 g
Distilled Water 1.0 gal
Oil of Winter-
green, or
Oil of Cinna-   3.5 g
mon, or
Oil of Orange

Color to suit, using any
pure-food coloring.

No. 2

Benzalkonium
Chloride 10% 6.00 cc

Saccharin
Sodium 0.45 g
Amaranth 0.20 g
Cinnamon Oil 2.00 cc
"Tween" 20 11.90 cc
Distilled Water
To make 1000.00 cc

Add the saccharin and amaranth to about 900 cc of the
water. Mix the cinnamon oil
and "Tween" and add this to
the aqueous solution. Add the
other ingredients and dilute to
volume with water. Mix well
and filter. When diluted with
two parts of water, this makes
a 1:5000 solution of benzalkonium chloride.

No. 3

Sodium
N-Lauroyl
Sarcoside 0.100-0.200
Ethyl Alcohol 10.000
Flavor 0.150
Soluble Saccharin .012
Distilled Water
To make 100.000

No. 4

(Pyorrhea)

British Patent 691,246

Lactic Acid 12
Sodium Lactate 12

Peppermint Oil   1
Calcium Phosphate   0.5
Water   To suit

No. 5

a Cetyl Pyridinium
    Chloride, USP   1.0
    Citric Acid, USP   1.0
b Cinnamon Oil, USP   0.5
    Peppermint Oil,
    USP   1.0
    "Tween" 60   3.0
    Ethanol, USP   100.0
c "Sorbo"   200.0
    Water   q.s.ad 1000.0
    Color   q.s.

Dissolve *a* in a sufficient quantity of water. Add all of *b*, adding alcohol slowly with agitation. Slowly mix *a* and *b*, and add *c* with agitation. Add enough water, in which the desired color has been previously dissolved, to make the total volume 1000 ml.

Cosmetic Tooth Lacquer
Formula No. 1

Austrian Patent 167,113

Amyl Acetate   40-50
Pyroxylin   20-30
Zinc Oxide   5-10
Acetone   8-12

Castor Oil   4-7
Pigment   To suit

No. 2

French Patent 1,270,663

Pyroxylin   4
Cellulose Aceto-
    butyrate   1
Titanium Dioxide or
    other pigment   2-5
Hydrogenated
    Abietic Acid   0.5-2
Ethyl Acetate
        To make 100

Discolored Teeth Cleaner

French Patent 943,987

Monoethanolamine   3
Triethanolamine
    Lactate   10
Water   87

Denture Cleaner

Sodium
    Perborate   240 g
Sodium
    Chloride   480 g
Exsiccated
    Magnesium
    Sulfate   30 g
Calcium
    Chloride   30 g

Anhydrous Sodium Carbonate 30 g
Methyl
Salicylate 1 minim
Menthol 2 g
Peppermint Oil 12 minim

To use, dissolve a small portion in water, and allow the dental plate to remain in the solution overnight. It should be noted that dentures of cellulose acetate are decomposed by alkaline substances, but that those of formaldehyde and phenol-formaldehyde withstand most chemical agents.

Denture Adhesive

Formula No. 1

1) Starch   40
2) "Polyox" WSR-30   1-2
3) Glycerol   7
4) Water   52

Mix 1, 3, and 4, then add 2 with vigorous stirring.

No. 2

Starch   40 g
"Polyox" Resin
   WSR-301   1 to 2 g

Glycerol   7 g
Water   52 cc
Preservative   To suit

Using high turbulence and low shear, add "Polyox" resin to the mixture last, forming a smooth paste. The paste is stable to body and room temperatures. It is highly adhesive and holds the denture in place for prolonged periods of time.

No. 3

U.S. Patent 2,759,838

Petrolatum   44.0
Karaya Gum   50.0
Isopropyl Palmitate   4.0
Aluminum
   Distearate   0.5
8-Hydroxyquinoline   0.1
Color and Flavor   To suit

Toothache Drops

Clove Oil   2 dr
Chloroform   3 dr
Camphor   4 dr
Phenol   4 dr
Turpentine   1 oz
Alcohol   1 oz

Antiseptic Denture Adhesive

U.S. Patent 2,574,476

Petrolatum	36.0
Mineral Oil	12.5
8-Hydroxyquinoline	0.2
Red Color	1.5
Karaya Gum	48.8

Permanently Antiseptic Dental Filling

German Patent 936,060

Zinc Phosphate Cement Powder	1 kg
Cupric Oxalate, Dry	50 g

Pulverize in a ball mill.

Dental Pulp-Capping Composition

U.S. Patent 2,516,438

a	Calcium Hydroxide	100 g
b	Eugenol	100 cc
	Linseed Oil	15-30 cc

Mix *a* and *b* and work together before use.

Topical Dental Solution

Benzocaine	8
Benzyl Alcohol	10
Methyl Salicylate	20

Dental Desensitizing Paste

Sodium Fluoride C. P.	10
Kaolin	10
Glycerol	10

Ointment Base

No. 1
Polyethylene Glycol Ointment USP XV

"Carbowax" Polyethylene Glycol 4000	40-50
"Carbowax" Polyethylene Glycol 400	50-60

Warm the ingredients together on a water bath to 167°F (75°C), remove heat, stir and cool rapidly until congealed.

The base is homogeneous, white, semi-solid, and similar to petrolatum in consistency, and is completely water-soluble. The consistency of the base can be changed by altering the ratio of liquid and solid polyethylene glycols. When solid medicaments are added, more of "Carbowax" poly-ethylene glycol 400 may be needed; with liquid medicaments, more of "Carbowax" polyethylene glycol 4000 should be used.

Stability tests at 104°F (40°C) for six weeks indicate that this base is a satisfactory vehicle for the following therapeutic agents:

Sulfur
Ammoniated Mercury
Benzalkonium Chloride, dissolved in 1 cc of water for 100 g of base
Resorcinol
Aqueous Solutions (max 3%)
Phenol
Mercuric oxide
Ichthammol
Boric acid
Tannic acid
Balsam Peru
Benzocaine

Sulfa drugs are compatible with this base, but a yellow color frequently develops. Agar cup tests show that this discoloration does not in any way impair the antibacterial action of the ointments.

No. 2
(Anhydrous)

"Carbowax" Polyethylene Glycol 4000	47.5
"Carbowax" Polyethylene Glycol 400	47.5
Cetyl Alcohol	5.0

Warm the ingredients on a water bath to 158°F (70°C). Remove heat and stir until congealed.

This base is white, homogeneous, and semi-solid, firmer than No. 1, but with better slip when applied to the skin. It is especially recommended for mixing with aqueous or alcoholic solutions, or for salicylic acid and its derivatives. Aqueous solutions or water can be added to the extent of 10 per cent; alcoholic solutions should not exceed 5 per cent of the formula. This ointment with 6 per cent of salicylic acid will become softer, but will not separate after prolonged storage at 99°F (37°C).

The base is also a vehicle for zinc oxide or sulfur if they are first triturated to a paste with a small amount of propylene glycol, glycerol or "Carbowax" polyethylene glycol 400. Direct introduction of these powders into the base gives a finished ointment which is too firm for smooth application.

Bases No. 1 and No. 2 are anhydrous mixtures of inert substances of suitable consistency. Water may be included to solubilize medicaments such an phenacaine hydrochloride.

No. 3
(Hydrous)

"Carbowax" Polyethylene Glycol 4000	50
"Carbowax" Polyethylene Glycol 4000	40
"Span" Emulsifier 40	1
Water	9

Heat and mix the components on a water bath to 158°F (70°C) until homogeneous, remove heat and stir until congealed.

This hydrous base is glossy, white, semi-solid and more like a cosmetic cream than the anhydrous preparations. It can be kept at 109°F (43°C) without separating and it loses little or no water at room temperature in an open container after several weeks. It maintains stability after the addition of up to 10 per cent of water or an aqueous solution.

It is compatible with the following medicaments:

Zinc Oxide
Coal Tar
Ammoniated Mercury
Ichthammol
Sulfur
Balsam Peru
Salicylic Acid (6% or less)
Phenol

No. 4
(Anhydrous)

"Carbowax" Polyethylene Glycol 4000	42.5
"Carbowax" Polyethylene Glycol 400	37.5
1,2,6-Hexanetriol	20.0

Heat "Carbowax" polyethylene glycol 4000 with the 1,2,6-hexanetriol on a water bath at 140-158°F (60-70°C). Add this melt to "Carbowax" polyethylene glycol 400 at room temperature with vigorous stirring. Stir occasionally until solidification takes place. The melting point of this base is 125.1°F (51.7°C).

This modified ointment base has a good appearance, little tendency to become granular, and is compatible with all important ointment ingredients.

The solvent power of the 1,2,6-hexanetriol increases the compatibility of "Carbowax" polyethylene glycols with drugs and improves the stability of the ointments. The low melting point and high viscosity of the triol lead to a softer preparation at all temperatures and prevent separation or bleeding of the liquid components. The triol, on 90-day feeding tests, is comparable to glycerol in toxicity. Patch tests have demonstrated that it is not a primary irritant nor a sensitizer.

The following ointments made with base No. 4 were stable at 25°C for at least three months, and at 40-42°C for at least two weeks:

MEDICAMENT

Ammoniated Mercury..	5
Benzoic Acid 4% and	
Salicyclic Acid......	4
Boric Acid...........	10
Calamine.............	15
Coal tar, 3%; Zinc oxide, 25%; and Starch....	25
Ethyl Aminobenzoate..	5
Iodine 4% and Potassium Iodide.........	4
Icthammol............	10
Mercuric Oxide, Yellow	1
Phenol...............	3
Resorcinol...........	10
Sulfur, Precipitated....	15
Tannic acid 20% and Anhydrous Sodium Sulfite..............	0.2
Zinc Oxide...........	20
Zinc Oxide 25% and Starch.............	25

The tannic acid and coal tar ointments were stable but somewhat sticky. Their consistency was improved by increasing the proportion of "Carbowax" polyethylene glycol 400.

If more than 10 per cent of water is to be added to base No. 4, the consistency will be improved by increasing the 4000 component or decreasing the 400 glycol.

FORMULATIONS

A. Whitfield's Ointment
B. Chlortetracycline ointment
C. Bacitracin ointment
D. Benzocaine anesthetic ointment
E. Chloramphenicol ointment
F. Cortisone ophthalmic ointment
G. Peroxide ointment
H. Salicylanilide ointment
I. Streptomycin ointment
J. Nitrofurazone ointment
K. Athlete's foot ointment (NF, X)
L. Athlete's foot ointment
M. Athlete's foot ointment
N. Athlete's foot ointment
O. Hydrocortisone ointment.

	A	B	C	D	E	F	G	H	I	J	K	L	M	N	O
"Carbowax" polyethylene glycol															
300				30		50			to 10	498					
400	49.5	10					100								
1500			5.5	15	30	200		94	5.5			19	5.5		39
1540										450					
4000	49.5	2		15			100		2			29.6	22.2	35	19
Polyethylene glycol Ointment D.S.P. to 1000											750				
Propylene glycol												10	7.8	5	31
Dibenzothioindigo (red)												0.4			
Triethanolamine												3	3.3		
Water, distilled												15	55.6	35	5
Zinc stearate														5	6
Bacitracin			1000 units												
Benzocaine				20											
Benzoic acid	60														
Cetyl pyridinium chloride						0.2									
Chloramphenicol					0.275										
Chlortetracycline	1														
Cortisone					0.25										
"Hyamine" 1622								1							
Hydrocortisone															1
Nitrofurazone											2				
Propanol													10		
Propionic acid													3.6		
Salicylanilide							5								
Salicylic acid	30														
Sodium propionate													16.4		
Streptomycin								0.5							
Sulfadiazine													3.6		
Undecylenic acid											50	5	5.55		
Zinc peroxide						10									
Zinc undecylenate											200	18			

Pharmaceutical Ointments & Creams

All these preparations have withstood a shelf test of at least four months before approval for inclusion here. Quantities are parts by weight.

Formula No. 1
Acriflavine

Acriflavine	0.023
Liquid Paraffin	25.000
"Polawax"	2.500
Water	75.000

No. 2
Benzoyl Peroxide

Benzoyl Perox. Anhyd.	10.0
Liquid Paraffin	37.5
Petroleum Jelly	37.5
"Polawax"	25.0

No. 3
Calamine Liniment

Calamine	3.5
Lanolin	2.5
Olive Oil	2.0
"Polawax"	2.5
Water	71.5

No. 4
Calamine Lotion

Calamine	42.0
Liquid Paraffin	37.5
"Polawax"	3.1
Water	62.1
Zinc Oxide	3.1

No. 5
"Eusol"

Beeswax	1
"Eusol"	50
"Polawax"	49

No. 6
Insect Repellant

Dimethyl Phthalate	10.0
Oleic Acid	2.7
"Polawax"	0.5
Triethanolamine	0.9
Water	10.0

No. 7
Methyl Salicylate

Methyl Salicylate	12.5
Petroleum Jelly	10.0
"Polawax"	5.0
Water	72.5

No. 8
Dibucaine

5-Aminoacridine Hydrochloride	0.1
Glycerol	18.0
Liquid Paraffin	44.0
Dibucaine Hydrochloride	0.1
Petroleum Jelly	24.0
"Polawax"	14.0

No. 9
Phenyl Mercuric Acetate

Liquid Paraffin	20.000
Petroleum Jelly	15.000
"Polawax"	15.000
Phenyl Mercuric Acetate	0.125
Water	60.000

No. 10
Proflavine-Urea

Liquid Paraffin	12.200
Petroleum Jelly	4.050
"Polawax"	4.050
Proflavine Sulphate	0.300
Sodium Phosphate	0.135
Sodium Chloride	0.597
Urea	3.750
Water	70.900

No. 11
Sulphanilamide

Castor Oil	25
Glycerol	10
"Polawax"	10

Sulphanilamide 10
Water 45

No. 12
Turpentine

Camphor 5
"Polawax" 15
Turpentine 50
Water 150

No. 13
Acetic Turpentine

Acetic Acid 7.5
"Polawax" 5.0
Turpentine 2.5
Water 17.0

No. 14
Sulphacetamide

Sulphacetamide
Soluble 20
"Polawax" 12
Liquid Paraffin 9
Water 66

No. 15
Cupro-Zinc

Copper Sulphate 1.00
Zinc Sulphate 1.50
Camphor 0.25
"Polawax" 12.00
Liquid Paraffin 9.00
Water 72.00

No. 16
"Eusol Alternative"

Chloramine 4
"Polawax" 12
Liquid Paraffin 9
Water 72

No. 17
D.D.T. (Scalp)

D.D.T. Cream 2.0
"Polawax" 4.0
Liquid Paraffin 22.5
Water 80.0

No. 18
Camphor (Skin)

Camphor 10
"Polawax" 20
Petroleum Jelly 20
Liquid Paraffin 20
Water 40

No. 19
Proflavine

Proflavine Mono-
hydrochloride 0.1
"Polawax" 3.0
Liquid Paraffin 36.0
Water 58.0

No. 20
Cod Liver
(Bedsore)

Cod Liver Oil 10
"Polawax" 20

Lanolin	5
Glycerol	10
Water	75

No. 21
Penicillin

Penicillin	
(Ca or Na)	q.v.
"Polawax"	7.0
Paraffin Wax	5.0
Liquid Paraffin	41.0
Chlorocresol	0.1
Water	47.0

No. 21A
Aureomycin Cream Base

"Polawax"	8
Propylene Glycol	18
Liquid Paraffin	40
Hard Paraffin	4
Distilled Water	30

No. 22
Ammonium Alum

"Polawax"	15
Liquid Paraffin	10
Water	70
Ammonium Alum	5

No. 23
Atropine Sulphate

"Polawax"	15
Liquid Paraffin	10
Water	74
Atropine Sulphate	1

No. 24
Balsam of Peru

"Polawax"	15
Liquid Paraffin	10
Water	65
Balsam of Peru	10

No. 25
Benzocaine

"Polawax"	15
Liquid Paraffin	10
Water	71
Benzocaine	4

No. 26
Benzoic Acid

"Polawax"	15
Liquid Paraffin	10
Water	65
Benzoic Acid	10

No. 27
Betanaphthol

"Polawax"	15
Liquid Paraffin	10
Water	70
Betanaphthol	5

No. 28
Boric Acid

"Polawax"	15
Liquid Paraffin	10
Water	65
Boric Acid	10

No. 29
Burows Solution

"Polawax"	15
Liquid Paraffin	10
Water	65
Burows Solution	10

No. 30
Chlorobutanol

"Polawax"	15
Liquid Paraffin	10
Water	74
Chlorobutanol	1

No. 31
Chloral Hydrate

"Polawax"	15
Liquid Paraffin	10
Water	65
Chloral Hydrate	10

No. 32
Coal Tar

"Polawax"	15
Liquid Paraffin	10
Water	70
Coal Tar	5

No. 33
Ephedrine Hydrochloride

"Polawax"	15
Liquid Paraffin	10
Water	74
Ephedrine Hydrochloride	1

No. 34
Gentian Violet

"Polawax"	10.0
Liquid Paraffin	10.0
Water	79.9
Gentian Violet	0.1

No. 35
Iodoform

"Polawax"	15
Liquid Paraffin	10
Water	65
Iodoform	10

No. 36
Lead Sub-acetate

"Polawax"	15
Liquid Paraffin	10
Water	70
Lead Sub-acetate	5

No. 37
Lime Water

"Polawax"	15
Liquid Paraffin	10
Water	65
Lime Water	10

No. 38
Menthol and Methyl Salicylate

"Polawax"	15
Liquid Paraffin	10
Water	55

Menthol	10
Methyl Salicylate	10

No. 39
Mercuric Oxide
(Red)

"Polawax"	15
Liquid Paraffin	10
Water	65
Mercuric Oxide, Red	10

No. 40
Mercuric Oxide
(Yellow)

"Polawax"	15
Liquid Paraffin	10
Water	74
Mercuric Oxide	1

No. 41
Methyl Salicylate

"Polawax"	15
Liquid Paraffin	10
Water	65
Methyl Salicylate	10

No. 42
Oleum Rusci

"Polawax"	15
Liquid Paraffin	10
Water	65
Oleum Rusci	10

No. 43
Phenol

"Polawax"	15
Liquid Paraffin	10
Water	73
Phenol	2

No. 44
Picric Acid
(3 months only)

"Polawax"	15
Liquid Paraffin	10
Water	74
Picric Acid	1

No. 45
Potash Alum

"Polawax"	15
Liquid Paraffin	10
Water	70
Potash Alum	5

No. 46
Potassium Iodide/Sodium
Thiosulphate

"Polawax"	15
Liquid Paraffin	10
Water	64
Potassium Iodide	10
Sodium Thiosulphate	1

No. 47
Pyrogallol
(3 months only)

"Polawax"	15
Liquid Paraffin	10

Water	73
Pyrogallol	2

No. 48
Sulphadiazine

"Polawax"	15
Liquid Paraffin	10
Water	70
Sulphadiazine	5

No. 49
Sulphur Créam

"Polawax"	15
Liquid Paraffin	10
Water	60
Sulphur	15

No. 50
Zinc Oxide

"Polawax"	15
Liquid Paraffin	10
Water	65
Zinc Oxide	10

No. 51
Cresantol No. 3

"Polawax"	15
Liquid Paraffin	10
Water	72
Cresantol, No. 3	3

No. 52
Aminoacridine
Hydrochloride

"Polawax"	15.0
Liquid Paraffin	10.0
Water	74.9
Aminoacridine Hydrochloride	0.1

No. 53
Aminoacridine
Hydrochloride/
Sodium Bicarbonate

"Polawax"	15.0
Liquid Paraffin	10.0
Water	74.8
Aminoacridine Hydrochloride	0.1
Sodium Bicarbonate	0.1

No. 54
Lauryl Thiocyanate

"Polawax"	15
Liquid Paraffin	10
Water	70
Lauryl Thiocyanate	5

No. 55
"Adrenaline"

"Adrenaline"	0.01
Tartaric Acid	0.10
Sodium Meta-bisulfite	0.10
Preservative	0.10
Liquid Paraffin	45.00
Distilled Water	40.00
"Polawax"	15.00

No. 56
Pyrilamine Maleate (I)

Pyrilamine Maleate	1.5
White Wax	9.0
"Polawax"	9.0
Liquid Petrolatum	37.0
Distilled Water	q.s. 100.0

No. 57
Pyrilamine Maleate (II)

Pyrilamine Maleate	1.50
White Wax	5.00
"Polawax"	10.00
Glycerol	10.00
"Methocel" 1500	0.05
Distilled Water	q.s. 100.00

No. 58
Chlorocresol

"Polawax"	20.00
Lanolin	10.00
Chlorocresol	0.05
Water	100.00

No. 59
Antazoline

Antazoline Hydrochloride	2.00
"Polawax"	15.00
Petroleum Jelly	25.00
Liquid Paraffin	10.00
Chlorocresol	0.10
Water	100.00

No. 60
Proflavine

Proflavine Hemisulphate	0.1
"Polawax"	3.0
Liquid Paraffin	36.0
Water	58.0

Chloramphenicol

Cetyl Alcohol	12
Liquid Paraffin	11
Sodium Lauryl Sulfate	1
Chloromycetin	1
Water	75-77
Propyl p-Hydroxy-benzoate	0.1

Melt the cetyl alcohol and warm to about 95°C. Add the sodium lauryl sulphate with stirring, then add 6 parts of water with continuous stirring and hold at 110°C until frothing ceases. Cool to about 50°C, with stirring; add the liquid paraffin and the propyl p-hydroxybenzoate (Nipasol); and finally stir in a premixed solution of 1 part chloramphenicol (BP) in 69 parts of water. Cool with stirring. All parts given are by weight.

For Haemorrhoids

"Dehydag"

Wax SX	10.0
"Cetiol" V	6.0
Lanolin	20.0
Petrolatum, White	20.0
Procaine Hydro-chloride	1.0
Menthol	0.2
Bismuth Sub-gallate, Basic	5.0
Zinc Oxide	5.0
Rice Starch	5.0
Witch Hazel Extract, dist.	10.0
Water	17.8
Preservatives	As desired

Heat "Dehydag" Wax SX, "Cetiol" V, Lanolin, petrolatum, and bismuth gallate on the water bath to 70-80°C (158-176°F). Dissolve Procaine, Witch Hazel Extract, and preservative in the water, likewise heated to 70-80°C (158-176°F), and stir these into the molten fats.

Emulsification takes place while stirring until cold. Into resultant ointment incorporate the zinc oxide and rice starch at room temperature. Finally pass the finished ointment through a roller mill and fill into tubes or other suitable containers.

Antibiotic Vehicle

U.S. Patent 2,713,019

Formula No. 1

"Carbowax" 4000	32.26
"Carbowax" 1540	32.20
"Carbowax" 1000	32.26
Polyethylene Glycol 400W	103.00

No. 2

Polyethylene Glycol 400 Dilaurate	60
Polyethylene Glycol 400 Distearate	10
White Petrolatum	10
Liquid Petrolatum	10
"Glycowax" 5932	10

Acne Cream

a	"Veegum"	1.75
	Sodium Carboxymethyl Cellulose (type 7 MSP)	0.40
	Water	34.45
b	Glycerol	5.00
	Allantoin	0.25
	Resorcinol	3.00
	"Triton" X100	0.20
	Water	29.45

c "Nytal" 300   16.00
  Titanium Dioxide   2.90
  Iron Oxides   1.10
  Sulfur   5.00
d "Vancide" BN
  Solution (50%)   0.50

Dry blend "Veegum" and CMC and add to the water slowly, agitating continually until smooth. Add *b* to *a*. Pulverize *c* and add to *a-b*. Add "Vancide" BN solution to preceding.

Ointment for Boils

Tetracycline   2.4
Zinc Oxide and
  Talc   36.0
Glycerol   24.0
Bentonite   12.0
Water   To make 240.0

Glucose Ointment

Glucose   10
"Dehydag" Wax SX   15
"Cetiol" V   15
Water   50
Preservative   As desired

Melt "Dehydag Wax" SX and "Cetiol" V in a water-jacketed vessel at a temperature of 176-194°F. Dissolve the glucose and preservative in the water, add them to the fatty compound at 176-194°F, and stir until cool.

The ointment thus obtained can be immediately filled into tubes which have been coated on the inside with a corrosion-proof lacquer.

Chest Rub

Nicotinic Acid
  Methyl Ester   0.5
Rectified Oil of
  Turpentine   5.0
Pine Needle Oil   5.0
Oil of Thyme   1.0
Rosemary Oil   1.0
Eucalyptus Oil   1.0
Oil of Juniper   0.5
Oil of Sage   0.5
Flowers of Camphor   3.0
"Emulgade" F   18.0
"Cetiol" V   5.0
Liquid Paraffin   5.0
Peanut Oil,
  Hydrogenated   5.0
Dist. Water
    To make 100.0
Preservative

Massage Stick

Petrolatum   39.5
Lanolin   40.0

Paraffin Wax   16.0
Beeswax   4.0
Perfume   0.5

Melt the first four ingredients on a water bath. Then add the perfume and pour the product into molds.

Suppository Base

A. "Carbowax" 1000   96
   "Carbowax" 4000   4
B. "Carbowax" 1540   94
   Hexanetriol-1,2,6   6
C. "Carbowax" 1000   75
   "Carbowax" 4000   25
D. "Carbowax" 4000   88
   Hexanetriol-1,2,6   12
E. "Carbowax" 1540   70
   "Carbowax" 6000   30
F. "Carbowax" 6000   50
   "Carbowax" 1540   30
   Water & Medication   20

Base A, a low-melting base, is useful if rapid disintegration is desired, or if the suppository is to be used in a reasonably short period of time. This base may require refrigeration during the summer months.

Bases B and C are more stable formulations and may be used if the suppository is to be subjected to storage at higher temperatures or when a slow release of the medicament is advisable. Base B, incorporating hexanetriol-1,2,6, is especially useful for the preparation of suppositories by compression methods, and with active ingredients that have no effect on the melting point of the "Carbowax" compounds.

Base D, a higher-melting base incorporating hexanetriol-1,2,6, is recommended for preparation of suppositories by compression methods and where active ingredients slightly lower the melting point of the "Carbowax" compounds.

Base E was formulated to accommodate those medicinal substances that lower the melting point of the "Carbowax" compounds.

Base F, a base containing water, facilitates the incorporation of water-soluble, polyethylene glycol insoluble substances.

Vaginal Suppositories

"Cetrimide"   50 mg
Diiodohydroxy-
   quinoline   200 mg

Suppository

    Base   To make 100 g

Tablet for Vaginal Douche

"Cetrimide"	50 mg
Boric Acid	q.s.

Use: one tablet to one quart of water; To give 100 mg after neutralization by Sodium Carbonate q.s. for effervescence.

To Stop Nosebleed

A special hemostatic ointment used as a practical method for stopping spontaneous nosebleeds.

White Gelatin	25
Zinc Oxide	10
Calcium Chloride	10
Glycerol	5
Sulfachrysoidine	5
Distilled Water	50

At normal room temperature, the ointment is whitish, firm, and pliant, and without odor. Before use, it is heated gently to soften. Apply freely on a small piece of cotton, and leave it in the nasal cavity for one to two minutes. After the first application, the bleeding decreases markedly and stops completely after the second or third application. The ointment has no toxic after-effects and is not irritating to the mucous membrane.

Contraceptive Gel

7-Chloro-4-Indanol	0.1
Sodium Chloride	10.0
Ricinoleic Acid	1.0
Sodium Dodecyl Sulfate	0.2
Water	To make 50-100

Dressing for Burns

Glycerol	18.0
Procaine HCl	1.2
Magnesium Sulfate Heptahydrate	20.0
Magnesium Oxide	6.0
Water-Soluble Base	54.8

Poison Ivy Protectant

Monostearin, Self-emulsifying	10.0
Glycerol	6.0
Triethanolamine Ricinoleate	1.0
Wool Alcohols	6.0
Oleyl Alcohol	3.0
Chlorocresol	0.2
Zirconium Oxide	4.0
Distilled Water	To make 100.0

(Stable) Calamine Lotion

	Formula No. 1	No. 2
Calamine	8.00	8.00
Zinc Oxide	8.00	8.00
Glycerol	2.00	3.00
Sodium Bitartrate	1.00	—
"Veegum"	1.00	0.50
Sodium Citrate	0.25	—
CMC Type 70H	—	1.00
Dioctyl Sodium Sulfosuccinate	0.03	0.05
"Span" 20	0.10	—
Distilled Water To make	100.00	100.00

No. 3
(High Viscosity)

A thicker-than-normal calamine lotion can be formulated from the following ingredients:

Calamine	80
Zinc Oxide	80
Talc	100
Glycerol	250
Corn Starch	100
Magma Bentonite	
To make	1000

To Preserve Fish Liver

Methyl	
"Parasept"	1.75 lb
Butyl	
"Parasept"	2.25 lb

Isopropanol	33.00 gal
Water	67.00 gal

One volume of mixture is used for 10 volumes of fresh livers. This preserves them for about two weeks at 70°F. The isopropanol is later removed from the livers or oil by washing with hot water and centrifuging.

To Remove Callus

Salicylic Acid	100
Lactic Acid (80%)	100
Lanolin, Anhydrous	100
Petrolatum, Yellow	695

Melt together gently and mix. While mixing to cool add

Lavender Water Oil	5

Skin Fungicides

Formula No. 1

Calcium Propionate	15.0
Zinc Propionate	5.0
Polychloro-Copper-Phthalocyanine	0.5
Talc, USP	79.5

No. 2

Zinc Undecylenate	20.0
Talc, USP	76.0
Undecylenic Acid (Grade AA)	2.0
Dibenzo Thioindigo (Red)	2.0

No. 3

Salicylic Acid (USP Powder)	2.0
Boric Acid (USP Powder)	6.0
Zinc Stearate, USP	3.0
Talc, USP	86.5
Dibenzo Thioindigo (red)	2.0
Polychloro Phthalocyanine (green)	0.5

No. 4

Undecylenic Acid, Grade AA	5.0
Triethanolamine	3.0
Zinc Undecylenate	18.0
Propylene Glycol	10.0
"Carbowax" 1500	19.0
"Carbowax" 4000	29.6
Water, Distilled	15.0
Dibenzo Thioindigo (Red)	0.4

No. 5

Undecylenic Acid	5
"Kessco" Wax A-21	4
Stearic Acid	19
White Mineral Oil (65-75 visc)	3
Propylene Glycol	5
Sodium Alginate Solution (1%)	64

Heat all of the ingredients together to 90°C. Allow the mixture to cool to 35°C with continual stirring. The preparation remains liquid at this temperature and may be poured into containers. It sets to a firm cream in approximately 16 hours.

Antiseptic Lubricant

"Cetrimide"	0.1
Gum Tragacanth	2.5
Glycerol	20.0
Ethanol	2.5
Water	To make 100.0

Antiseptic Jelly

"Cetrimide"	1.0
Gelatin	2.5
Water	To make 100.0

This formula gives a solid gel of the suppository type; more fluid preparations of the hand-jelly type can be prepared by using alternative thickening agents such as cellulose ethers.

Antibiotic-Antiseptic Solution

"Cetrimide"	0.05
Tyrothricin	0.05
Ethanol	0.50
Water and	
Color	To make 100.00

Germicide

Chloroxylenol	50 g
Terpineol	100 ml
Ethanol (95%)	200 ml
Ricinoleic Acid	50 g
Potassium Hydroxide Solution	q.s. to neutralization
Distilled Water	To make 1000 ml

Mix together the ricinoleic acid and 50 ml of the ethanol and reserve 3 ml of the solution. To the remainder gradually add with constant stirring the solution of potassium hydroxide until a few drops of the resulting soap solution gives a pink color with phenolphthalein, and then add reserved solution. Dissolve the chloroxylenol in the remainder of the ethanol, mix with terpineol, add the mixture to the soap solution, and finally add sufficient distilled water to produce the required volume.

The final solution is a yellow-to-amber colored liquid with a terpineol odor and is soapy to touch. When diluted with nineteen times its volume of water it gives a white stable emulsion.

For Surgical Instruments

"Cetrimide"	0.20
Sodium Nitrite	0.05
Sequestrant	0.05
Sodium Sesquicarbonate	4.70
Dry Vehicle	To make 100.00

Flavorings for Drugs

Analgesics and Antipyretics: Grenadine.

Antibiotics: Banana, butterscotch, chocolate, citrus, maple, mint/spice, orange, peach, peach/anise, pineapple, vanilla, wild cherry.

Antihistamines: Anise/birch, anise/mint, blackcurrant, cherry, chocolate, citrus, coconut, coconut/vanilla, raspberry, rum/peach, spice/vanilla, wild cherry.

Barbiturates: Anise/citrus, blackcurrant, cherry, lime, mint, concentrated orange juice, raspberry, spearmint, spice, strawberry, vanilla, wine/raspberry.

Bitters: Liqueur flavours, port, raspberry, salt/sugar, salt/sugar/anise, vanilla.

Bromides: Apricot, peach, peach/anise.

Cough Syrups: Apricot, blackberry, blackcurrant, butterscotch, gooseberry, loganberry, peach, peach/anise, peach/orange, peach/rum, pineapple.

Digestive Preparations: Chocolate, maple, maple/butterscotch, mint/anise, orange, peppermint, raspberry, vanilla, wild cherry.

Emulsions: Black currant, butterscotch.

Expectorants: Black currant.

Geriatric Preparations: Banana, coffee, pineapple, port, vanilla.

Iron Preparations: Aspic, aspic/sodium glutamate, black currant, loganberry, strawberry.

Laxatives: Aspic, aspic/sodium glutamate, butterscotch, butterscotch/maple, pineapple, pineapple/orange, strawberry.

Liver Preparations: Black currant, sodium glutamate.

Paediatric Preparations: Banana, butterscotch, chocolate, coconut, coconut/vanilla, grenadine, pineapple, vanilla.

Phosphates: Port, raspberry, vanilla.

Piperazine: Strawberry.

Protein Hydrolysates: Aspic, aspic/sodium glutamate.

Sulphonamides: Pineapple, strawberry.

Vitamins: Aspic, aspic/sodium glutamate, banana, butterscotch, butterscotch/maple, caramel, cream soda, maple, maple/honey, maple/vanilla, pineapple, strawberry, vanilla.

Yeast: Black currant, strawberry.

Flavors with wide applications were lemon/lime, lime, and passion fruit.

In a series of specific tests, quinine hydrochloride was found best flavoured with cacao, aromatic eriodictyon, licorice or synthetic coconut syrups; additions of salt and increased amounts of citric acid improved the disguising power of the syrups. Incorporation of carboxymethylcellulose to increase viscosity also proved an advantage.

For chlorotetracycline hydrochloride (0.05-34.63 mg/4 mils dose), simple syrup was found best, followed by root beer and cream soda. Oxytetracycline was best masked by cacao syrup.

Low-Caloric Drug Syrup

Methylcellulose, USP	5.0 g
Cyclamate Sodium, NF	27.0 g
Glycerol, USP	2.0 cc
Imitation Strawberry Flavor	9.8 cc
Amaranth Solution, USP	10.0 cc
Distilled Water	q.s. ad. 1.0 liter

Heat 400 ml of water to 70 to 80°C. Add the methylcellulose and stir until hydration is effected. Cool to 12°C to ensure complete solution of the methylcellulose. Add remaining ingredients and water to volume and stir until homogenous.

This syrup has been improved by adding methyl and propyl parabens USP in the usual concentrations to prevent microbial contamination. The taste may be improved with a combination of artificial sweeteners. Reduce the 27 g of cyclamate sodium to 13.4 g and add 1.34 g of saccharin sodium. When high concentrations of artificial sweeteners are required, it is preferable to use a combination of cyclamate and saccharin.

Cough Syrup
Formula No. 1

Dihydrocodeinone Bitartrate	0.32
Pyrilamine Maleate	2.40
Potassium Guaiacol Sulfonate	32.00
Sodium Citrate	16.00
Citric Acid	16.00
Sorbic Acid	0.50
Water	160.00
Honey	1150.00

The solid ingredients are dissolved in about 140 ml of water with the aid of gentle heat. The solution is filtered, the filter is washed with the remainder of the water, and the filtrate and washings are mixed with the honey.

No. 2

Codeine Sulfate	2.10
Promethazine Hydrochloride	1.05
Potassium Guaiacol Sulfonate	8.40
Sodium Citrate	37.50
Citric Acid	12.50
Ipecac Fluidextract	2.30
Chloroform	4.10
Water	155.00
Glycerol	190.00
Honey	940.00

The water solution of the solid ingredients is prepared as for No. 1. This solution is added to a mixture of the fluidextract, glycerol and honey, and when it has cooled to room temperature the chloroform is incorporated. Filtration of the product, under pressure and using filter aid, is desirable.

Nasal Spray

"Cetrimide"	0.025
Phenylephrine Hydrochloride	0.500
Neomycin (as Sulfate)	0.300
Water To make	100.000

Throat Spray

"Cetrimide"	0.05
Tyrothricin	0.02
Diperodon Hydrochloride	0.05
Water To make	100.00

Throat Lozenge

"Cetrimide"	5 mg
Tyrothricin	2 mg
Diperodon Hydrochloride	1 mg
Dry Vehicle To make	100 g

Hematinic Syrup

Cyanocobalamin, USP	400.00 mcg
Ascorbic Acid, USP	20.00 g
Ferrous Gluconate, USP	17.00 g
Flavor	q.s.
Sodium Citrate, USP	q.s.
"Sorbo"	q.s. ad

(Sorbitol   1000.00
Solution,   ml
USP)

Dissolve ferrous gluconate in "Sorbo" at 70°C. Cool the solution to 45°C and dissolve the ascorbic acid. Cool to room temperature and add the vitamin B_{12}. Adjust the pH to 4.0 with sodium citrate.

Wild Cherry Syrup

Wild Cherry   q.s. (about
  Flavor Con-   30 ml)
  centrate
Citric Acid,
  USP   2.0 g
Methylparaben,
  USP   1.2 g
Propylparaben,
  USP   0.2 g
Sucrose, USP   654.5 g
"Sorbo" Sorbi-
  tol Solution,
  USP   200.0 ml
Water   q.s. ad 1000.0 ml

Heat 300 ml of water to boiling and dissolve the parabens in it. Add the sucrose with constant mixing. Allow the mixture to cool to 40°C and add the citric acid, mixing thoroughly. When the mixture has come to room tempera-

ture, mix in the "Sorbo" and add sufficient quantity of water to make total volume 1000 ml.

Antihistamine Elixir

Chlorpheniramine
  Maleate   0.4 g
Benzaldehyde   0.1 ml
Vanillin   0.2 g
Amaranth
  Solution   1.0 ml
Ethanol   100.0 ml
"Sorbo"   450.0 ml
Water   To make
                 1000.0 ml

Vitamin C Syrup

Sodium   100 mg of
  Ascorbate   act. per
                    5.0 ml
Wild Cherry Fluid
  Extract, NF   q.s.
Citric Acid, USP   10.0 g
Color   q.s.
Preservative   q.s.
"Sorbo"   q.s. to
  Sorbitol   1000.0 ml
  Solution,
  USP

In about 900 ml of the "Sorbo," mix and dissolve all ingredients. Then add sufficient "Sorbo" to make the total volume 1000 ml.

Multiple Vitamin Syrup

Nicotinamide, USP	2.00 g
Riboflavin-5′-Phosphate Sodium	0.33 g
Thiamine Hydrochloride, USP	0.30 g
Pyridoxine Hydrochloride, USP	0.10 g
Ascorbic Acid, USP	8.00 g
Vitamin B_{12} Oral Powder	q.s.
Vitamin A Palmitate	q.s.
Vitamin D	q.s.
"Tween" 80	10.00 g
Sodium Carboxymethylcellulose, USP	10.00 g
Preservative	q.s.
Flavors	q.s.
Saccharin Sodium, USP	0.32 g
Cyclamate Sodium, NF	3.33 g
"Sorbo" Sorbitol Solution, USP	600.00 g
Water	q.s. ad 1000.00 ml

(*a*) Heat 575 g of "Sorbo" to 85°C, and dissolve in it the nicotinamide, riboflavin-5′-phosphate, synthetic sweeteners, and preservatives. Cool to 25°C and mix well under inert gas. Then add and dissolve the thiamine and pyridoxine hydrochlorides with thorough mixing. Add and dissolve the ascorbic acid.

(*b*) Slurry the vitamin B_{12} in 10 ml of distilled water and add to the main batch (*a*), mixing well.

(*c*) Heat 300 ml of distilled water to boiling and dissolve the sodium carboxymethylcellulose, using high-shear equipment. Cool to 25°C and add to the main batch (*a*), mixing well.

(*d*) Mix the vitamins A and D with sufficient "Tween" 80 to obtain clarity (about 10 g). Add 25 g of "Sorbo." Then add 50 ml of distilled water, mix well, and homogenize. Rinse the container with small quantities of water to a total of 25 to 50 ml. Add to the main batch (*a*), and mix well.

(*e*) Add a sufficient quantity of water to make the total volume 1000 ml.

Liquid Antacid Drug

a	"Veegum"	1.0
	"CMC" (type 7	
	MSP)	0.5
	Water	71.4
b	Saccharin	0.1
	Water	6.0
c	Aluminum Hydrox-ide gel F2000	12.0
d	Magnesium Trisilicate	9.0
	Preservative	q.s.

Dry-blend *a* adding to the water slowly, agitating continually until smooth. Mix *b* together and add to *a*. Add small portions of *a-b* to *c*, triturating until smooth. Add *d* and mix until smooth.

Liquid antacids are particularly difficult to formulate as they tend to settle with hard packing, or gel on aging. This formula is very stable on aging and is easily restored to its original viscosity by shaking.

Kaolin-Pectin Suspension

a	"Veegum"	0.88
	"CMC" (type 7	
	MSP)	0.22
	Water	79.12
b	Kaolin	17.50
c	Pectin	0.44
	Saccharin	0.09
	Glycerol	1.75
d	Flavor*	q.s.
	Preservative	q.s.

Mix ingredients as for preceding.

This preparation remains stable on long aging. It appears to thicken but can easily be returned to its original viscosity by shaking.

Pentobarbital Elixir

Pentobarbital Sodium, USP	4 g
Sweet Orange Peel Tincture, USP	30 ml
Ethanol, USP	200 ml
Propylene Glycol, USP	100 ml
Diluted Hydro-chloric Acid, NF	6 ml
Caramel, NF	2 g
"Sorbo" Sorbitol Solution, USP	600 ml
Water	q.s. ad 1000 ml

Dissolve the pentobarbital sodium in 60 ml of water, then add the "Sorbo," ethanol, propylene glycol, sweet orange peel tincture, and caramel.

Mix thoroughly, and add the hydrochloric acid and sufficient water to make the total volume 1000 ml. Mix well, and filter if desirable.

This elixir remains clear at 25°C and at 4°C. The pH is 4.6 to 4.8, and there is no need to add preservative. The pentobarbital content is similar to that of the official preparation.

___

Phenobarbital Elixir

Phenobarbital, USP	4.00 g
Orange Oil, USP	0.25 ml
Amaranth Solution, USP	10.00 ml
Ethanol, USP	200.00 ml
Propylene Glycol, USP	100.00 ml
"Sorbo" Sorbitol Solution, USP	600.00 ml
Water q.s. ad	1000.00 ml

Dissolve the phenobarbital in the ethanol, then add the propylene glycol, orange oil, "Sorbo," and amaranth. Mix thoroughly, and add sufficient water to make the total volume 1000 ml. Mix well and filter, if desirable.

This mixture remains clear even at temperatures as low as 4°C. The pH is 6.2 to 6.4 and bacteriological testing has indicated that there is no need for additional preservatives. Tasting tests with this formula indicated no significant difference when compared to the official elixir.

___

Compound Glycerophosphates Elixir

Sodium Glycerophosphate	35 g
Calcium Glycerophosphate	16 g
Ferric Glycerophosphate	3 g
Manganese Glycerophosphate	2 g
Quinine Hydrochloride	875 mg
Strychnine Nitrate	125 mg
Citric Acid	600 mg
Lactic Acid	20 ml
Compound Cardamom Spirit	2 ml
Ethanol	125 ml
"Sorbo"	458 ml
Water To make	1000 ml

Dissolve the calcium and sodium glycerophosphates in

100 ml of water containing the lactic acid. Dissolve the ferric and manganese glycerophosphates and the citric acid in 50 ml of water with the aid of heat, and add to the first solution. Dissolve the quinine hydrochloride in the alcohol containing the compound cardamom spirit, and add the "Sorbo" to the solution. Mix the three solutions, and add sufficient water to make the product measure 1000 ml. Filter if necessary.

Iron, Quinine and Strychnine Elixir

Soluble Ferric Phosphate	35 g
Quinine Phosphate	5 g
Strychnine Phosphate	250 mg
Orange Oil	1 ml
Ethanol	250 ml
"Sorbo"	392 ml
Water	To make 1000 ml

Dissolve the soluble ferric phosphate in 217 ml of water by cold maceration, and add 98 ml of "Sorbo." Dissolve the strychnine phosphate in the ethanol and add the orange oil, the quinine phosphate, and the remainder of the "Sorbo." Shake until thoroughly mixed; then add the ferric phosphate solution and enough water to bring the volume to 1000 ml. Allow the mixture to stand 24 hours, shaking repeatedly until the quinine phosphate is dissolved. Filter, if necessary, using 10 grams of purified talc.

Aspirin Elixir

Aspirin	2.50
"Sucaryl"	3.50
Soluble Orange Oil	0.25
Ethanol	20.00
Water	15.00
Polyethylene Glycol 400	To make 100.00

Estrone Solvent, Stable

U.S. Patent 2,822,316

Ethyl Lactate	1 pt
Castor Oil	1 pt

Bronchodilator Amine Formulations
Aerosol

No. 1

Isoproterenol HCl	0.25
Water	1.50

Ethanol (Absolute) 33.25
Dichloro-
    difluoromethane
    (Propellant 12)   65.00

No. 2

Isoproterenol HCl   0.25
Water   1.50
Ethanol (Abs)   33.25
Dichlorotetra-
    fluoroethane
    (Propellant 114) 40.00
Dichlorodi-
    fluoroethane
    (Propellant 12)   25.00

No. 3

Isoproterenol HCl   0.20
Water   2.00
Ethanol (Abs)   37.80
Trichloromono-
    fluoromethane
    (Propellant 11)   30.00
Dichlorodi-
    fluoromethane
    (Propellant 12)   30.00

No. 4

Epinephrine USP   0.25
3N Solution of Hy-
    drochloric Acid   0.50
Ascorbic Acid   0.15
Water   1.00
Ethanol (Abs)   33.10

Dichlorodi-
    fluoromethane
    (Propellant 12)   25.00
Dichlorotetra-
    fluoroethane
    (Propellant 114) 40.00

No. 5

Ephedrine   15.00
Ethanol 95%   26.00
Dichlorodi-
    fluoromethane
    (Propellant 12)   59.00

No. 6

Isoproterenol HCl   0.25
Water   1.50
Ethanol (Abs)   33.25
Dichlorotetra-
    fluoroethane
    (Propellant 114) 45.50
Dichlorodi-
    fluoromethane
    (Propellant 12)   19.50

Cardiovascular Drugs

Formula No. 1

Narcotine   1.00
Water   0.65
Ethanol (Abs)   12.35
Chloroform   20.60
Dichlorodi-
    fluoromethane
    (Propellant 12)   34.80

Dichlorotetra-
fluoroethane
(Propellent 114)   30.60

No. 2

Nicotine	1.00
Ethanol 95%	34.00
Dichlorodi- fluoromethane (Propellent 12)	25.00
Dichlorotetra- fluoroethane (Propellent 114)	40.00

No. 3

Octyl Nitrite	0.1
Ethanol 95%	20.0
Dichlorotetra- fluoroethane (Propellent 114)	49.2
Dichlorodi- fluoromethane (Propellent 12)	30.7

No. 4

Octyl Nitrite	0.1
Ethanol 95%	9.9

Dichlorotetra-
fluoroethane
(Propellent 114)   55.4
Dichlorodi-
fluoromethane
(Propellent 12)   34.6

No. 5

Octyl Nitrite	0.1
Diethyl Ether	14.9
Dichlorotetra- fluoroethane (Propellent 114)	45.0
Dichlorodi- fluoromethane (Propellent 12)	40.0

Antispasmodic Formulation

Atropine	0.1
Ethanol 95%	9.9
Dichlorodi- fluoromethane (Propellent 12)	45.0
Dichlorotetra- fluoroethane (Propellent 114)	45.0

Vitamin A Aerosol Foam

Concentrate:

Vitamin A (as Palmitate in Corn Oil)	2000 u/g

Vehicle:

	% wt
Double-Pressed Stearic Acid	2.6
Acetylated Lanolin Alcohols	1.5

Vehicle (cont.)

Glyceryl Monostearate		3.0
Polyoxyethylene Sorbitol Monolaurate USP		3.0
Triethanolamine USP		2.0
Propylene Glycol USP		3.0
Preservative		0.1
Perfume		0.2
Water	To make	100.0

Fill: Concentrate, 90.0; Propellent 12/114 (40/60) 10.0
Package: 6-oz Crown Spratainer (lacquer-lined); Clayton
Foam Valve

Vitamin A Stability:	*Potency Retained* (Based on Theoretical)
Initial	100
3 mo , 43°C	95
6 mo , R.T.	95

Neomycin, Polymyxin, Hydrocortisone Aerosol

Concentrate	*Per Gram*
Neomycin, USP	3.5 mg
Polymyxin, USP	4000 units
Hydrocortisone, USP	10.0 mg

Vehicle		*% Wt*
Anhydrous Lanolin, USP		0.45
Stearic Acid, USP		6.86
Cetyl Alcohol		0.45
"Span" 80		0.45
"Tween" 60		2.75
"Sorbo"		4.90
Water	To make	100.00

Fill: Concentrate, 90.0; Propellent 12/114 (15/85), 10.0
Package: Wheaton S-32F1 plastic-coated glass bottle (10 ml practical fill); Risdon 4935-28 gold-anodized valve with 4935-EF4 actuator

Stability	*Potency Retained, based on theoretical %*			
		2 mo,	6 mo,	15 mo,
	Initial	43°C	RT	RT
Hydrocortisone	125	102	105	85
Neomycin	110	108	89	82
Polymyxin	107	91	106	76

Neomycin, Polymyxin, Hydrocortisone Nitrosol

Concentrate	*Per Gram*
Neomycin, USP	3.5 mg
Polymyxin, USP	4000 units
Hydrocortisone, USP	10.0 mg

Vehicle	*% Wt*
Polyethylene Glycol 400	
Monostearate USP	2.5
Isopropyl Palmitate	2.5
Stearic Acid USP	3.0
Propylene Glycol, USP	10.0
Sodium Benzoate, USP	0.2
Water To make	100.0

Fill: Concentrate 50.0 g; Nitrogen 90.0 psi g
Package: 2¼-oz Peerless Aluminum Can (Epon-lined); Precision 6B Valve with 6B actuator

Stability	*Potency Retained, based on theoretical %*	
	Initial	2 mo, RT
Neomycin	103	89
Polymyxin	81	81
Hydrocortisone	105	92

Procaine Penicillin Suspension

U.S. Patent 2,897,120

A preparation containing 300,000 units penicillin/ml, which gives the minimum of difficulty from settling and caking.

		mg/ml
a	"Plasdone," Polyvinylpyrrolidone	5.0
	Sodium Carboxymethyl Cellulose (Low viscosity)	5.0
	Sodium Citrate Granular USP	5.7
	Methylparaben USP (sterile)	2.63
	Water for injection	q.s.

		units/ml
b	Procaine Penicillin G milled (sterile bulk, 4% Lecithin-coated)	157,500
	Procaine Penicillin G micronized (sterile bulk, 1.25% Lecithin)	157,500

To prepare the vehicle, mix the sodium citrate, sodium carboxymethylcellulose, and "Plasdone" in the water for injection and sterilize by heating to 120°C for 30 min. When cool, add the sterile methylparaben. Mix the milled and micronized penicillins and add to a under aseptic conditions. Pass the suspension through a colloid mill and fill aseptically into sterile containers.

Antacid Suspension

Magnesium Hydroxide	40.00
Aluminum Hydroxide	40.00
Sodium Carboxymethylcellulose, USP	7.50
Magnesium aluminum silicate	10.00
Preservative	q.s.
Flavors	q.s.
"Cyclamate" Sodium, NF	25.00
"Sorbo" Sorbitol Solution, USP	200.00
Water q.s. ad	1000.00

Heat the water to boiling and dissolve the preservative in it. Split this volume in half. To one half, add and dissolve the CMC. In the other half, disperse the magnesium aluminum silicate. Mix the two phases slowly with agitation. Mix in the "Sorbo." Add and dissolve the cyclamate sodium. Add and disperse the two hydroxides, mixing well. Mill the preparation if necessary to disperse insolubles. Add flavors and then sufficient quantity of water to make the total volume 1000 ml.

Antibiotic Suspension

Formula No. 1

Calcium Dioxytetracycline	q.s.
Sodium Metabisulfite, USP	0.983 g
Cyclamate Sodium, NF	0.833 g
Methylparaben, USP	0.899 g
Propylparaben, USP	0.099 g
Color	q.s.
Flavor	q.s.
"Sorbo" Sorbitol Solution, USP	q.s. ad 1000.000 ml

Citric Acid, USP (60% w/v solution)	q.s.

a) Blend together all solids, except the antibiotic calcium dioxytetracycline and the color, in a twin-shell blender. Pulverize these solids through a No. 60 screen and reblend.

b) Add a to about 800 ml of "Sorbo" with rapid stirring, followed by the flavor.

c) Pulverize the antibiotic through a No. 60 screen and add to ab. Then add the color.

d) Add sufficient "Sorbo" to bring the final volume to 1000 ml, and run the preparation through a mechanical homogenizer.

e) Adjust pH to 7.5 with a 60% w/v solution of citric acid, and saturate the preparation with nitrogen.

No. 2

Antibiotic (Insoluble)	q.s.
Magnesium Aluminum Silicate	9.5
Sodium Carboxymethylcellulose, USP	2.5
Sodium Citrate, USP	25.0

Flavor q.s.
Color q.s.
Methylparaben, USP 0.9
Propylparaben, USP 0.2
"Tween" 80 1.0
"Sorbo" Sorbitol
Solution, USP 500.0
Water q.s. ad 1000.0

a) Heat 200 ml of water to boiling, and dissolve in it one half of the parabens. Cool to about 70°C, then mix in the "Tween" 80. Sprinkle in the silicate, stirring until a uniform smooth suspension results.

b) Heat an additional 200 ml of water to boiling, and dissolve in it the remainder of the parabens. Disperse the CMC in this until a smooth gel results. Mix in the "Sorbo." Then dissolve the sodium citrate.

c) Add b to a slowly, with constant stirring. Cool the mixture to 25°C. Add the antibiotic, flavor and dye, mixing thoroughly. Add sufficient quantity of water to make the total volume 1000 ml.

Cottonseed Oil,
USP 600.00
Butylated
Hydroxyanisole 0.05
"Arlacel" 161 (Non-
self-emulsifying
glyceryl mono-
stearate 30.00
"Tween" 60 30.00
Lactose, USP 140.00
Sodium Benzoate,
USP 0.2
Flavors q.s.
Color q.s.
Water q.s. ad 1000.00

Mix the cottonseed oil, the anti-oxidant (butylated hydroxyanisole), "Arlacel" and "Tween," heating to 65°C so that all are melted together. Dissolve the lactose and sodium benzoate in water and heat to 67-70°C. Add the water phase to the oil phase with thorough agitation. Allow the emulsion to cool and then add the flavor and color, mixing well. Mix in the sulfisoxazole to obtain a uniform dispersion, and homogenize.

Sulfisoxazole Emulsion

Sulfisoxazole,
USP 200.00

Paraldehyde Emulsion

Paraldehyde, USP 333.3
Mineral Oil, USP 366.3

"Tween" 60   41.7
"Span" 80   41.7
Saccharin Sodium,
  USP   2.5
Peppermint
  Water,
  USP   q.s. ad 1000.0

Mix the mineral oil, paraldehyde, "Tween" and "Span." Dissolve the saccharin in the water and slowly add this solution to the oil mixture with constant stirring.

Pharmaceutical Vegetable Oil Emulsion

Cottonseed Oil,
  USP   500.0
"Tween" 60   50.0
"Span" 80   50.0
Sucrose, USP   50.0
Methylparaben,
  USP   1.0
Propylparaben, USP   1.0
Flavors   q.s.
"Sorbo" Sorbitol
  Solution, USP   50.0
Water   q.s. ad 1000.0

Mix the "Span" and "Tween" emulsifiers and the cottonseed oil and heat to 65°C. Dissolve the sucrose and parabens in water heated to 70°C. Add the water phase to the oil phase with stirring. Add the "Sorbo." Allow the emulsion to cool and add the flavor.

Antiseptic Dusting Powder

Iodine   1
Boric Acid   99

Bedsore Powder

Bismuth
  Subgallate   120 gr
Powdered Benzoin 75 gr
Starch   To make 1 oz

Sachet Powders

Formula No. 1

Talc   60
Light Precipitated
  Chalk   25
Zinc Stearate   5
Perfume Oil   10

No. 2

Talc   50
Light Precipitated
  Chalk   20
Zinc Stearate   5
Colloidal Kaolin   15
Perfume Oil   10

The high concentration of perfume oil in sachets causes the powder to cake slightly in the jar so that it does not flow freely.

Rubbing Alcohol (Lubricating)

"Solulan" 98	2.0
70% S.D. Ethanol or 70% Isopropanol	} 98.0

Cigarette-Smoking Deterrent

Danish Patent 86,468

Silver Acetate	1-3
Cocarboxylase	0.03-0.70
Ammonium Chloride	3-6
Licorice	10-15
Lactose	35-40
Starch	35-40
Peppermint Oil	To suit

Form into tablets.

Cigarettes smoked up to 3-4 hours after taking a tablet leave a very bad taste.

Anti-Seborrhea Shampoo

"Cetrimide"	15.6
Cetyl Alcohol	15.6

Lanolin	1.0
Water and Perfume To make	100.0

Scalp Protective against Chemicals

Lecithin (14% in Isopropanol)	88
Castor Oil	6
Oleic Acid	6

Dietary Salt Substitute

U.S. Patent 2,742,366

Calcium Glutamate	1-10
Dextrose	0-25
Tartaric Acid	0-5
Calcium Chloride	0-10
Potassium Chloride	40-90
Sucrose	5-10
Calcium Stearate	0.5-1

For Athlete's Foot (Powder)

Undecylenic Acid	5
Zinc Undecylenate	10
Salicylanilide	10
Talc	75

(Ointment)

a	"Vancide" 26	2.90
	Terpineol	0.05

Lanolin	9.67
Stearic Acid	19.30
b Triethanolamine	1.30
Water	58.03
c Diethylene Glycol	
Monoethyl Ether	4.35
Propylene Glycol	4.35
Ethyl Salicylate	0.05

Heat *a* to 75°C; heat *b* to 70°C; add *a* to *b*. Then add *c*, one at a time and mix all until cool.

Antiseptic Spray
(Foot and Shoe)

Concentrate

a "Vancide" 89RE	0.2
b Isopropanol	97.8
c Isopropyl Myristate	2.0

Packaging: Concentrate 80-50; Propellent 12/11 (30/70), 20-50.

Add *a* to *b* and heat to 60°C. Cool to room temperature and add *c*. Fill concentrate into appropriate containers and pressurize with propellent.

Aerosol Foot Powder

Concentrate

| "Nytal" 400 | 86.0 |

Magnesium	
Carbonate	5.0
Kaolin	5.0
Zinc Stearate	3.0
Methyl "Tuads"	0.5
Isopropyl Myristate	0.5

Packaging: Concentrate, 10; Propellent 12/11 (50/50), 90.

Sanitizing Foot Bath

Use 2 oz lithium hypochlorite to 1 gal water.

Skin Shield against
X-Rays

| Bismuth Subnitrate | 84 |
| Lanolin | 16 |

Rub together until uniform and apply to the skin. This has 62% of the efficiency of lead.

Ephedrine Sulfate Jelly

Ephedrine	
Sulfate	10.0 g
Sodium Alginate	40.0 g
Methyl	
Salicylate	0.1 cc
Eucalyptol	1.0 cc
Sodium Benzoate	2.0 g
Glycerol	100.0 g
Distilled Water	857.0 cc

Triturate the sodium alginate with the glycerol in a mortar until a smooth paste is formed. Dissolve the ephedrine sulfate and sodium benzoate in the distilled water and add the solution, in divided portions, to the first mixture, triturating after each addition. Finally add the oils, mix well, and allow all to stand about 3 hours.

Elixir Aspirin

Aspirin	4 g
Sodium Citrate	12 g
Sucrose	30 g
Tincture of Orange	12 cc
Distilled Water	25 cc
1% Amaranth Solution	2 drops
Orange Color	3 drops
Glycerol	To make 100 cc

Effervescent Aspirin

U.S. Patent 3,024,165

Acetylsalicylic Acid USP (100 Mesh)	10
Citric Acid, Anhydr (Gran)	25
Sodium Bicarbonate USP (Gran)	45
Calcium Acid Phosphate, Anhydr (Powd)	1/6
Thiamine Mononitrate, Anhydr (Powd)	1/65

Mix all at 70°F and relative humidity of 10-15%. Compress into tablets at 6 tons pressure. Wrap and seal tablets in foil.

Electrode Jelly
(For Electrocardiograph)

Formula No. 1

Sodium Chloride	6.5 lb
Pumice Powder	8 lb
Gum Tragacanth	8 oz
Potassium Bitartrate	4 oz
Glycerol	24 oz
Carbolic Acid	1 oz
Water	2 gal

Heat the gum tragacanth and glycerol in half the volume of water for 6 hours. Dissolve the potassium bitartrate and sodium chloride in the remaining volume of water and add to preceding. Stir the resulting mixture thoroughly with an electric mixer and heat it again for 1 hour. Then add the carbolic acid and pumice together with more water if necessary, and mix the whole again to a creamy consistency.

No. 2

U.S. Patent 3,048,549

Saturated Sodium Chloride Solution	87.60
Sodium Carboxy-methylcellulose	3.50
Mineral Oil	8.76
Sodium Benzoate	0.10

Burow's Solution Emulsions

Formula No. 1

Burow's Solution	10
Hydrous Lanolin	20
Zinc Oxide Ointment	30

No. 2

Burow's Solution	4
Zinc Oxide	8
Talc	8
Peach-Kernel Oil	25
Bentonite	2
Lime Water To make	100

No. 3

Phenol	0.6
Menthol	1.2

Zinc Oxide	20.0
Starch	20.0
Burow's Solution	80.0
Peanut Oil	160.0
Lime Water To make	600.0

No. 4

Anhydrous Lanolin	30
Olive Oil	120

Mix with heating.

Zinc Oxide	30
Talc	30

Mix and add:

Burow's Solution	8
Lime Water To make	300

Sterilizing Solution For Storage of Catgut

Isopropanol	70.0 g
Formaldehyde	2.5 cc
Sodium Bicarbonate	0.1 g
Sodium Nitrite	0.1 g
Distilled Water To make	100.0 g

This solution is used for the storage of tubes of catgut before use. The tubes sink in this solution and the exterior of the tubes remains sterile. The effectiveness of the solution may be lost if it is exposed to evaporation. The solution so exposed should be removed at regular intervals.

Neutral Emollient Lotion

Glyceryl Monostearate	2.0
Spermaceti	1.0
"Carbowax" 400	2.0
Olive Oil	5.0
Sodium Lauryl Sulfate	0.4
Gelatin	0.3
Distilled Water	To make 100.0

Combine the waxes and oil and melt on a water bath. Dissolve the sodium lauryl sulfate and gelatin in 85 cc of distilled water and heat the resulting solution to 75°C. Add the melted wax-oil mixture to the hot-water phase with constant stirring. Cool with constant stirring to room temperature.

The emulsion is of the oil-in-water type and particle size is 3-25 microns. The pH is 7.8 to 7.09 at 20°C.

The addition of two drops of oil of rose geranium and amaranth to 100 cc of this lotion, produces a pleasing, non-greasy, water-soluble hand lotion. Addition of 10% or more olive oil provides a general body lubricant. From 0.5-1% menthol is incorporated for an antipruritic effect.

The addition of 5% *liquor carbonis* detergent and 1% resorcinol to the base results in a product which is effective for the treatment of cradle cap and seborrheic dermatitis. To make an antibiotic lotion, one 500-milligram neomycin tablet is crushed and mixed with 10 oz of the base in a Waring blender. The final preparation is 12 fluid oz because of the incorporated air. Among other substances successfully incorporated in this emollient lotion in various percentages are: estrogens in oil, resorcinol, salicylic acid, vitamin A, water-soluble hormones, and cycloform.

Corticosteroid-Antibiotic Topical Lotion

Hydrocortisone (0.5 w/v)	50
Neomycin Sulfate (0.5 w/v)	50
Polyethylene Glycol 400 Monostearate	250
Propylene Glycol	1000
Isopropyl Palmitate	130

Stearic Acid	150
Sodium Benzoate	
(preservative)	20
Distilled	
Water, q.s.	10 liters

Dissolve the hydrocortisone with stirring and heating (65°C) in the propylene glycol. Cool the propylene glycol solution to 50°C, add with stirring the stearic acid, PEG-400 monostearate, and isopropyl palmitate, and filter. Heat 6.5 liters of distilled water to 50° and dissolve with stirring the sodium benzoate and neomycin sulfate, and filter. While maintaining both solutions at 50°C, add with stirring the aqueous solution *to* the propylene glycol solution, and bring to final volume with water. Pass the mixture through a colloid mill at 40°C and conduct the filling operation while continuously stirring.

Acne Lotion

Resorcinol	2.4-9.6
Precipitated	
Sulfur	2.4-12.0
Zinc Oxide	15.0
Talc	15.0
Glycerol	10.0

Ethanol	40.0
Water	40.0

Contraceptive

Polyethylene Oxide	1.50
8-Quinolinol Sulfate	0.03

Rubbing Alcohol Compound

Ethyl alcohol, or ethanol, the compound used almost exclusively in massage treatment, has certain disadvantages. It irritates the skin by rapid evaporation at room temperature. Because of this, there is inadequate time for proper massage. There is also lack of lubricity in most formulations with ethanol.

The following formulation provides a superior rubbing alcohol compound:

"Polyox" Resin	
WSR-301	1 g
Water	262 cc
Ethanol	1000 cc

Using high turbulence and low shear, disperse the "Polyox" resin in the ethanol and then slowly add the water to dissolve the resin. "Polyox" resin contributes a small amount of viscosity and needed lubricity. The solution does not irritate the skin and it dries relatively slowly.

Liniment

Formula No. 1

Wormwood Oil	5 cc
Menthol	1 g
Thymol	0.50 g
Iodine Crystals	0.25 g
Cajeput Oil	2 cc
Tyrolese Pine Oil	2 cc
Ethanol	25 cc

No. 2

Spirits of Camphor	20
Spirits of Ammonia	5
Wintergreen Oil	3
Rubbing Alcohol	72

No. 3

Turpentine	10
Light Mineral Oil	25
Wintergreen Oil	25
Camphor	1
Mustard Oil	3
Sodium Lauryl Sulfate	1
Distilled Water	35

Mix and dissolve all the ingredients except the water; then add the water in small quantities, with constant stirring. This can all be done at room temperature. A very small quantity of this preparation can be made by shaking directly in a bottle.

Anesthetic Lubricant

Formula No. 1

"Intracaine" Hydrochloride	2.00 g
Methylcellulose Powder (4000 cp)	1.00 g
Butyl p-Hydroxy-benzoate	0.015 g
Methyl p-Hydroxy-benzoate	0.13 g
Distilled Water	100.00 cc

Dissolve the preservatives in 75 cc of the water which has been heated to boiling. Use 25 cc of this solution, while hot, to wet the methylcellulose. Cool the remainder of the preservative solution, add it to the wet methylcellulose, shake this, and place it in a refrigerator until a clear solution of the methylcellulose is obtained. Add the "Intracaine" hydrochloride (previously been dissolved in the remaining 25 cc of distilled water) and mix thoroughly. Autoclave for 15 minutes at 121°C in 30-cc wide-mouth tightly capped bottles or in an Erlenmeyer flask covered with glassine paper and gauze. Pour into sterile jars.

No. 2

Dibucaine Hydrochloride	0.5 g
Powdered Tragacanth	1.5 g
Ethanol	5.0 cc
Glycerol	20.0 cc
Benzalkonium Chloride, (12.8%)	0.1 cc
Distilled Water	To make 100.0 cc

Suspend the tragacanth in the ethanol in a large conical graduate and add to this suspension rapidly the |dibucaine" and glycerol dissolved in 70 cc of the water. Stir vigorously for several minutes, dissolve the benzalkonium chloride, and add sufficient water to make 100 cc. Stir again and allow to stand for about one hour. Viscosity can be controlled by varying amount of glycerol.

Pediatric Antispasmodic

a Syrup	80.00
Tartaric Acid	0.24
Citric Acid	0.24
Ethanol	21.00
Distilled Water	To make 120.00
Grape Flavor	0.60

Mix first four ingredients, add purified talcum, and filter; add the water and then grape flavor to suit.

b Atropine Sulfate	6/250 gr
Phenobarbital Sodium	1.5 gr

Add b to a.

Each teaspoonful or 5 cc of this liquid will contain 1/1000 gr of atropine sulfate and 1/16 gr of phenobarbital sodium.

Bromoform Emulsion for Whooping Cough

Bromoform	6 oz 3 dr
Irish Moss	5 oz
Water	5 pt
Olive Oil	1.5 pt
Syrup	To make 1 gal

Boil the Irish moss and water for 10 minutes; cool, strain, and add water to make 5 pt. Then add the other ingredients and homogenize.

Anti-Infective-Analgesic
Ear Drop

Tyrothricin (0.1% w/v)	10 g
Benzocaine (5% w/v)	500 g
Chlorobutanol (0.5% w/v)	50 g
Polyethylene Glycol-400, q.s.	10 liters

Triturate the tyrothricin in portions with small quantities of the polyethylene glycol-400. Add more polyethylene glycol-400 and heat to 50°C. Dissolve with stirring the benzocaine and chlorobutanol. Bring up to volume with polyethylene glycol-400 and filter solution through kieselguhr.

Sodium-Free Dietary Salt

U.S. Patent 2,601,112

Potassium Chloride	60
Ammonium Chloride	15

Choline Dihydrogen Citrate	8
Choline Chloride	2
Magnesium Stearate	3
Corn Starch	12

Antihistamine Ointment

White Beeswax	5
Liquid Paraffin	15
Stearic Acid	12
Glycerol	5
Triethanolamine	2
Antazoline Hydrochloride	2
Distilled Water	To make 100 cc

Melt the beeswax, paraffin, and stearic acid over a water bath; dissolve the triethanolamine and glycerol in 30 cc water; warm, and add, with constant stirring, to the melted waxes. Dissolve the antazoline hydrochloride in the remainder of water; warm, and add, with constant stirring, to the first mixture. Stir gently until cool.

Ophthalmic Vehicle

Formula No. 1

Petrolatum	80
Mineral Oil	20

No. 2

(Emulsifiable with Water)

Petrolatum	80
Mineral Oil	15
Glyceryl Monooleate	5

Eye Wash

"Methocel" (4000 cp)	0.1
Normal Saline Solution	30.0

Mentholated Talcum Powder

Menthol	3 dr
Charcoal	4 dr
Powdered Alum	1.5 dr
Carmine	4 dr
Boric Acid	1.5 oz
Perfume	As desired
Talcum	20 lb
Yellow Food Color	To suit

Fungicidal Skin Coating

U.S. Patent 2,576,987

Tincture of Benzoin USP	70 cc
Salicylanilide	4 g
Isopropyl Alcohol	To make 100 cc

Water-Removable Skin Paste

Talc	6.5
Zinc Oxide	6.5
Haldane's Emulsion Base	17.0

Haldane's emulsion base is a pleasant, creamy-white water-washable base; composition: Sodium Lauryl Sulfate, 2; Glyceryl Monostearate, 9. The proportion of this emulsion base to be used can be altered for the consistency desired.

Appetite Satisfier

U.S. Patent 2,631,119

Sodium Glutamate	1767
Sodium Chloride	550
Protein Hydrolysate	257
Glutamic Acid	137

Take it 15 to 30 minutes before eating a meal.

Gastric Sedative

Sodium Bromide	10
Chloral Hydrate	5
Spirit of Anise	1
Chloroform Water	To make 60

Peptic Ulcer Remedy

U.S. Patent 2,638,433

Cottonseed Flour	50
Dextrose	40
Magnesium Trisilicate	5
Calcium Carbonate	5

Improved-Taste Phenobarbital Solution

Formula No. 1

Phenobarbital	0.20
"Tween" 80	4.00
Sodium "Sucaryl"	0.50
Lemon Oil	0.25
Citric Acid	0.20
Glycerol	4.00

Methylparaben	0.10
Water	50.00

No. 2

Phenobarbital	0.2
"Tween" 80	5.0
Raspberry Syrup	50.0
Citric Acid	0.5
Glycerol	5.0
Methylparaben	0.1

No. 3

Phenobarbital	0.20
"Tween" 80	4.00
Sodium "Sucaryl"	0.50
Peppermint Oil	0.25
Glycerol	5.00
Methylparaben	0.10
Water	50.00

Formulations for Tablets

	Sodium Bicarbonate, g	Aspirin, g	Ascorbic Acid, g	Magnesium Carbonate, g	Sodium Salicylate, g
Medicinal	100.0	85.0	75.0	100.0	100.0
Lactose	—	—	30.0	10.0	—
Corn Starch	10.0	15.0[a]	—	15.0	—
Binder	q.s.	q.s.	q.s.	q.s.	q.s.
Lubricant	1.0	0.5	1.1	1.25	2.2
"Silene" EF	—	—	—	—	10.0
Talc	—	—	—	—	2.2

[a] 10 g incorporated into wet granulations; remainder added just before compression.

General procedure for manufacture of granulations and tablets:

1. Medicinal, diluent, and disintegrant weighed and sifted through No. 40 mesh screen.
2. Powder mass moistened by binders.
3. Wet mass forced through screen of predetermined mesh to establish size of granules.
4. Dried for a pre-determined time in hot air oven.
5. Dried granules sized through appropriate screen. Granules passing through No. 40 mesh screen considered fines.
6. Granules mixed with experimentally determined amount of lubricating agent.
7. Compressed on F. J. Stokes single-punch, Model E tablet machine.

Drug Granulations

	Erythromycin-Triple Sulfa (Tablets)	Ox Bile-Dehydrocholic Acid (Tablets)	Aluminum Aspirin (3) with Caffeine, Acetophenetidin, and Methapyrilene HCl, (Granules for Encapsulation)
Total Weight dry ingredients	640	22	26
"Plasdone"	4.4	0.5	0.3
Solvent	alcohol	alcohol-acetone	alcohol
Concentration (%) "Plasdone" in granulating solution	2.9	4.9	4.9
"Plasdone" (%) on dry weight other ingredients	0.7	2.3	1.2

The "Plasdone" solution is added to the premixed dry ingredients in the usual manner. The granulation is screened, dried at 120°F, blended with lubricants, and compressed into tablets or ground for filling into capsules.

APC Granules

U.S. Patent 2,820,741

Aluminum Aspirin	14.64
Acetophenetidin	7.64
Caffeine	1.53
Methapyrilene	
Hydrochloride	1.72
Talc	0.25
Ethanol SD3A	6.50
PVP	0.32

Mill ingredients. Mix caffeine, methapyrilene, and talc. Charge into a mass mixer along with aluminum aspirin and acetophenetidin. Wet the dry mixture with the solution of PVP in ethanol and mix thoroughly. Granulate through ¼-in. mesh screen, dry at 120°C, and grind to suitable size.

Tablet Coating

Formula No. 1

Polyvinyl-	
pyrrolidone	5% w/v
Polyethylene	
Glycol 600	2% v/v

Ethanol, 70	
per cent	q.s.

The required amount of polyvinylpyrrolidone is weighed and transferred to a mortar. Twenty cc of 70 per cent ethanol is added and the mixture triturated until solution is clear. Polyethylene glycol 600 is added and the solution diluted to 100 cc with the ethanol.

No. 2

Polyvinyl-	
pyrrolidone	5% w/v
Polyethylene	
Glycol 600	2% v/v
Acetylated	
Mono-	
glyceride	5% w/v
Ethanol, 70	
per cent	q.s.

The solution of polyvinylpyrrolidone is prepared as for No. 1. Instead of diluting to 100 cc, the polyvinylpyrrolidone-polyethylene glycol-600 solution is diluted to 75 cc and warmed on a water bath. The acetylated monoglyceride is melted on a water bath and poured into the warm ethanolic solution of polyvinylpyrrolidone. This solution is diluted to 100 cc with the ethanol.

Approximately 500 tablets are placed in the coating pan and compressed air is applied to remove any dust or loose powder. No sub-coat is applied. Nine cc of coating solution No. 1, containing 10 per cent polyvinylpyrrolidone is applied to the tablets just to wet them. The tablets are stirred by hand to distribute the solution, then the pan is allowed to rotate. At the end of a 2-minute interval warm air is applied and the tablets are allowed to rotate another 2 minutes. This is followed by a rest period of 1 minute during which the pan is turned off and the warm air continued. Three successive applications, of 6, 3, and 2 cc respectively, are then made using the same time sequence as outlined above. Although the finished coat is uniform, smooth and hard it is also brittle. Varying amounts of polyethylene glycol-600 are added to the coating solution as a plasticizing agent. 2 per cent polyethylene glycol-600 in the coating solution decreases the brittleness of the finished coat. Coating solutions containing from 2-10 per cent polyvinylpyrrolidone and 2 per cent polyethylene glycol 600 are applied to uncoated tablets according to the procedure outlined. This stage in the coating procedure requires about 30 minutes. Tablets coated with the 5 per cent polyvinylpyrrolidone solution are uniform, smooth, hard, non-tacky, and non-brittle. The average thickness of the coat was 0.0017 inch.

Tablets are coated with solution No. 2 containing 5 per cent polyvinylpyrrolidone, 2 per cent polyethyleneglycol 600, and varying amounts of acetylated monoglyceride. The technique and procedure outlined on No. 1 is used with one modification. The uncoated tablets and the coating solution are warmed before the coating solution is added. This ensures complete solution of the acetylated monoglyceride and production of a uniform coat.

Color is applied to the tablets by dissolving or dispersing the coloring agent in the coating solution and applying successive portions (3-6 cc) to the coated tablets. The concentration of color is varied from 0.02 per cent to 0.1 per cent (based on volume of coating solution used) and the solutions applied in the order of ascending intensity of color. The intensity can be varied by increasing the number of coatings or by increasing the concentration of coloring agent in the coating solution. The application of sufficient coats of color to produce uniformly colored tablets requires about one hour.

The final step in the coating process is the polishing of the tablets. Five hundred coated tablets are placed in a canvas-lined 8-inch pan and 9 cc of warm wax solution is added. The pan is allowed to rotate thirty minutes after which a small amount of cosmetic grade talc is dusted on the tablets and rotation is continued for fifteen minutes. Another 6 cc of wax solution is added to the tablets while they are tumbling and rotation is continued for one hour. The resulting tablets have a high gloss.

Drug-Tablet Coating

U.S. Patent 2,693,436

Sugar	200
Water	100
Water-Soluble Hydroxyethyl-cellulose	2-7
Titanium Dioxide (for opacity)	0.5-3

Enteric Coating

Formula No. 1

Butyl Stearate	70
Carnauba Wax	30

No. 2

Butyl Stearate	45
Carnauba Wax	30
Stearic Acid	25

No. 3
(For Gastric Tablets)

Cellulose Acetate Phthalate	20
Hydroabietyl Alcohol	10
Acetone	30
Ethanol	35
N-Propanol	5

This coating resists simulated gastric fluid for 4 hr and intestinal fluid one half hr.

Tablet Granulation Binder

Zein G200	100
Propyleneglycol	10
Stearic Acid	10

Aqueous Ethanol	200
Water	10

Heating this alcoholic solution of zein to 40°C will cause the stearic acid to dissolve. For lower solids, tablet makers can dilute this formulation with 90% ethanol. Tablets can be colored by introducing the proper alcohol-soluble dye.

Aspirin Compound Tablet

(*Composition for 100,000*)

Aspirin (3.5 gr to tablet)	22.68 kg
Phenacetin (2.5 gr tablet)	16.20 kg
Caffeine, anhyd. (0.5 gr tablet)	3.24 kg
Dried Corn Starch	4.33 kg
Magnesium Stearate	0.108 kg

Colorants:

Gray: Activated Charcoal	108 g
Green: Emerald Green Dye	38 g
Yellow: Tartrazine Dye	66.5 g
Pink: Rose Pink Dye	64.6 g

For gray tablet, blend charcoal with corn starch and dry in oven. For other colors, dissolve dye in 950 ml of water, blend into corn starch, and dry in oven. For all colors: Add colored starch to aspirin, phenacetin, and caffeine. Mix and pass through 30-mesh screen on Fitzpatrick comminuting machine (impact forward). Mix 10 minutes, slug on Stokes D-3 rotary press, and break down through 8-mesh screen on granulator. Add magnesium stearate through 30- to 100-mesh screen, mix well, and compress using 13/32-inch punch and die. Weight of

10 tablets is 72 grains. Batch weight for 100,000 tablets is 102 lbs, 8 oz.

Antihistamine Tablet

For 100,000

Pyranisamine Maleate (30 mg to tablet)	3000 g
Confectioner's Sugar	10,250 g
Corn Starch	2560 g
Stearic Acid	171 g
Magnesium Stearate	57 g
Tartrazine Dye	25 g

Dissolve dye in 240 ml of water and blend well into the corn starch. Add colored starch to pyranisamine maleate and confectioner's sugar and mix well. Pass through a 30-mesh screen on Fitzpatrick comminuting machine (impact forward). Mix for 15 minutes and granulate with 20% acacia solution. Force through 8-mesh screen on granulator, or 3A screen on comminuting machine (knives forward). Place on trays, and dry in oven at 90°-100°F. When dry, force through 12- to 14-mesh screen on granulator. Add stearic acid and magnesium stearate (screened 30-100 mesh), mix well, and compress using 9/32-inch punch and die. Weight of 10 tablets is 25 grains. Batch weight for 100,000 tablets is 35 lbs, 12 oz.

Antihelminthic Wafer

For 10,000:

Piperazine Citrate (Equiv. 500 mg Piperazine Hexahydrate each wafer)	5520 g
Confectioner's Sugar	6160 g
Cyclamate Sodium	518 g
Corn Starch	180 g
Acacia	180 g
Magnesium Stearate	135 g
Stearic Acid	68 g
1:1 Lime/Mint Flavor	55 g
Emerald Green Dye	11.7 g

Dissolve dye in 70 ml of water and blend into corn starch. Mix with piperazine citrate, confectioner's sugar, cyclamate sodium, and acacia, and pass through Fitzpatrick comminuting machine (impact forward). Mix 15 minutes and granulate with 50% glucose solution. Force through 8-mesh screen on granulator, or 3A screen on comminuting machine (knives forward). Spread as thin as possible on trays and dry in oven at 100°F. When dry, force through 14-mesh screen. Screen out approximately 7 oz of 30-mesh fines and mix intimately with magnesium stearate, stearic acid, and flavor (all screened 30-mesh). Mix with granulation 10 minutes and compress using ¾-inch punch and die. Weight of 10 wafers is 200 grains. Batch weight for 10,000 wafers is 28 lbs, 8 oz.

Anorexigenic Chew-Tablet

For 100,000:

d-Amphetamine Sulfate (5 mg each tablet)	500 g
Methylcellulose (Extra dense)	10,000 g

Confectioner's Sugar	31,400 g
Cocoa	2,500 g
Vanillin	200 g
Magnesium Stearate	1,200 g

Blend and mix all ingredients with 14 liters of *anhydrous* methanol. Force through 8-mesh screen and dry in oven for at least 12 hours at 120°F until granulation is free of methanol. When dry, transfer from tray to drum. Do not screen because fragile granule will be destroyed. Slug granulation and break down through 8-mesh screen. Mix granulation well and compress using 7/16-inch punch and die. Weight of 10 tablets is 70 grains. Batch weight for 100,000 tablets is 100 lbs.

Compression Coated Tablet

Formula No. 1

Core:

a β-Lactose	1000.00 g	
10% Gelatin Solution	8.50 g	(85 ml)
b "Carbowax" 6000	4.03 g	

Magnesium
  Stearate   1.01 g

Prepare a granulation of *a*. Dry and screen through a 25-mesh screen. Dry thoroughly and add the lubricant *b*.

Coating:

a Mannitol   1000.00 g
    10% Gelatin   18.50 g
      Solution   (185 ml)
b Stearic Acid   4.07 g
    Magnesium
      Stearate   4.07 g

Prepare a granulation of *a*, dry and screen through a 25-mesh screen. Dry thoroughly and add the lubricant *b*.

Preparation of Tablets

Core:

    Machine: Stokes Rotary B-2
    Speed: 550 tablets/minute
    Punches: ¼ inch standard
                concave
    Pressure: 400 pounds
    Die fill: 0.218 inch

Coating:

    Machine: Stokes #538 Press
                Coater
    Speed: 550 tablets/minute
    Punches: $1\frac{3}{32}$ inch standard
                concave
    Pressure: 1.5 tons
    Die Fill: 0.375 inch

Tablet Characteristics:

    Thickness: 0.206 in.
    Weight: 453 mg
    Coating Thickness: 0.040 in.
    Coating Weight: 353 mg
    Hardness: 9.0 kg
    Disintegration time:
      Coating   6 min
      Core   3 min
          Total   9 min

No. 2

Core:

a Mannitol   1000.00 g
    10% Gelatin   18.50 g
      Solution   (185 ml)
b Stearic Acid   2.55 g
    "Carbowax"
      6000   2.55 g

Prepare a granulation of *a*. Dry and screen through a #25-mesh screen. Dry thoroughly and add the lubricant, *b*.

Coating:

a Mannitol   1000.00 g
    10% Gelatin   18.50 g
      Solution   (185 ml)
b Stearic Acid   4.07 g
    Magnesium
      Stearate   4.07 g

Prepare a granulation of *a*. Dry and screen through a #40-mesh screen. Dry thoroughly and add the lubricant *b*.

Preparation of Tablets

Core:

Machine: Stokes Rotary B-2
Speed: 550 Tablets/min
Punches: ¼ inch standard
          concave
Pressure: 400 pounds
Die Fill: 0.218 inch

Coating:

Machine: Stokes #538 Press
          Coater
Speed: 550 tablets/min
Punches: 13⁄32 inch standard
          concave
Pressure: 1.3 tons
Die Fill: 0.375 inch

Tablet Characteristics:

Thickness: 0.218 inch
Weight: 465 mg
Coating Thickness: 0.046 inch
Coating Weight: 365 mg
Hardness: 8.5 kg
Disintegration time:
  Coating   5 min
  Core   5 min
        Total   10 min

No. 3
Multi-Layer Tablet

Upper Layer: (Fastest
dissolution)

a	Lactose	1000.00 g
	10% Starch	17.00 g
	Paste	(170 ml)
b	Stearic Acid	2.54 g
	Magnesium	
	Stearate	1.53 g

Prepare a granulation of *a*,
dry and screen through a #25-
mesh screen. Add lubricant *b*.

Middle Layer: (Medium-
speed dissolution)

a	Mannitol	1000.00 g
	2% "Methocel"	5.32 g
	Solution	(266 ml)
b	Stearic Acid	5.03 g
	Magnesium	
	Stearate	1.01 g

Prepare a granulation of *a*,
dry and screen through a #25-
mesh screen. Add lubricant *b*.

Lower Layer: (Slowest
dissolution)

a	Mannitol	1000.00 g
	10% Gelatin	18.50 g
	Solution	(185 ml)
b	Stearic Acid	5.09 g
	Magnesium	
	Stearate	1.02 g

Blend components of *a*, dry
and screen through a #40-
mesh screen. Add lubricant *b*.

Preparation of Tablets

Machine: Stokes #566
Speed: 600 tablets/min
Punches: 7⁄16-inch diam flat-
          face beveled edge
Pressure: 1.5-10 tons
Hardness: 5.5 kg

Die fill:

Upper layer	0.187 inch
Middle layer	0.093 inch
Lower layer	0.093 inch
Total	0.373 inch

Tablet Characteristics:

Thickness:

Upper layer	0.065 inch
Middle layer	0.042 inch
Lower layer	0.040 inch
Total	0.147 inch

Weight:

200 mg	200 mg
Middle layer	140 mg
Lower layer	122 mg
Total	462 mg

Disintegration time:

Upper layer	2 min
Middle layer	9 min
Lower layer	5.5 min
Total	9 min

Antacid Tablet

No. 1
Calcium Carbonate

Calcium Carbonate pptd	0.3300
Mannitol	0.3300
Saccharin, Soluble	0.0007
10% Starch Paste, *ca.* (as starch) 0.24 g	0.0240
Peppermint Oil	0.0007
Talc	0.0076
Magnesium Stearate	0.0070

Blend calcium carbonate and mannitol. Dissolve saccharin in starch paste and use this to granulate the medicament-mannitol blend. Dry at 140°F and screen through 16-mesh screen. Incorporate flavor, talc, and magnesium stearate; mix well. Age at least 24 hours and compress.

No. 2

Magnesium Trisilicate-Aluminum Hydroxide

Magnesium Trisilicate	0.5000
Aluminum Hydroxide, Dried gel	0.2500
Mannitol	0.3000
Saccharin, Soluble	0.0020
5% Starch Paste, *ca.* (as starch) 0.65 g	0.0325
Peppermint Oil	0.0010
Magnesium Stearate	0.0100
Corn Starch	0.0100

Mix the medicaments and mannitol. Dissolve saccharin in small quantity of distilled water, then combine this with starch paste. Granulate medicament-mannitol blend with starch paste. Dry at 140°F and screen through 16-mesh screen. Add flavor oil, magnesium stearate, and corn starch; mix well. Age at least 24 hours and compress.

Multi-Vitamin Tablet

No. 1
(Wet Granulation)

	Each Tablet
Vitamin A, Dry, stabilized	5000 USP units
Vitamin D, Dry, stabilized	1000 USP units
Vitamin C	60 mg
Vitamin B_1	1 mg
Vitamin B_2	1.5 mg
Vitamin B_6	1 mg
Vitamin B_{12}	2 microg
Calcium Pantothenate	3 mg
Niacinamide	10 mg
Saccharin, Soluble	0.003 g
Orange Flavor (Dry)	0.0070 g
Mannitol	0.2348 g
Acacia Powder	0.0065 g
Magnesium Stearate	0.0065 g
Talc	0.0070 g

Blend all vitamins, except A, D, and C, with 10% of the riboflavin, mannitol, and acacia. Dissolve saccharin in water (0.25 g/30 ml) and use the solution to granulate the blended dry ingredients. Dry at 120°F and pass through 16-mesh screen. Add the flavor oil and age 24 hours. Mix ascorbic acid with magnesium stearate, and the vitamins A and D with remainder of riboflavin. Add these and the talc to the previous mixture, mix well, and compress.

No. 2
(Dry Granulation)

	Each Tablet
Vitamin A, Dry, stabilized	5000 USP units
Vitamin D, Dry, stabilized	1000 USP units
Vitamin C	60 mg
Vitamin B_1	1 mg
Vitamin B_2	1.5 mg
Vitamin B_6	1 mg
Vitamin B_{12}	2 microg
Calcium Pantothenate	3 mg
Niacinamide	10 mg
Saccharin, Soluble	0.0011 g

Orange Oil	0.0015 g
Mannitol	0.2362 g
Corn Starch	0.0166 g
Magnesium	
Stearate	0.0066 g
Talc	0.0100 g

Treat the ascorbic acid crystals with a light coat of pharmaceutical glaze. Then mix all the ingredients except the corn starch, magnesium stearate, and talc. Slug, then pass through a 20-mesh screen. Add the corn starch, magnesium stearate, and talc; mix well, and compress.

Sustained Release
Medicinals

U.S. Patent 2,793,979

Formula No. 1

d-Amphetamine Sulfate (100%), 60-mesh	5.0
Lactose, USP 60 mesh	57.5
Sucrose, USP (aqueous solution: 85% (wt) sucrose to vol water	15.0
Glyceryl Distearate	25.0

The glyceryl distearate is melted. The *d*-amphetamine sulfate and lactose are mixed together and then added to the melted glyceryl distearate while stirring. After a thorough mixing of the glyceryl distearate, *d*-amphetamine sulfate and lactose, the mixture is cooled until it congeals to a hard mass, the stirring being continued as long as possible. This mass is ground and sieved through a #35-mesh screen. The sucrose syrup is added to the powder thus obtained and thoroughly mixed with it to mass the powder. The resulting product is granulated through a #10-mesh screen. The granules are dried overnight at 37°C, and sieved through a #14-mesh screen.

No. 2

Chloroprophen-pyridamine Maleate, 50-mesh	5
Terra Alba, 60-mesh	45
Sucrose, USP (aqueous solution: 75% (wt) sucrose to vol water)	15
Cetyl Alcohol, NF	10
Stearic Acid, NF	5
Glyceryl Trilaurate	20

The cetyl alcohol, stearic acid, and glyceryl trilaurate

are melted together. The chloroprophenpyridamine maleate and terra alba are added to the melted mixture while stirring. After a thorough mixing of the cetyl alcohol, stearic acid, glyceryl trilaurate, and chloroprophenpyridamine maleate, the mixture is cooled until it congeals to a hard mass, the stirring being continued as long as possible. The formed mass is ground and sieved through a 30-mesh screen. The sucrose syrup is added to the powder thus obtained and thoroughly mixed with it. The resulting product is ground through a 14-mesh screen. The granules are dried 10 hours at 37°C and sieved through an 18-mesh screen.

Protective Film for Skin (Aerosol)

PVP/VA E-335 (100% Solids)	5.0
"Carbowax" 600	0.2
Ethanol (SDA-40 anhydr)	24.8
"Genetron" Propellent 142b/114a (30/70)	70.0

Aerosol Bandage

No. 1
Neomycin-Hydrocortisone

Concentrate:

Neomycin Micronized (sulfate, USP)	3.5 mg/g solids content
Hydrocortisone USP	2.5% solids content

Vehicle: Wt %

PVP/VA E-535	40.0 (as solids)
"Carbowax" 600	0.4
Ethanol (anhydr)	59.6

Fill:

Concentrate	25.0
Propellent 11/12 (70/30)	75.0

This product was packaged in 6-ounce Crown Spra-Tainer (lacquer-lined cans fitted with Precision NN Valve and a standard spray actuator). Studies of stability showed no loss of potency even after 15 months at room temperature or 12 months at 43°C.

No. 2
Neomycin

Concentrate:

Neomycin, Micronized (sulfate, USP)	3.5 mg/g solids content

Vehicle: Wt %

PVP/VA E-335	16.7 (as solids)
"Citroflex" A-2	0.8
Ethanol (anhydr)	82.5

Fill:

Concentrate	30.0
Propellent 12/11 (30/70)	70.0

This formula was packaged in 2¼-ounce Peerless Aluminum Can (Epon-lined with Precision NN valve and a standard spray actuator). Studies of stability were conducted for 6 months at 43°C and 9 months at room temperature without any loss of potency.

Poultry Wormer

	No. 1	No. 2	No. 3
Piperazine	16.9	—	—
Piperazine Hexahydrate	—	38.3	—
Piperazine Citrate, Hexahydrate	—	—	46.9
Water	100	100	100

Use 2 oz to gallon of water.

Formula No. 4

Phenothiazine	50
Dichlorophene	25
Piperazine	25

Use 1 lb to 100 lb mash. This is a toxic material not to be used on sick or weak birds.

Hog Scour

Sodium Arsanilate	9.13 g
Inert material	To make 100.00

Form into tablets. Use 1 tablet to 1 gallon drinking water.

Embalming Fluid

U.S. Patent 2,521,108

Formula No. 1

Sodium Hexameta- phosphate	1
Sodium Citrate	13
Sodium Carbonate	20
Sulfonated Castor Oil	10
Water	62
Formaldehyde	7.5-19

This remains stable for 3 years.

No. 2

Isopropanol	33.3
"Glucarine" B	8.3
Formalin	2.1
Phenol	6.7
Benzalkonium Chloride	0.1
Water	To make 100.0

Chapter V

DISINFECTANTS AND DEODORANTS

Lilac-Odor Disinfectant

Terpineol	80
Potash Soap	8
Isopropanol	4
Water	8

This disinfectant has a phenol coefficient of 5.

Soluble Cresylic Disinfectant

Cresylic Acid	50
Sodium Xylene Sulfonate	50

Iodine Disinfectant

No. 1
(Liquid)

"Idonyx" (16%)	12.5
"Neutronyx" 640	20.5
Hydrochloric Acid (37%)	10.0
Phosphoric Acid (85%)	12.0
Water	45.0

Provides 1.75% min. available iodine.

Dissolve "Idonyx" in warm (110-120°F) water. Add "Neutronyx" 640 and mix thoroughly. Cool, add hydrochloric acid slowly; add phosphoric acid and cool to room temperature.

No. 2
(Powder)

"Idonyx" (16%)	12.5
"Neutronyx" PX	50.0
Sulfamic Acid	30.0
Sodium Acid Pyrophosphate	7.5

Provides 1.75% min. available iodine

273

Mix all ingredients until uniform. Comminute, if a fine powder is desired.

Acid Cleaner Disinfectant (Liquid)

Detergent Concentrate (AL)	19.9
"BTC"-927	12.7
Hydroxyacetic Acid	17.0
Water	to 100.0

127 ppm active quaternary has hard water tolerance of 600 ppm; 200 ppm active quaternary, 700 ppm.

For Barber Shop Instruments

"BTC"-50 in Isopropanol	0.5
Isopropanol	10.0
"Marcol" JX	88.5
"Lorol" 5	1.0

Pine Oil Disinfectant (Phenol coefficient approx. 5)

Deod. Kerosine	66.0
Pine Oil	28.0
"BTC"-824	4.0
"Neutronyx" 600	2.0

Pine Oil Disinfectant (Concentrate)

Deod. Kerosine	55.0
Pine Oil	25.0
"BTC"-824	20.0

For Dishes and Glassware

"BTC"-1100	50.00
Ammonium Chloride	12.76
Sodium Bicarbonate	26.37
Inverted Sugar (gum, starch)	10.87

A tablet prepared from this composition weighing 2.27 g should be dissolved in 1½ gallons of water to yield 200 ppm quaternary. A tablet weighing 3 g should be dissolved in 2 gallons of water to yield 200 ppm quaternary.

Air Sanitizer

Formula No. 1

Essential Oils	3.5
Propylene Glycol	4.4
Triethylene Glycol	4.4
Isopropanol	22.7
Dichlorodifluoromethane	65.0

This air sanitizer is a high-pressure composition sold in the U.S.A. and Canada. As the pressure is too high for containers of tin-plate, it is packed in steel bulbs fitted with metering valves.

No. 2

A similar formulation but with a lower pressure is as follows:

Perfume	2
Propylene Glycol	3
Triethylene Glycol	3
Glycerol	2
Isopropanol	10
Dichlorodifluoro-methane	40
Trichloromono-fluoromethane	40

Room Deodorants (Aerosol)

Formula No. 1

Perfume	0.5
Odorless Kerosine	12.5
Isopropyl Myristate	2.0
Dichlorodifluoro-methane	42.5
Trichloromono-fluoromethane	42.5

No. 2

Perfume Oil	0.5
Metazene Concentrate (10%)	14.5
Dichlorodifluoro-methane	42.5
Trichloromono-fluoromethane	42.5

No. 3

Metazene (100%)	2.0
Light Perfume	0.3
"Emcol" E. 607 Cation	0.1
Isopropyl Myristate	0.3
Odorless Naphtha	17.8
Methylene Chloride	10.0
Propellent 11	24.0
Propellent 12	36.0
Isobutane	10.0

No. 4

Isopropanol	9.7
Triethylene Glycol	5.0
Propylene Glycol	5.0
"Hyamine" #1622	0.1
Perfume Oil	0.2
Propellent	80.0

Provided that the isopropanol is anhydrous, or very nearly so, there should be no corrosion with this last formulation in either tin-plate or aluminum. Good seams are essential, to prevent leakage and a resistant compound should be used for the lining.

Toilet Deodorant Block

British Patent 806,726

p-Dichloro-benzene	97.5-95.0
"Dantoin"	2.5-5.0

Swimming Pool Sanitizer

Use 2 oz lithium hypochlorite to 1000 gallons of water. Add 1 oz daily for each 5000 gal to maintain about 5 ppm chlorine.

Air Freshener or Deodorant

U.S. Patent 2,715,611

Trichlorofluoro-methane	2000.0
Dichlorofluoro-methane	2000.0

Cumene Hydro-peroxide	5.7

Pack in pressure cans.

Radioactive Decontaminating Detergent

Formula No. 1

Tetrasodium "Tetrine"	25
Anhydrous Sodium Orthosilicate	28
Sodium Tripoly-phosphate	35
"Pluronic" F68	10
Sodium Carboxy-methylcellulose	2

No. 2

Tetrasodium "Tetrine"	25
Sodium Hemi-phosphate	25
Sodium Carboxy-methylcellulose	2
40% Alkylaryl Sulfonate	48

CHAPTER VI

EMULSIONS

Hydrogen Peroxide Emulsions

The following formulas range from thin to heavy lotions:

Increasing Viscosity

Formula	No. 1	No. 2	No. 3	No. 4	No. 5
"Arlacel" 83	5.7	11.4	12.0	18.0	17.0
"Tween" 40	—	—	8.0	12.0	—
"Tween" 80	4.3	8.6	—	—	13.0
Hydrogen Peroxide (35%)	23.0	23.0	23.0	23.0	23.0
Water	57.0-67.0	57.0	57.0	47.0	47.0
Stabilizer	(Type and quantity as required by formulator)				

Mix surfactants, add H_2O_2, and blend. Add water and blend.

WARNING: Extreme care is essential in formulating with hydrogen peroxide because rapid release of oxygen can occur under heat or contamination, resulting in explosion of containers. Any composition should be adequately stabilized to avoid danger.

No. 6

A white pourable cream can be prepared as follows:

"Arlacel" 83	20.0
"Tween" 80	15.0
H_2O_2 (35%)	23.0
Water	42.0
Stabilizer	As required

As usual with hydrogen peroxide products, formulas should be stabilized and adequate studies should be made to establish the keeping quality, emulsion stability, and freedom from liability to dangerous decomposition.

Polyethylene Emulsions

Formula No. 1

"A-C" Polyethylene 629	40
"Tergitol" NPX	10
20% KOH Solution	3
Water	147

"Tergitol" NPX can be substituted by "Triton" X100, "Renex" 690 or 697, or "Poly Tergent" B-300.

Melt the "A-C" Polyethylene 629 and the "Tergitol" NPX together at 257°F (max). When completely melted, cool to 230°F and slowly add the KOH solution with stirring to break up the foam that forms.

NOTE: Care should be exercised in this operation. If the temperature is too high or the solution is added too rapidly, the material will foam over the top of the heating kettle.

While the "A-C" Polyethylene 629 is being melted, heat the water to boiling (turn off heat just before using).

With rapid stirring (but below speed at which air is incorporated), slowly add the melt to the water (1-3 minutes). During the addition, the temperature of the melt should be maintained at 230°F, and the water maintained at 203-210°F (not boiling). Add the melt at a steady rate to the top of the vortex formed by the stirring; it should not hit the side of the vessel or the shaft of the stirrer, but spiral down the vortex and be emulsified. If stirring is too slow or rate of addition too rapid, the melt with accumulate in the vortex and impair the quality of the finished product.

Cover, and with moderate stirring allow mixture to cool to 104-122°F; add water to make up water lost by evaporation.

The procedure as outlined should be studied and practiced before modifications are attempted. Fine-particle, translucent emulsions should be obtained which show no tendency to "cream." This procedure is easily adapted to commercial production.

With the KOH and "Tergitol" NPX system, a wide variety of concentrations is possible. Using 20% "A-C" Polyethylene 629 solids, completely stable products can be formed using 3 to 10% "Tergitol" NPX and 0.15-0.3% KOH based on weight of emulsion. The pH of the concentrate (20% "A-C" Polyethylene) is from 7.5 to 10.

Below 3% "Tergitol" NPX and 0.15% KOH good emulsions are obtained, but these will not remain stable after freezing.

With concentrations of "Tergitol" NPX above 5% and KOH above 0.3%, translucent-like emulsions of excellent stability are obtained, but with pH greatly above 9. The best overall properties are found in the range of 3-5% "Tergitol" NPX and 0.15 to 0.30% KOH for 20% "A-C" Polyethylene 629 emulsion concentrates.

Non-Ionic Polyethylene Emulsions
(Wax-to-Water Method)

"A-C" Polyethylene 629	40	40	40	40	40	40	40	40	40
"Michelene" D. S.	8	—	—	—	—	—	—	—	—
"Alrosol"	—	8	—	—	—	—	—	—	—
"Monamine" ACO-100	—	—	5	—	—	—	—	—	—
"Pluramine" S-100	—	—	—	4	—	—	—	—	—
"Permaline" A-100	—	—	—	—	8	—	—	—	—
"Nopcogen" 14-L	—	—	—	—	—	8	—	—	—
"Hyonic" FA-40	—	—	—	—	—	—	8	—	—
"Ninol" 737	—	—	—	—	—	—	—	8	—
"Ninol" 2012E	—	—	—	—	—	—	—	—	8
Water	q.s. for desired concentration								

Non-Ionic
(Non-Nitrogen)

"A-C" Polyethylene 629	40	40	40	40	40
"Tergitol" NPX	10	—	—	—	—
"Renex" 697	—	10	—	—	—
"Triton" X-100	—	—	10	—	—
"Polytergent" B-300	—	—	—	—	10
"Igepal" CO 530	—	—	—	10	—
20% KOH Solution	3	3	3	3	3
Water	147	147	147	147	147

There are other non-ionic emulsifiers similar to those above
which are equally satisfactory. Borax, sodium hydroxide, triethanolamine and other amines may be used in place of the
potassium hydroxide.

Anionic Polyethylene Emulsions

"A-C" Polyethylene 629	40	40	40	40	40	40
Oleic Acid	4	6	5	5	6	7
3-Methoxy propylamine	3	—	—	—	—	—
2-amino, 2-methyl, 1-propanol	—	3	—	—	—	—
Morpholine*	—	—	7	—	—	—
Monoethanolamine	—	—	—	4	—	—
Diethanolamine	—	—	—	—	5	—
Triethanolamine	—	—	—	—	—	7
Water	q.s. for desired concentration					

Other fatty acids that may be used are: linseed, myristic,
palmitic, soya, stearic, tall oil.

Cationic Polyethylene Emulsions

"A-C" Polyethylene 629	40	40	40	40	40	40
"Armac" T	8	—	—	—	—	—
"Armac" HT	—	8	—	—	—	—
"Armac" C	—	—	8	—	—	—
Acetic Acid	—	—	—	2	1	1
"Ethomeen" 18-12	—	—	—	8	—	—
"Alro" Amine O	—	—	—	—	7	—
"Alro" Amine S	—	—	—	—	—	6
Water	To make up to desired concentration					

NOTE: When morpholine or other very low boiling amines are used in the emulsifier, the kettle should be covered and care should be taken to avoid loss of amine due to excessive temperatures. Local ventilation may be desirable.

Emulsions of "Vistanex" LM

Formula No. 1

"Vistanex" LM	100
Stearic Acid	3
Triethanolamine	2
Water	80

Mix at about 200°F to obtain smooth emulsion.

No. 2

"Vistanex" LM	100
"Staybelite" Ester No. 3 or 10	24
Stearic Acid	4
Triethanolamine	2
Water	90

Keep temperature at about 180°F. Staybelite Ester increases adhesive strength.

No. 3

"Vistanex" LM	100
Potassium Stearate	5
Water	70

Add potassium stearate slowly at about 160°F to ensure complete dispersion.

No. 4

"Vistanex" LM	100
Light Mineral Oil	6
Stearic Acid	3
Triethanolamiue	2
Water	50

Hold temperature at about 160°F. This formulation has excellent tack.

These emulsions are easily prepared with heavy-duty mixing equipment, such as a Baker-Perkins-type mixer with steam or hot-water jacket. Dispersion of emulsifier in "Vistanex" should be complete before water is added, and is facilitated by heat. The water should be added slowly at the start to ensure a uniform emulsion. Stability of such emulsions is often improved by addition of 0.5 to 1% of alkali metal salts or hydroxy compounds.

Preparation of emulsions of "Vistanex" MM is somewhat more difficult, mainly because of the elastic properties of these polymers. Plasticization with well-broken-down rubbers, thermoplastic resins, oils, solvents, etc. greatly aids the process. Emulsions of MM grades of "Vistanex" can be prepared directly in a Banbury mixer, or by a two-step process, first mixing the "Vistanex" and emulsifiers with enough water to plasticize in a Banbury or on a rubber mill, then adding the rest of the water in an open mixer. Typical recipes follow.

Emulsions of "Vistanex" MM

Formula No. 1

"Vistanex" MM	
L-100	100
Light Mineral Oil	20
Oleic Acid	8
Potassium Hydroxide	2
Water	100

Add oil to "Vistanex" in Banbury, followed by oleic acid and potassium hydroxide in 50% water solution. Transfer to open mixer and add rest of water.

No. 2

"Vistanex" MM	
L-100	100
Casein	10
Triethanolamine	3
Water	80

Add casein, triethanolamine, and 10 parts water to "Vistanex" in Banbury, then add 20 parts water containing 1% wetting agent. Add remainder of water in open mixer.

No. 3

"Vistanex" MM L-100	100
Carbon Tetrachloride	360
Ammonium Linoleate	10
Ammonium Stearate	5
Ammonium Hydroxide	6
Water	110

Mix in an open heavy-duty mixer. Keep temperature at about 70°F.

No. 4

"Vistanex" MM L-100	100
Masticated Pale Crepe Rubber	20
Oleic Acid	8
Potassium Hydroxide	2

Mix rubber, "Vistanex," and oleic acid in a Banbury or two-roll mill, then add potassium hydroxide as 45% water solution in a mixer. Add water as desired.

"Oronite" (Polybutene) Emulsions (O/W)

	Nonionic	Cationic	Anionic
Emulsifying Agent	"Oronite" NI-W	"Oronite" ADE-50	Oleic Acid + Triethanol amine
Temperature	120-130°F		
Equipment	High Speed Laboratory Stirrer		

Components	Weight Percent		
Emulsifier	3.4	3.0	1.8 Oleic Acid
			1.2 Triethanolamine
Polybutene 32	55.0	55.0	54.0
Water	41.6	42.0	43.0

A mixture of the emulsifying agent and polybutene is heated to temperature and stirred until it becomes homogeneous. Preheated water is added in small increments under agitation until inversion from water-in-oil to oil-in-water emulsion takes place. The rest of the water can then be added more rapidly. In the anionic emulsion, triethanolamine is added to the polybutene-oleic acid mixture as a solution in a small amount of water.

"Oronite" polybutenes form a wide range of odorless, creamy white emulsions that are easy to handle. These emulsions combine the chemical resistance and low permeability to gas and moisture of the base polymer with a high degree of chemical and mechanical stability.

"Aroclor" Emulsions

There are several simple methods for making emulsions with "Aroclor"; the one used may be selected to suit the kind of "Aroclor" and type of formulation in which it will be used.

Viscous "Aroclors"

a	"Aroclor"	16 lb
	Stearic Acid	1 lb
b	Water	8 lb
	Triethanolamine	4 oz

Heat the "Aroclor" to a workable viscosity (180°F-plus) and stir in the stearic acid thoroughly. Heat the water to almost boiling (207°F) and stir in the triethanolamine thoroughly. Pour a *into* the water portion (b) agitating vigorously. Then combine a and b with a high-speed emulsifying stirrer, . . . or process through a colloid mill.

Liquid "Aroclors"

a	"Aroclor" 1254	100
	Oleic Acid	4
b	Water	92
	Ammonium Hydroxide (28%)	2
	"Lustrex" X-810	2

Mix the ammonium hydroxide and "Lustrex" X-810 thoroughly in the warmed water, using vigorous agitation. Mix the "Aroclor" 1254 and oleic acid, heat to 45°C and agitate vigorously. Maintain the 45°C temperature and agitation and add the water portion *slowly*. Continue agitation for one-half hour till inversion of phase is complete.

Emulsifiable Concentrated Stock Solutions of "Aroclors"

"Aroclor"	79
Toluene	16.70
Isopropyl Alcohol	3.55
"Sterox" CD (non-ionic emulsifier)	1.00
"Santomerse" #3 (anionic wetting agent)	0.75

This formulation is readily emulsifiable with water. If the more resinous "Aroclors" are used, increase the amount of toluene (or xylene) as needed to dissolve the "Aroclor" resin.

Vegetable Oil Emulsion (Pharmaceutical)

Cottonseed Oil, USP	500 g
"Tween" 60 Polyoxyethylene Sorbitan Monostearate (polysorbate 60)	50 g
"Span" 80 Sorbitan Monooleate	50 g
Sucrose, USP	50 g
Methylparaben, USP	1 g
Propylparaben, USP	1 g
Flavors	q.s.
"Sorbo" Sorbitol Solution, USP	50 g
Water	q.s. to 1000 ml

Dye-Carrier Emulsion

	Formula No. 1	No. 2
"Triton" X-400	1	1
o-Phenylphenol	8	—
Water	50	50
Methyl Salicylate	—	8

Lindane Emulsion

German Patent 1,066,047

Lindane	25
Polyglycol (400) Monooleate	8
Xylene	67

This emulsifies easily on addition of water.

Stable Asphalt Emulsion

Japanese Patent 12,771 ('61)

Asphalt	520
Fatty Acid Pitch (Acid No. 100, Sap. Value 170)	15

Heat to 130°C and pour slowly, with good agitation into caustic soda (0.3% solution) 465. (Heated >70°C.)

Soft Asphalt Emulsion

Water	4000.0
"Carbopol" 941	1.50
Sodium Hydroxide	
(10% solution)	4.50
"Ethomeen" C-25	0.75
Soft Asphalt	400

Carefully disperse the "Carbopol" 941 in the water. After thorough dispersion, neutralize with the sodium hydroxide followed by the "Ethomeen" C-25. Heat this mucilage to 65-70°C. (The water can be heated before "Carbopol" 941 and neutralizer are added as this will assist dispersion of the "Carbopol" 941.) Separately heat the asphalt to 85°C. Utilizing moderately high-shear mixing, slowly add the molten asphalt to the water mucilage. Add the asphalt no faster than the rate of dispersion. If the asphalt is added too rapidly, the emulsion will invert and separate.

When the asphalt has been added, continue mixing for a short time to ensure uniformity. Shock-cool the emulsion without mixing and allow it to remain undisturbed until its temperature falls to 30-35°C.

An emulsion prepared in the foregoing manner has a final pH of 6.5 and a viscosity of 5500 cps (Brookfield, 20 rpm). It is stable on aging and very smooth in appearance.

Tar Emulsion

Water	50.000
Tar	50.000
"Carbopol" 941	0.125
Sodium Hydroxide	
(10% solution)	0.250
"Ethomeen" C-25	0.063

Carefully disperse the "Carbopol" 941 in the water. When dispersion is complete, add the sodium hydroxide and then the amine. Lastly, slowly add the tar in a thin, continuous stream (so that the rate of addition is no greater than the rate of dispersion) with good mixing. Fairly high shear, such as that provided by a medium-speed Eppenbach homogenizer, forms the best, smallest droplet-size emulsion. Excessively rapid addition of the tar results in a grainy emulsion.

Creosote Emulsion
(Wood Preservative)

Water	200.00
"Carbopol" 934	0.60
Sodium Hydroxide	
(10% solution)	0.55
Creosote	200.00

Carefully disperse the "Carbopol" resin and then add the sodium hydroxide. Add the creosote in a slow stream while vigorously agitating the mix. Stir to uniformity. The resultant emulsion is ready for use.

When applied to wood, such an emulsion wets the surface well and penetration of creosote into the wood is excellent.

Silicone Oil Emulsion

DC-200 Silicone	
Oil	200.0
"Carbopol" 934	2.0
Dodecylamine	1.0
Sodium Hydroxide	
(10% solution)	6.0
Water	200.0

Carefully disperse the "Carbopol" 934 in the water and mix to a thin, cloudy dispersion without lumps. Add the sodium hydroxide and then the amine with mild agitation. The dispersion will gel immediately. Lastly, add the DC-200 in a slow, steady stream with good mixing. A white, stable emulsion forms immediately.

Tripropyl Phosphate Emulsion

"Emulphor" ELA	1
Tripropyl Phosphate	8
Water, warm	50

Polyglycol Monolaurate Emulsifier

Methyl Laurate	214 g
Polyethylene	
Glycol 400	400 g
Sodium Methylate	
Catalyst	3 g
(25% Solution	
in Methanol)	

The methyl laurate is simply charged into a 3-neck flask provided with a capillary nitrogen-inlet tube, stirrer, thermometer, and connection to a water aspirator. A vacuum of about 100 mm is applied, which starts the nitrogen slowly bubbling through, and the methyl laurate is heated to 125°C. Then 1.5% of alcoholic sodium methylate (based on the weight of the laurate) is added to the PEG 400, and this is slowly run into the flask from a dropping funnel in about 45 min. The temperature is brought back to 125°C and held for two to three hours to ensure completion of the reaction. By weighing the entire flask periodically it is possible to tell when all of the methanol has gone off, as the flask will come to constant weight. The pH can be adjusted to any desired value with an acid like phosphoric, and salts or traces of fatty acid are then filtered off.

Demulsifier for Petroleum Emulsions

Czechoslovakian Patent 89,390

Ammonium Dibutyl-naphthalene Sulfonate	90
9-Octadecenyl Alcohol-Ethylene Oxide Condensate	10

Powdered Steel Suspension

Methanol	14.00
"Carbopol" 941	0.50
Di-Normal-Butylamine	0.38
Steel Powder	85.00

Completely disperse the "Carbopol" 941 in the methanol and neutralize the dispersion with the amine. Stir in the powdered steel. The "Carbopol" resins are not generally useful as film forming agents or binders, but the "Carbopol" 941/di-*n*-butylamine salt is a very effective binder here. In this system, the di-*n*-butylamine is uniquely useful as the neutralizing agent to enhance the resultant binding ability and to improve the "Carbopol" film.

The suspension is poured into plaster casts, vibrated under vacuum to eliminate voids and dried at 180°F in an oven with circulating air. After drying, the molded parts exhibit unusually good green strength and excellent surface. No releasing agent is used, because the castings do not stick to the plaster mold. After sintering in a hydrogen atmosphere, the resulting casting maintains its excellent surface and displays very high specific gravity.

Molybdenum Suspension

Water	20.0
"Carbopol" 934	1.0
Sodium Hydroxide (10% solution)	4.2
Molybdenum Powder (2-10 Microns)	80.0

Carefully disperse the "Carbopol" 934 in the water and then neutralize with the sodium hydroxide. Add the metal and stir to uniformity. With so high a content of metal, the suspension is a stiff paste. No water separates on aging.

"Fiberglas" Suspension

Water	1 gal
"Carbopol" 941	3.8 g
Sodium Hydroxide (10% solution)	13.0 g
"Fiberglas", Chopped	0.5 lb

Carefully disperse the "Carbopol" 941 in the water, and then neutralize the dispersion with the sodium hydroxide. (If this small amount of sodium hydroxide is considered detrimental, ammonium hydroxide, or other base, can be substituted.) Lastly, disperse the "Fiberglas" in the 941 mucilage. The suspension will be stable yet easily pumped.

Graphite Suspension

Water	149.0
"Carbopol" 934	0.8
Sodium Hydroxide (10% solution)	4.2
Graphite No. 0399	50.0

Carefully disperse the "Carbopol" 934 in half the water and neutralize with the sodium hydroxide. Disperse the graphite in the remaining water and carefully stir this dispersion into the neutralized "Carbopol" 934 mucilage.

This suspension has a viscosity of 9,720 cps (Brookfield, 20 rpm) and displays no flocculation or settling on aging. It is important, however, that the mucilage be neutralized to at least pH 9 before the graphite dispersion is added, or flocculation and settling will occur.

Titanium Dioxide Suspension

"Titanox" AWD	60
Water	39.73

"Carbopol" 941   0.04
Sodium Hydroxide   0.20

Carefully disperse the "Carbopol" 941 in half the water, and then neutralize with the sodium hydroxide. Separately disperse the "Titanox" AWD in the remaining water and add this slowly to the "Carbopol" 941 mucilage. Stir to uniformity.

There is no settling on aging, but the suspension tends to form a weak gel which reverts to a fluid on very slight stirring.

Chapter VII

FARM AND GARDEN SPECIALTIES

Termite- and Decay-Proofing Fenceposts

Treatment is best within one week after posts are cut. They should be cut six in. longer than needed. They should not be allowed to dry out. Just before treatment bark should be peeled off.

Formula No. 1

To prepare Solution A, add 24 gallons of water to the first container. If a standard 50-gal wooden barrel is used, the water will be about 16 inches deep. Next add 18 lb (about 7 coffee cans) of copper sulfate crystals. Stir until the crystals dissolve. (Less stirring is required if the crystals are added to the water on the day before the solution is to be used.)

To prepare Solution B, add 26 gallons of water to the sec-
ond container. If a standard 55-gal oil drum is used, the water will be about 15 inches deep. Then add 18 lb (about 6 coffee cans) of powdered sodium chromate slowly while stirring. This chemical dissolves readily.

Stand the peeled or partially peeled green posts with the butt end down in solution A. Add enough posts so that the solution nearly reaches the top of the container. In this way the ground area of the post is surrounded by liquid. If crooked posts are being treated, the liquid may not rise high enough. In that case, prepare extra solution by adding about two-thirds of a coffee can of copper sulfate crystals to a 10-quart pail of water.

After the posts have stood in solution A for two days, remove them and stand them butt down in Solution B. If extra solution is needed, add about two-thirds of a coffee can of sodium chromate powder to a pail of water. After the posts have stood for one day, turn them over and let them stand top down in solution B for one day.

If necessary, the posts may be used at once, but it is better to pile close together to keep them from drying out, and to let them stay that way for several weeks. This rest period helps to distribute the chemicals more evenly throughout the posts.

Although it is not altogether necessary to rinse the posts after treatment, it is well to do so if they are to be used soon. If they are not rinsed and are used soon after treatment, the chemical remaining on the surface may irritate the skin. This can be avoided by wearing rubber gloves.

Before a second set of posts can be treated, more water and chemical must be added to the containers. Add enough water to bring Solution A to its original level. Then add 3 lb (1 heaping coffee can) of copper sulfate crystals.

After bringing Solution B up to its original level, add 3 lb (1 level coffee can) of sodium chromate. Stir each solution until the chemical dissolves. The solutions should be fortified in this way before each additional set of posts is treated.

No. 2

DDT	5
Kerosine	95

Wood Preservative (Fungicide) for Pressure Impregnation

	Formula No. 1	No. 2
Copper Naphthenate	12½	12½
#2 Fuel Oil	87½	—
Creosote	—	87½

No. 3

For surfaces to be painted

Copper Naphthenate	25
Spar Varnish	40
Mineral Spirits	35

Wait 48 hours before painting over this coating.

Insecticide Concentrates
(Emulsifiable)

Formula No. 1

Aldrin
(2 lb/gal)

Aldrin (93%)	26
Heavy Aromatic Solvent	71
"Gafac" RE-610	3

No. 2

BHC
(1 lb/gal)

BHC	29
Heavy Aromatic Solvent	68
"Gafac" RE-610	3

No. 3

Chlordane
(4 lb/gal)

Chlordane	46
Kerosine	50
"Gafac" RM-510	4

No. 4

DDT
(2 lb/gal)

| DDT | 25.0 |
| Heavy Aromatic Solvent | 72.0 |

| "Gafac" RM-510 | 1.8 |
| "Gafac" RM-710 | 1.2 |

No. 5

Dieldrin
(1.5 lb/gal)

Dieldrin	19
Heavy Aromatic Solvent	78
"Gafac" RE-610	3

No. 6

Lindane
(1.5 lb/gal)

Lindane	20
Xylene-Range Solvent	76
"Gafac" RE-610	4

No. 7

Malathion
(5 lb/gal)

Malathion	60
Xylene-Range Solvent	35
"Gafac" RM-510	5

No. 8

Toxaphene
(6 lb/gal)

Toxaphene	62
Kerosine	34
"Gafac" RE-610	4

Fly Spray Concentrate

Formula	No. 1 gal	lb	No. 2 gal	lb	No. 3 gal	lb	No. 4 gal	lb	No. 5 gal	lb	No. 6 gal	lb
"Crag" Fly Repellent	50.0	412.5	50.0	412.5	50.0	412.5	50.0	412.5	50.0	412.5	50.0	412.5
"Pyrenone" 50-5 or "Powco" 50-5 Special	14.0	110.6	—	—	10.0	79.0	10.0	83.0	—	—	—	—
"Pyrexcel" 50-5 Special	—	—	14.0	116.2	—	—	—	—	—	—	—	—
"Pyrocide" Booster Concentrate-M	—	—	—	—	—	—	—	—	10.75	83.9	—	—
90% Technical Methoxychlor Oil Conc.	—	—	—	—	—	—	—	—	—	—	→	46.0
Emulsifier ("Atlox" 1045 A)	13.6	116.7	13.6	116.7	13.6	116.7	13.6	116.7	8.90	76.5	5.9 ("Triton" X-188.	42.5
Aromatic Solvent ("Velsicol" AR-50)	—	—	—	—	—	—	—	—	—	—	to 100.0	332.0
Petroleum Distillate (Sp. gr. 0.768)	22.4	145.6	22.4	145.6	26.4	171.6	26.4	171.6	30.35	197.3	—	—

Aerosol Insecticide Spray

Formula No. 1
(For Plants)

		Carbon Dioxide	Nitrous Oxide
"MGK" Pyrocide Mix #5312		25.00	25.00
Pyrethrins	(0.020)		
"MGK" 264	(0.300)		
Rotenone	(0.100)		
Other Cube Resins	(0.200)		
Methoxychlor, Technical	(0.300)		
Dichlone	(0.120)		
"Karathane"	(0.113)		
Other Nitrophenols and Derivatives	(0.012)		
Petroleum Distillate	(0.115)		
Isopropanol		71.25	70.50
Propellent		3.75	4.50

Blend Pyrocide Mix #5312 with Isopropanol before filling.

No. 2
(For Cattle and Horses)

		Carbon Dioxide	Nitrous Oxide
"MGK" Pyrocide Mix #5662		30.00	30.00
Pyrethrins	(0.15)		
Piperonyl Butoxide	(0.30)		
"MGK" 264	(0.50)		
"MGK" Repellent 11	(0.20)		
Petroleum Distillate	(28.85)		
Petroleum Distillate ("Shellsol" 72)		67.00	65.70
Propellent		3.00	4.30

Blend Pyrocide Mix #5662 with petroleum distillate before filling.

No. 3
(For Roaches and Ants)

		Carbon Dioxide	Nitrous Oxide
"MGK" Pyrocide Mix #5601		75.00	75.00
Diazinon	(0.500)		
Pyrethrins	(0.050)		
Piperonyl Butoxide	(0.100)		
"MGK" 264	(0.166)		
Petroleum Distillate	(74.184)		
"Shellsol" 72		20.30	21.88
Propellent		4.70	3.12

Blend Pyrocide Mix #5601 with petroleum distillate before filling. Intermediate Mix #5680 may also be used. This intermediate is 20 times as concentrated as #5601. To use, blend 3.75% of Intermediate Mix #5680 with 91.25% petroleum distillate before filling.

No. 4
(For Plants and Pets)

"MGK" Pyrocide Mix #5951		2.00
Pyrethrins	(0.056)	
Rotenone	(0.008)	
Other Cube Resins	(0.016)	
Pine Oil	(0.090)	
Petroleum Distillate	(0.406)	
Water		96.72
Nitrous Oxide		1.28

Make a filling stock of the concentrate and water. Use two parts concentrate with 93 parts water. (Avoid using water of extreme hardness.) Mix thoroughly and continue agitation during the filling process to be sure filling stock is always uniform. Do not make up more filling stock than will be used in one day.

Rotenone Insecticides

Flea	Rotenone	1.0	20	Cube powder 5% rotenone
Powder	Other cube extractives	2.0		
	Inert	97.0	80	Pyrophyllite or talc
Garden	Rotenone	1.0	20	Cube powder 5% retenone
Dust	Other cube extractives	2.0		
	Sulfur	25.0	25	Dusting sulfur
	Inert	72.0	55	Pyrophyllite or talc
Garden	Rotenone	0.50	10	Cube powder 5% rotenone
Dust	Other cube extractives	1.00		
	Pyrethrins	0.03	5	Pyronyl dust concentrate
	Piperonyl butoxide	0.30		
	Inert	98.17	85	Pyrophyllite or talc
Emulsifiable	Rotenone	2.00	6.0	Cube resin 33.3% rotenone
Garden	Other cube extractives	4.00		
Spray	Pyrethrins	0.40	7.6	"Pyronyl 50-5" emulsion concentrate
	Piperonyl butoxide	4.00		
	Petroleum distillate	3.20	76.4	"Velsicol AR 50" or "Sovacide 544C"
	Alkylated naphthalene	76.40		
	Inert	10.00	10.0	Emulsifier, heat to 80-90°C. to dissolve and filter.
Agricultural	Rotenone	5.0	15	Cube resin 33.3% rotenone
Emulsifiable	Other cube extractives	10.0		
Rotenone 5%	Xylene	82.0	82	Petroleum xylene
	Inert	3.0		
	Warning — Flammable! Keep away from heat and open flame		3	Emulsifier, heat to 90°C to dissolve and filter.
Emulsifiable	Rotenone	5.0	15	Cube resin
Rotenone	Other cube extractives	10.0		33.3% rotenone
5% for Fish	Inert	85.0	80	"Velsicol AR50"
Control			5	Emulsifier, heat to 90°C to dissolve and filter.
Garden	Rotenone	0.5	1.5	Cube resin, 33.3% rotenone
Aerosol	Other cube extractives	1.0		
	Pyrethrins	0.1	1.9	"Pyronyl 50-5" oil concentrate
	Piperonyl butoxide	1.0	2.0	Acetone
	Petroleum distillate	10.4	9.6	"Sovaspray 100"
	Inert	87.0	10.0	Methylene chloride
			32.5	"Freon" 11
			42.5	"Freon" 12
Pressurized	Rotenone	0.75	2.25	Cube resin, 33.3% rotenone
Garden	Other cube extractives	1.50		
Spray	Methoxychlor	3.00	10	Acetone
	Inert	94.75	34.42	Isopropanol
	Warning — Flammable!		25	"Freon 11"
	Do not spray near fire or flame.		25	"Freon 12"

Aerosol Mothicide—Mildewcide

"Vancide" 89RE	0.2
DDT	10.0
"Chlorothene"	89.8

Mix the components together until uniform.

Aerosol Packaging Procedure:

Above concentrate: 50%

Propellent 11/12: 50%

To formulate a mildewcide use:

"Vancide" 89RE   0.2%
Chlorothene   98.8%

These are simple, yet effective preparations. "Vancide" 89RE, bactericide-fungicide, is stable on long aging in these formulas.

To Attract Male Gypsy Moth

U.S. Patent 3,018,219

12-Acetoxy-1-hydroxy-9-octadecene is used as a bait.

Grain Fumigant

Formula No. 1

Carbon	
Tetrachloride	27.0
Ethylene Dichloride	63.0
Sulfur Dioxide	2.9
Ethylene Dibromide	7.1

No. 2
(Tablet)

Aluminum Phosphide	70
Ammonium	
Carbamate	26
Paraffin Wax	4

This must be kept dry; admission of moisture generates phosphine.

Use 120-180 g to 1000 bushels.

Herbicide Concentrates (Emulsifiable)

Formula No. 1

Butyl 2,4-D Ester	
(2.67 lb acid equivalent/gal)	
Butyl 2,4-D	42.0
Kerosine	55.0
"Gafac" RM-510	0.9
"Gafac" RM-710	2.1

No. 2

Butyl 2,4-D Ester	
(4 lb acid equivalent/gal)	
Butyl 2,4-D	58.4
Heavy Aromatic	
Solvent	38.6
"Gafac" RM-710	3.0

No. 3

Butyl 2,4-D Ester	
(6 lb acid equivalent/gal)	
Butyl 2,4-D	78.0
Heavy Aromatic	
Solvent	17.0
"Gafac" RM-710	5.0

No. 4

Isooctyl 2,4-D Ester	
(4 lb acid equivalent/gal)	
Isooctyl 2,4-D	68.0
Heavy Aromatic	
Solvent	27.0
"Gafac" RM-510	3.5
"Gafac" RM-710	1.5

No. 5

Isopropyl 2,4-D Ester	
(3.34 lb acid	
equivalent/gal)	
Isopropyl 2,4-D	45.0

Heavy Aromatic	
Solvent	52.0
"Gafac" RM-510	0.6
"Gafac" RM-710	2.4

No. 6

Isooctyl 2,4,5-T Ester	
(4 lb acid equivalent/gal)	
Isooctyl 2,4,5-T	67.0
Heavy Aromatic	
Solvent	28.0
"Gafac" RM-510	3.5
"Gafac" RM-710	1.5

Plant Leaf-Shine Spray
(Aerosol)

PVP/VA Copolymer 1-535 has been successfully used as a base for leaf-shine sprays. The following is a basic formulation in which a suitable non-phytotoxid bactericide (such as "Dithane"), and/or insecticide (such as pyrethrin) may be incorporated. If bactericides or insecticides are to be added resultant product should be checked for compatibility, i.e. the formation of a clear and flexible film.

PVP/VA 1-535	3.00 to 5.00
"Carbowax" 1500	0.20 to 0.35
Isopropanol	31.80 to 29.65
Propellents 12/11 (40/60)	65.00

Dissolve the copolymer in isopropanol. Add "Carbowax." Fill under pressure.

Agricultural Antifreeze for Plants

PVP/VA emulsion or solid copolymers may be sprayed on plants to prevent excessive transpiration and to prevent freezing during the winter months. Such formulations may also serve to prevent wilting of foliage during transplanting of ornamental plants, such as evergreen azaleas, rhododendron, and mountain laurel. A typical formulation consists of 1-5% (solids basis) PVP/VA S-630 or PVP/VA W-464 in water.

Preservative for Cut Flowers

U.S. Patent 2,923,094

Magnesium Sulfate	0.10
Borax	0.04
Citric Acid	5.00
Succinic Acid	5.00
Silver Nitrate	0.10
Water	1000.00

Tree Wound Dressing

British Patent 839,789

Water	131.0
Bentonite	6.5
Potassium Hydroxide	1.8
Ligno Sulfonic Acid	5.0
Bitumen (60-80 penetration)	217.0
Stearic Pitch	17.0
Copper 8-Quinolinolate	5.2

For Plant Chlorosis

German Patent 1,114,833

Graham's Salt	77
Iron Sulfate $7H_2O$	23
Water, to make a 2% solution.	

To Suppress Sprouting of Potatoes

U.S. Patent 2,999,746

In a closed bin introduce 1 pound of diproparyl ether for each 1000 pounds of potatoes.

To Control Land Crabs

Rice Bran	10 lb
Lindane (50% wettable Powder)	1 lb
Water, q.s. to make paste.	

This quantity is sufficient for 700 crab holes.

Bee Repellent

An absorbent pad, impregnated with propionic anhydride is placed in a special fume chamber of the beehive. This chamber is then fitted over a section of the hive called the *super*, from which honey is harvested by beekeepers. Air forced with a bellows into the fume chamber drives the bees into other supers, or into the brood chambers below the supers.

When propionic anhydride is used, bees become gentle and easy to work with. They are repelled from sealed honey supers in one to two minutes, in shade as well as in sun. There is no change in flavor or odor of the honey.

Repellent Coating for Seeds

Formula No. 1
(Birds)

"Drinox"	0.5-1
Spreader-Sticker	1
Water, to make	100

No. 2
(Birds, Rodents)

U.S. Patent 2,900,303

Iron Oxide	33-75
Calcium Carbonate	25-66
Copper Oxalate	2-10

To Rat-Proof Packaging Materials

Treat with a dilute solution of zinc dimethyldithiocarbamate-cyclohexylamine complex. It repels rats because of its taste.

Dog Shampoo

Formula No. 1

"Stepanol" WA Paste	36.0
Ethylene Glycol Monostearate Pure	2.0
"Stepan" T-6-B	4.5
Lanolin	0.5
"G-11"	0.5
Water	56.5

Mix "Stepanol" WA Paste, water and "G-11" (hexachlorophene) until uniform. Add other ingredients with stirring and heat to 175°F. Adjust pH to 7.4 with citric acid.

No. 2
(Aerosol)

a	Stearic Acid	100
	Myristic Acid	60
	Oleic Acid	100
	Cetyl Alcohol	3
	Propylene Glycol	607
	"Pyrenone"	22
b	Triethanolamine	150

PVP K-30	2
"Versene" Regular	8
Triethanolamine	
Lauryl Sulfate	1448
Water (Distilled)	485
Perfume	15

Heat a and b separately to 70°C until all ingredients are dissolved. Add b to a. Stir until cool and then add perfume. Charge 90% with 10% propellents 12/114 (40/60).

Chapter VIII

FOOD PRODUCTS

Brown 'N Serve Rolls

Flour	100 lb
Water	68 lb
Nonfat Milk	
Solids	4 lb
Salt	2 lb
Sugar	12 lb
Shortening	12 lb
Eggs	6 lb
Yeast	2 lb
Yeast Food	4 oz
Calcium Propionate	
(Mold Inhibitor)	4 oz

Dissolve the yeast in some of the water. Place the milk solids, salt, sugar, shortening, eggs, yeast food, mold inhibitor, and the balance of the water into the machine and stir together. Add the flour, then the yeast solution, and mix until a uniform dough is obtained.

This dough should have a temperature of 90°F when mixed. Ferment at 80°F.

To alter consistency of dough, water may be reduced to 66 or 64 lb.

Glaze for Rolls

Granulated Sugar	20 lb, 1 oz
Cellulose Gum, CMC-7HSP	
(amount depending on	
consistency desired)	1.75 to 2.25 oz
Water	80 lb

Blend sugar and cellulose gum together by sifting. Add the water to the dry blend and stir until smooth.

303

Potato Bread

Potato bread has become more popular since the development of a good potato flour, which eliminates the preparation of boiled potatoes. The crumb has a silky softness and fine taste. It is agreeably moist and keeps soft for a long time. Considering that 1 lb of potato flour is equal to about 4.5 lb of cooked mashed potatoes, a bread that contains 5% of potato flour is in every sense real potato bread and can be labeled as such.

Primary Ferment

Yeast	2 oz
Potato Flour	10 oz
Water (85°F)	3 lb

Beat mixture well and let it stand until it rises and breaks, i.e., about 30 minutes.

Dough

Bread Flour	12 lb	8 oz
Cold Water	5 lb	
Yeast		2 oz
Salt		5 oz
Sugar		8 oz
Shortening		8 oz
Malt Extract		1 oz
"Breadlac" Nonfat		
Dry Milk Solids		8 oz

Mix in the ferment and make a dough of medium stiffness at 80°F. Let it come to a full rise and punch. Take to bench in 20 minutes. Round and mold as usual. The yield should be 22 pounds.

Dairy Bread
(Straight Dough)

Formula No. 1

Flour	100 lb
Water	70 lb
"Breadlac"	6 lb
Sugar	5 lb
Salt	2 lb
Shortening	4 lb
Diastatic Malt	12 oz
Yeast	2 lb

The temperature of dough should be 80°F. First rise is 2½ to 3 hours; second rise, 45 minutes. Scale and round the pieces of dough. Let them rest 10 minutes after rounding; mold, and proof with moderate steam for about 45 minutes. Bake at 425°F for 30 to 45 minutes. The yield should be 189 lb, 12 oz.

No. 2

Flour	40 lb
Water	26-28 lb

"Breadlac" 6 lb
Sugar 5 lb
Salt 2 lb

Use 81°F temperature; at least 20 minutes floor time. Proof 50 to 60 minutes. The temperature of oven should be 425°F, and the yield, 187 lb. The dough should come from the mixer soft, but thoroughly mixed.

No. 3

(Sponge Dough)

Flour 60 lb
Water 40-42 lb
Shortening 4 lb
Diastatic Malt 4 oz
Yeast Food 4 oz
Yeast 1 lb, 8 oz

Work at 76°F and ferment for 4½ hours.

Cinnamon Bread

There are several methods of preparing cinnamon bread, most of which consist of the use of a mixture of cinnamon and sugar that is sprinkled on the surface of the flattened dough, then rolled to form a band, often with added raisins or nuts, that follows the curl of the mold. A very superior cinnamon loaf may be prepared as follows:

Scale and round a piece of standard sweet dough, about 12 to 15 oz, and allow it to rest on the bench until soft and plastic. Meanwhile arrange to have some batter for layer cake—either yellow or white—in the refrigerator. Mix into the batter sufficient plain cinnamon to color the mixture a rich brown shade.

Pin out the pieces of sweet dough to a thin sheet, the width of the pan, and about 14 inches long. Make it square at both ends. Spread the cinnamon-cake batter evenly over the entire surface. Roll loosely to shape a loaf. Pan and proof.

Just before baking, cut with a sharp knife to about 1 inch in depth. The oven heat will expand the cinnamon batter mix to make an attractive wide band that is very appealing when sliced.

School Lunch Bread

Formula No. 1
(Sponge)

Flour About 70 lb
Water 49 lb

"Breadlac"	6 lb
Shortening	2 lb
Diastatic Malt	8 oz
Yeast Food	4 oz
Yeast	1 lb, 12 oz

Dissolve the ingredients by running the mixer at slow speed; then add the flour and dissolved yeast. Mix the sponge until smooth at 77°F for 4 hours.

No. 2
(Dough)

Flour	30 lb
Water	25 lb
"Breadlac"	6 lb
Sugar	4 lb
Salt	2 lb, 2 oz
Shortening	2 lb

The dough should have at least 20 minutes floor time before going to the divider. Make up and proof normally. Bake at 400 to 410°F for 30 minutes for 1-lb loaves. The yield should be 193 lb, 10 oz.

This formula was made part of a governmental program designed to promote more nutritious foods for children. Great emphasis has been placed on a high content of milk solids.

Cracked-Wheat Hearth Bread

a	Water	3 lb
	Honey	6 oz
	Cracked Wheat	2 lb
	Whole Wheat	2 lb
b	Clear Flour	9 lb
	Water	5 lb
	Yeast	4 oz
	Salt	4 oz
	Shortening	6 oz
	"Breadlac"	3 oz

Mix a to a dough at 78°F and let it stand for 3 hours. (This is not a sponge, but a soaking dough.)

Add b to the first mix and make a dough at 78°F. First rise is 2 hours; second rise, 50 minutes. Divide, round, and let stand 15 minutes.

Mold as rye bread, place on mealed boards, and give medium proof. Cut on the peel as rye bread and bake on the hearth with steam at 425°F. The yield should be 22 lb, 9 oz.

Resembling rye bread in appearance, though somewhat darker, this loaf has a delicious flavor that is derived from the blend of flours with honey.

Salty Icebox Rye Bread

First Clear	
Flour	8 lb, 8 oz
Dark Rye	
Flour	4 lb, 8 oz
Yeast	6 oz
Salt	10 oz
Rye Sour	
(Optional)	1 lb
Water	8 lb
Caramel	
Color	As desired
Shortening	4 oz
Whole Caraway	
Seeds	10 oz
Malt Extract	2 oz

Mix thoroughly at 80°F. Ferment to full rise in 2 to 3 hours. Punch and stand 30 minutes. Scale 10-oz pieces (more or less, as desired) and mold into short lengths. Proof 20 minutes; then mold into tight loaves with square ends, 12 to 15 inches long. Proof on mealed boards or pans for only 5 minutes. Then bake on the hearth with much steam, keeping the loaves well spaced to avoid cracking. A further precaution against cracking is to wash the loaves with warm water just before baking. Bake at 400 to 450°F.

This is a popular bread for parties and special occasions. Sliced thin, it makes tasty snacks and provides an excellent base for hors d'oeuvres. It has long keeping qualities if refrigerated and has an excellent, tangy flavor.

Stable Ferment Process Bread

Water	70 lb
Nonfat Dry Milk	6 lb
Corn Sugar	4 lb
Salt	2.25 lb
Yeast	2.5 lb
Yeast Food	0.5 lb

The water is metered into a stainless-steel tank. The other ingredients are dispersed mechanically. This ferment in the tank is agitated slowly at a constant temperature (96 to 100°F) for 5 to 6 hours until properly conditioned for use.

After cooling, the ferment will remain stable for several hours. After the day's run, it may be refrigerated for use the following day.

The required ferment is metered into the dough mixer

and the conventional mixing of dough begins.

Bread Dough Mix

Flour	100 lb
Sugar	5 lb
Lard	4 lb
Yeast	0.5 lb
Yeast Food	0.25-0.5 lb

This dough is mixed 12 to 13 minutes at 84 to 85°F; floor time: 30 minutes.

In the ferment, spray or roller milk of good quality should be used. Certain measures are necessary to keep the roller type from foaming.

Strict sanitary control is needed because of the danger of infestation with bacteria.

For this reason, stainless-steel tanks with double walls and temperature controls are necessary. With this type of equipment, the pipes can be easily taken apart for cleaning at regular intervals.

Although the cost of ingredients is higher with this system than in the conventional sponge method, the advantages more than offset this. The following economies result: saving in floor space; no setting of individual sponges and the resulting variations which consume time; a full day's production of ferment can be made at one time; less waste, less evaporation; closer control of uniformity in the bread.

Protein Bread

	Dough (g or lb)	Sponge (%)*
Patent Flour	40.0	60.0
Water	29.4	34.6
Yeast	—	3.5
Malt	—	1.25
Shortening	—	3.5
Salt	1.75	—
Sugar	3.5	—
Non-fat Milk Solids	3.5	—
"Promine"-R or"Soyalose"105	5.0	—

* The percentage figures are based on the total flour weight in the finished mix.

Yeast food may be added to the sponge at a level of 0.25 to 0.50, based on the total flour in the finished mix, depending on the nature of the flour used.

To prepare sponge: Sift flour and add to mixing bowl. While mixing using dough hook at low speed, slowly add water, followed by the addition of yeast, malt (yeast food, if used) and shortening. Mix at medium speed until well mixed (dough will pull from bowl). Temperature of sponge should be about 75-78°F when finally mixed. Add sponge to trough and proof at 75-80°F for 3-4 hours and punch. The sponge may be placed in a retarding box at 40°F overnight for use the following day or immediately mixed with the dough for finishing.

To prepare dough: Mix milk solids, salt, sugar and soy flour (or "Promine"-R) with remaining water (warm), flour, and sponge. Mix about 10 minutes at low speed and 5 minutes at medium speed. The temperature after mixing should be about 78°F. Let dough rest 30 minutes. Divide, mold, pan, and proof for 30-45 minutes. Bake at 410-425°F for about 25 minutes.

Vienna Bread

High-Gluten Flour	6 lb
Patent Flour	7 lb
Water	8 lb
Yeast	4 oz
Salt	4 oz
Sugar	4 oz
Shortening	7 oz
Malt	3 oz
"Breadlac"	5 oz

The temperature of the dough should be 80°F; full rise 2 hours; fold over. Second rise 1 hour. Rest 30 minutes more and take to bench. Round up and let rest 20 minutes; then mold with great care.

Proof on mealed boards or in cloths until double in size. Transfer to peel and cut lengthwise, using two strokes of the blade: the first stroke from the tapered end to the middle, slightly to the left of center. Start the second stroke at the center and cut to the other tapered end. This gives a feathered effect that adds greatly to the appearance of the loaf.

Bake on the hearth at 425°F with the loaves well spaced, with plenty of steam. Cut off the steam when the loaves start to color. The yield should be 22 lb, 11 oz. This bread has a fine flavor and a crust that, though somewhat brittle, is very tender.

Rope (Mold) Preventive for Bread

Spanish Patent 197,203

Acetic Acid	50.0
Tartaric Acid	1.0
Lactic Acid	2.5
Hydrochloric Acid	0.3
Distilled Water	46.2

Cocoa-Whey Cookies

Granulated Sugar	95
Shortening	30
Butter	30
Whole Eggs	22
Salt	2
Cocoa	19
Cake Flour	100
Baking Powder	4
Dried Whey	12
Cinnamon	⅓
Water (Variable)	20

Scale the sugar, shortening, butter, eggs, and water into the mixer. Sift the cake flour, cocoa, baking powder, salt, cinnamon, and whey powder together and add to the other ingredients. Mix approximately 3 minutes at low speed. Deposit by spoonfuls on lightly greased pans. Bake 6 to 8 minutes at 360 to 380°F.

Oatmeal-Coconut Whey Cookies

Brown Sugar	120
Shortening	43
Butter	22
Cake Flour	100
Soda	2
Cinnamon	0.25
Salt	2
Whole Eggs	5
Ground Oatmeal	23
Ground Raisins	23
Macaroon Coconut	23
Dried Whey	12
Water (Variable)	10

Scale the sugar, shortening, butter, eggs, oatmeal, raisins, coconut, and water into the mixer. Sift the cake flour, soda, cinnamon, salt, and whey powder, and add to the other ingredients. Mix at low speed for approximately 3 minutes. Deposit by spoonfuls on lightly greased pans. Bake 10 to 12 minutes at 375 to 400°F.

Yeast-Raised Whey Doughnuts

Shortening	15
Sugar	15
Honey	3
Salt	1
Whole-Egg Powder	15
Mace	1
Coriander	1
Bread Flour	75
Cake Flour	25
Yeast	6
Dried Whey	6
Warm Water (Variable)	50

Cream the shortening, sugar, eggs, honey, salt, and spices thoroughly. Sift the flours together. Dissolve the yeast in the warm water and add to the mixture with the sifted flour. Mix until smooth. Let the dough rise for 90 minutes, punch, and give 30 minutes on bench. Make up by hand, proof light, and fry at 350 to 380°F.

Whey Sweet Dough

Sugar	17
Flavoring	As desired
Salt	1
Whole-Egg Powder	6
Shortening	12
Bread Flour	100
Yeast	5
Dried Whey	6
Warm Water (Variable)	55-58

Cream the shortening, sugar, eggs, salt, and flavoring. Dissolve the yeast in the warm water, add with the flour to the mixture, and mix until smooth. Temperature of dough should be approximately 82°F. Fermentation: 75 minutes; 30 minutes; 25 minutes. Make up, proof light, and bake at 375 to 400°F, for approximately 25 minutes.

Yeast-Raised Doughnut Mix

Spring Wheat Flour	2370
2X Sugar	91
Corn Sugar	210
Lard	270
"Aldo" 40	30
Salt	30
Dry Egg Yolk	10

Skimmed Milk	
Powder	5
Malt Flour	2
Wet Yolks	20

This is a straight dough for rings, bismarks, sticks, and pershings.

Use 100 lb of this mixture, after sifting, with 4 to 5 lb yeast, dissolved in 45 to 47 lb water. The temperature of the dough should be 78 to 80°F.

Mix in a vertical or horizontal mixer for 1 minute at low speed (40 to 50 rpm) and for 15 minutes at second or high speed (70 to 80 rpm) to the maximum mixing.

Let the dough set for 1 hour to 1 hour and 15 minutes at 80°F.

Fry the doughnuts at 375 to 380°F, 1 to $1\frac{1}{4}$ minutes on each side if at surface, or $2\frac{1}{2}$ minutes if submerged.

Lard has been found to be the best shortening but lard alone will not maintain the desired tenderness. Therefore, about 10% "Aldo" 40 is used on the lard. The quantity may vary slightly with the flour. "Aldo" 40 may be used up to 15% on the lard without increasing the absorption of fat.

Cake Doughnut

Soft Flour*	1000
Fine Granulated Sugar (Cane or beet)	550
Corn Sugar	50
Liquid Vegetable Oil (Soybean, cottonseed, peanut)	77
S-1098 (Mixed Monoglycerides)	3 to 4
Egg Yolk	70
Skim-Milk Powder	100
Sodium Acid Pyrophosphate	34
Soda	23
Salt	29
Full-Fat Soya Flour	25

* Protein, 8.5–9.0; moisture, 13.0–13.5; ash, 0.38–0.40.

Use 5 lb of this mix with 2 lb to 2 lb, 2 oz of water at a temperature of 74 to 78°F, for the dough. Flavor with lemon and vanilla or nutmeg and mace.

Babka
(Polish Cake)

Yeast	2 oz
Lukewarm Water	1 pt
Scalded and Cooled Milk	1 pt

Sugar (Sucrose or Dextrose)	12 oz
Egg Yolks	1½ pt
Melted and Cooled Butter	1 lb
Salt	⅔ oz
Flour	3 lb, 6 oz
Almond Extract	½ oz
Ground Walnuts	1 lb
Cinnamon	⅓ oz
Apple Jelly	12 oz

Dissolve the yeast in the water. Combine the milk, one half of the sugar, egg yolks, salt, and butter. Add two thirds of the flour and the yeast solution, and mix until smooth.

Let this stand about 2½ hours then add the remaining sugar, flour and the almond extract and mix until smooth. Then combine the cinnamon and walnuts and work these into the dough just enough to distribute them.

Place in greased round 9-inch cake pans, 4 inches deep. Brush the batter with beaten egg white and allow it to rise until double in bulk. Bake in slow oven at 325°F, for about 1 hour. Remove, cool, melt the apple jelly, and brush over the cake.

Helpful Hints for Baking Sweet Goods

Quality ingredients ensure success. Every baker knows he can purchase cheaper flour, shortening, spices, and other ingredients. It is a temptation to cut costs when prices soar, but repeated sales are not built on mediocre quality. The difference in unit-cost between a fine product and a poor one is little, but the difference in taste-appeal is tremendous. And it is the taste that tells.

Use sugar of fine granulation as it will dissolve more readily. Undissolved sugar tends to result in coarse-grained cakes.

Use a bowl of the proper capacity for the size of the batch to be made. The sugar and shortening should occupy about a quarter of the bowl area before creaming.

Avoid overgreasing the pans. A well conditioned pan requires very little grease. The use of pan liners is recommended. They conserve grease, reduce cripples, tend to produce better volume, and eliminate peaking of the cakes.

Watch baking temperatures. Goods should be baked

in the shortest possible time so that excessive drying out will be avoided. Fruit and pound cakes are baked at 300 to 350°F; loaf cakes at 350 to 375°F, and layers and sheets, at 375 to 425°F.

Altitude Alters Formulas

Varying altitudes alter the results in tested formulas. A recipe may give perfect results at sea level and turn out poorly at an elevation of 5,000 feet unless changes are made. The reason for this is that as the altitude increases the atmospheric pressure decreases. As a result, there is a greater expansion of steam and gases in the mix. To balance this, it becomes necessary to reduce the creaming time or the amount of leaven used.

Yellow Layer Cake

Formula No. 1

a	125% Granulated Sugar	12 lb, 8 oz
	Butter	2 lb, 8 oz
	Emulsifier Shortening	2 lb, 8 oz
	Vanilla Extract	1 oz

b	High-Grade	
	Cake Flour	10 lb
	Baking Powder	8 oz
	"Parlac" (Dry Whole Milk)	1 lb, 9 oz
	Salt	5 oz
c	Water	7 lb, 8 oz
d	Whole Eggs	5 lb

Cream ingredients of a until light and fluffy.

Sift b together, add to a, and cream until the sugar and shortening are well distributed. Add 4 lb of water gradually; work until smooth. Add eggs slowly. When the eggs are well incorporated, add rest of the water and mix smooth and glossy.

Batch weight: 42 lb, 7 oz. Scale 7 oz to 7-inch layer tin; 9 oz to 8-inch layer tin. Bake at 420°F. The cakes should be quite soft out of the oven.

No. 2

"Parlac"	1 lb, 2 oz
Water	8 lb, 6 oz

Place the "Parlac" on top of the water and dissolve with a hand whip. Hold ready until needed.

High-Grade
 Cake Flour 10 lb
Emulsifier
 Shortening 4 lb, 12 oz
Powdered
 Orange
 Juice 12 oz

Cream 5 to 8 minutes at low speed.

140% Granulated
 Sugar 14 lb
Salt 6 oz
Baking Powder 10 oz
Liquid Milk
 ("Parlac") 4 lb
Vanilla Extract 2 oz

Sift the dry ingredients and add them alternately with the milk. Scrape the bowl. Cream 3 to 5 minutes at low speed.

Whole Eggs 6 lb
Liquid Milk
 (balance of
 "Parlac") 5 lb, 8 oz

Stir with a hand whip. Add in three stages and smooth out the mix.

 Batch weight: 55 lb, 10 oz. Scale 6 oz to 5-inch tin; 9 oz to 8-inch tin. Bake at 360°F.

Milk-Chocolate Cake

Cake Flour 10 lb
High-Grade
 Shortening 4 lb

Cream 5 minutes at low speed

"Parlac" Dry
 Whole Milk 6 lb, 8 oz
Soda 1 oz
Salt 6 oz
Baking Powder 4 oz

Sift and add to the first batch.

150% Granulated
 Sugar 15 lb

Add immediately.

Water 5 lb

Add gradually. Mix 5 minutes at low speed and scrape the bowl.

Whole Eggs 6 lb
Water 6 lb, 8 oz
Vanilla Extract 2 oz

Mix together with a hand whip. Add gradually in several stages.

Melted Bitter
 Chocolate 2 lb, 8 oz

When smooth, incorporate at low speed.

Batch weight: 56 lb, 5 oz. Scale 7 oz in 7-inch layer tins; 9 oz in 8-inch layer tins. Bake at 375°F.

Chocolate-Fudge Icing

"Parlac"	1 lb, 8 oz
Confectioners' Sugar	20 lb
Salt	1 oz
Vanilla Extract	2 oz

Mix the dry ingredients together and place them in a mixing bowl.

Water	3 lb, 4 oz
Granulated Sugar	4 lb
Corn Syrup	4 lb
Butter and/or High-Grade Shortening	4 lb, 4 oz
Shaved Bitter Chocolate	4 lb, 4 oz

Melt the chocolate and butter on low heat. Bring the sugar, syrup and water to the boiling point. Add the butter and chocolate and cut off the heat. Place in a bowl with the confectioners' sugar and mix.

Beat with a creaming paddle until smooth and glossy at medium speed. Add a little more water if needed to obtain the desired consistency.

Cuban Rum Cake

Prepare Turk-head or angel-cake pans by greasing heavily with a mixture of 1 lb shortening and 10 oz cake flour. Then line the pans with finely chopped nuts or a crunch. This mix makes twenty-eight cakes when scaled at 20 oz each.

"Parlac" Dry Whole Milk	13 oz
Luke-Warm Water	5 lb, 6 oz

Place the "Parlac" on top of the water and dissolve with a hand whip. Hold ready.

Cake Flour	4 lb, 8 oz
High Grade Shortening	3 lb

Cream 3 minutes at low speed.

125% Granulated Sugar	9 lb, 6 oz
Cake Flour	3 lb
Baking Powder	6 oz
Salt	3 oz
Liquid Milk "Parlac"	3 lb

Sift the dry ingredients and add. Cream 3 minutes. Scrape the bowl.

Whole Eggs   4 lb, 2 oz
Vanilla Extract   2 oz

Add in three portions, mixing 3 minutes each.

Crushed Pineapple   3 lb
Liquid Milk
   "Parlac"   3 lb, 2 oz

Add and mix smooth.

Batch weight: 40 lb. Bake at 360°F. When cakes are cool, have syrup hot; dip cakes and drain them on screens.

Rum Syrup for Cake

Dissolve 6 lb granulated sugar, 2 lb glucose, and 0.5 oz cream of tartar in 3 lb, 4 oz of water and bring to a boil. When cooled to 110°F or below, add rum extract to taste. The syrup should be hot and the cakes cool to obtain the best penetration. A 1-lb cake should absorb about 4 oz of syrup.

White Layer Cake

"Parlac"   1 lb, 3 oz
Water   8 lb, 5 oz

Place the "Parlac" on top of the water and dissolve with a hand whip.

High-Grade
   Cake Flour   10 lb
Emulsifier
   Shortening   4 lb, 12 oz

Cream 4 minutes at low speed.

130% Granulated
   Sugar   13 lb
Salt   5 oz
Baking Powder   10 oz
Cream of Tartar   2 oz
Liquid Milk
   (from the Above) 4 lb

Add and incorporate. Cream 4 minutes at low speed and scrape the bowl.

Egg Whites   6 lb, 12 oz
Vanilla Extract   3 oz
Liquid Milk
   (Balance of
   Above)   5 lb, 8 oz

Mix together with a hand whip. Incorporate in three stages and smooth out.

Batch weight: 54 lb, 8 oz. Scale 8 oz to a 7-inch layer tin; 11 oz to an 8-inch layer tin. Bake at 350°F.

Angel Food Cake

Egg Whites (Liquid)	41.55
Granulated Sugar	20.79
Cream of Tartar	0.65
Salt	0.33
Vanilla (Pure Extract)	0.31
Powdered Sugar	20.79
Cake Flour	15.58*

* When using "Starbake"-100 Powdered for optimum quality of cake, substitute this for 30% of the flour (reducing flour content to 10.91%). Tests indicate additional quality by adding one-half of the "Starbake" (4.67%) with salt and granulated sugar, the remainder with powdered sugar and flour.

Place freshly thawed egg whites in the 12-quart mixing bowl with whip of a Hobart C-100 mixer and whip them to a "wet peak" on second speed (approximately 3.5-4.5 minutes).

Sift cream of tartar, salt, and granulated sugar together and add them to the mixer. Then add vanilla and continue mixing on second speed until properly whipped. This is determined by cutting the batter with a spatula—the batter should go together in 15-20 seconds. Mixing time is approximately 2 minutes.

Sift powdered sugar and cake flour together 3 times, and add this mixture in approximately one-third portions, folding in with a flat egg beater 15 times with each portion.

Place six hundred grams of batter in a standard aluminum cake pan 8½-inch bottom diameter, 9½-inch top diameter, 4-inch depth.

Spread the batter evenly in the pan with a 1-inch spatula by cutting the batter and bake for 35 minutes at 375°F.

Invert pan and allow cake to cool 90 minutes before removing it.

Pie Crust

Flour	70
"Starbake"-100 Powdered	30
Shortening	50
Salt	1
Water, Approx.	25

Mix flour and "Starbake" together and blend in shortening and salt. Add water to proper consistency. The control crust should contain 100% flour and 60% shortening.

Lemon Pie Filling

a	Water	41.0
	Granulated Sugar	13.0
b	Starch	4.0
	Water	13.5
c	Granulated Sugar	20.0
	Cellulose Gum,	
	CMC-7HP	0.5
d	Egg Yolks	4.5
	Butter	0.75
	Lemon Juice	2.75

Mix a and heat just to boiling. Slurry b together and pour into a with constant stirring. Stir until thick and smooth; bring back to boil. Dry-mix c by sifting together several times. Add slowly to a-b with constant stirring. Stir until smooth and thick. Mix egg yolks with a portion of preceding mix, then stir into entire mix and cook on steam heat for about 2 minutes. Stir in butter; then stir in lemon juice.

Cherry Pie Filling

a	Drained Cherry Juice	10 lb	8 oz
	Water	10	—
b	Water	2	—
	Starch (variable)		
	For heavy syrup	—	10.5
	For medium firmness	—	15.0
c	Granulated Sugar	1	—
	Cellulose gum,		
	CMC-7HSP	—	3
d	Granulated Sugar	5-6	—
	Salt	pinch	—
e	Cherries (Drained)	19	8

Bring a to boil. Dissolve b, add to boiling mix, and cook until thick and clear. Mix c together; add to hot mixture and stir until smooth. Add d to hot mixture and bring back to boil. Turn off heat. Finally add cherries, mixing gently.

Cream Pie Fillings

	Large Batch		Test Batch
Water	6 lb	0 oz	272 g
Skim Milk Powder	0	12	34
Sugar	2	8	114
Salt	0	1	2.8

Dissolve, heat to boiling

Water	2 lb	6 oz	110 g
Whole Eggs	1	0	46
"Starbake"-100, Powdered	0	12	34

Mix slowly, add to boiling mixture. Cook until thick.

Shortening (Butter, Margarine)	0	8	23
Vanilla Flavor	0	¼	0.7

Ice-Cream Pie

Stale Macaroons	4 lb
Butter	4 oz
Sugar	⅔ oz
Chocolate Ice Cream	½ gal
Coffee Ice Cream	½ gal
Shredded Coconut	12 oz

Melt the butter and add it to the crushed macaroons and sugar. Stir well and press well to the sides and bottom of pie plates. Refrigerate for 15 minutes. Form ice-cream balls with a scoop and roll these in the coconut. Pile them into the macaroon crust, sprinkle all over with coconut, and refrigerate, or wrap and store in a freezer.

High-Grade Non-settling Custard Pie

Water	2 lb, 4 oz

Bring to a boil.

Cornstarch	3 oz
Water	1 lb

Stir together to make a smooth cream. When the first water is boiling, stir in the starch cream. Remove from the fire at once without further cooking.

"Breadlac"	1 lb, 2 oz
Granulated	
Sugar	1 lb, 6 oz
Salt	⅜ oz
Nutmeg	1/16 oz

Mix together dry.

| Whole Eggs | 2 lb, 4 oz |

Rub into this the dry mix until smooth.

| Water | 5 lb, 8 oz |

Stir until smooth and then stir in the cooked starch.

This makes ten 9-inch pies. Fill into shells and bake at 425 to 450°F with strong bottom heat. If there is a flash heat, leave the oven door open. Remove pies before they are entirely set.

For coconut-custard pies, blend equal parts of short-thread coconut and custard mix and deposit 2 to 3 oz of coconut mix in each shell before filling.

Sour Cream for Cheese Cake

Determine the fat content best suited using the following table:

Percentage	Whole Milk (Gallons)	Butter		Shortening	
0.10	3.5	2 lb	6 oz	2 lb	2 oz
0.20	3.5	4	4	4	4
0.30	3.5	6	6	6	6
0.40	3.5	8	8	8	8

Heat slowly until the fats are melted. Pasteurize at 180°F for 10 minutes with constant agitation.

Homogenize and cool to 70°F without agitation. When the cream has reached 70°F, stir in 2 qt of sour cream or bacteria-cultured milk. Allow mixture to ripen for 12 to 15 hours.

Do not pasteurize under 160°F; 180°F is preferable as higher heat treatment lowers the oxidation-reduction potential to a point where prompt growth of the bacteria is possible.

After the cream has ripened, retain 2 qt to inoculate the next batch.

Good results are obtained with half whole fresh milk and half reconstructed whole milk.

If a very heavy cream is desired quickly, add 10 drops of commercial rennet extract with the inoculating starter.

Fig Newtons

Powdered Sugar	9 lb
Corn Syrup	1 lb
Shortening	4 lb
Salt	3 oz
Vanilla Flavor	To suit
Molasses	8 oz

Cream thoroughly.

| Eggs | 1 lb, 6 oz |

Add gradually while creaming.

| Water (Variable) | 4.75 lb |
| Ammonia | 1 oz |

Mix together to dissolve; then add, just stirring in.

Cookie Flour	20 lb
Cornstarch	1 lb
Baking Soda	2.25 to 2.5 oz

Add and mix until smooth.

Fig Paste

| Twice-Ground Figs | 5 lb, 6 oz |

Granulated Sugar	2.25 lb
Invert Syrup	2.25 lb
Corn Syrup	12 oz

Mix all to a smooth paste.

Fig Cookies

Shortening	3.25 lb
Ground Figs	2.25 lb
Brown Sugar	10 lb, 6 oz
Salt	3 oz
Flavor	To suit

Mix thoroughly.

| Eggs | 2.25 lb |

Add while creaming.

| Liquid Fresh or Reconstituted Nonfat Dry Milk Solids | 3 lb |

Add, just stirring in.

| Bleached Cake or Pastry Flour | 10 lb, 8 oz |

Add and mix only until smooth.

Roll out, cut as desired, place on greased and dusted pans, and bake at 375 to 390°F.

Filled Fig Cookies

Sugar	3 lb
Shortening	4.5 lb
Pastry Flour	9 lb
Baking Powder	1 oz
Salt	1 oz
Eggs	1 pt
Butter Flavor	To suit

Cream sugar and shortening together. Add eggs gradually. Sift salt, baking powder, and flour and add at low speed. Pin out dough in strips 4 inches wide and ¼-inch deep, to the width of the baking sheet. Fill the baking sheet with these strips. Crimp the sides. Cover the center liberally with fig jam. Cut strips of dough 4 inches long and ¼-inch wide, criss-cross these over the jam and press into the edges of the dough. Bake at 400°F. When baked, cut into strips about 1½ inches wide.

Walnut-Fudge Bars

Granulated Sugar	2 lb, 8 oz
"Primex"	1 lb
"Cinacoa"	8 oz
Glucose	12 oz
Salt	1 oz
Pastry Flour	1 lb, 8 oz
Whole Eggs	12 oz
Walnut Flavor	To suit
Water	4 oz
Black Walnuts	1 lb

Scale all ingredients into a mixing bowl. Mix at moderate speed to a smooth dough.

Scale approximately 9 lb into a greased and papered bun pan. Bake light at 375 to 400°F. Cool and cut into squares or bars.

Fruit-Nut Bars

Cake Flour	3.75 lb
Sugar	2.5 lb
Shortening	1.25 lb
Molasses	12 oz
Cake Crumbs	1 lb, 8 oz
Ammonium Bicarbonate	0.2 oz
Soda	1 oz
Salt	0.6 oz
Eggs	4 lb
Milk	12 oz
Raisins	1 lb

Nuts	12 oz
Cinnamon	0.5 oz

Cream the sugar, shortening, molasses, salt, eggs, and cinnamon. Sift the cake crumbs, flour, and soda together, and add to preceding. Dissolve the ammonium bicarbonate in the milk, add to preceding mixture and mix until uniform. Finally add the raisins and nuts.

Deposit the strips on sheet pans, flatten, wash with an egg wash or milk and egg, and bake. Cut into bars when cool.

To make the bars, roll the dough into strips of sufficient length to cover a sheet pan. Flatten the strips and, to obtain a glossy surface, apply an egg wash before baking. After baking, allow the cake to cool sufficiently so that the bars can be cut either directly across the pans or at an angle.

Nut Wafers

a	Granulated Sugar	1.5 lb
	"Primex"	1.5 lb
	Pastry Flour	1.5 lb
	Salt	0.5 oz
	Ground Walnuts	
	or Pecans	2.0 lb

	"Cinacoa"	0.5 oz
b	Egg Whites	1 lb
	Granulated Sugar	0.5 oz
	Flavor	To suit

Cream a together to a smooth paste. Beat b to a stiff meringue and fold into this paste. Run out with a bag and tube on greased and dusted pans, allowing 4 to 6 oz for a dozen cookies. Bake light at 400°F.

Butterscotch Squares

"Cinacoa"	4 oz
Granulated Sugar	2.5 lb
"Primex"	1 lb
Salt	0.5 oz
Chopped Pecans	5 oz
Baking Powder	0.25 oz
Pastry Flour	1.5 lb
Whole Eggs	1 lb
Water	4 oz
Flavor	To taste

Scale all ingredients into a mixing bowl and mix at medium speed to a smooth dough. Spread into a well-greased paper-lined bun pan and bake at 375°F. When cool, cut into squares.

Honey Jumbles
(For Wire-Cut Machine)

Shortening	10 lb
Sugar	10 lb
"Proflo"	5 lb
Salt	1 lb
Whole Eggs	8 lb
Honey	60 lb
Invert Sugar	30 lb
Soda	5 oz
Ammonia	2 lb
Water	52 lb
Flour	196 lb
Flavor	As desired

Cream the shortening, sugar, "Proflo," flavor, and salt. While creaming, gradually add the eggs and mix until thoroughly incorporated. Then add the honey and invert sugar and mix until smooth. Dissolve the soda and ammonia in the water. Add this solution, along with the flavor, to the creamed mix. Then mix again until smooth.

30% Protein Cookie
(For Teething)

"Promine"-D	150 lb
Wheat Flour	100
"Soyabits" #20 Cooked and Toasted	80
Arrowroot Starch	25
Sugar, Baker's Fine	160
Shortening	100
"Promolip" *	10
Salt	2
Ammonium Bicarbonate	5
Vanilla Sugar	5
Water	120

* A protein-lecithin combination.

Cream sugar and shortening together, add Soyabits #20, "Promolip," "Promine"-D, Arrowroot starch, salt, ammonium bicarbonate and vanilla sugar and mix. Add wheat flour and water and mix well. Bake at 375°F for 20 minutes.

Protein Snack

Cracker Meal	125 lb
Bread Flour	85
Sunflower Meal	85
"Promine"-D	15
Gum Arabic	3
Onion Salt	2
Celery Salt	1
Water—approximately	80

Mix all dry ingredients and add sufficient water to make a slurry that will just flow.

Extrude in ribbons into deep fat at 375°F and fry until light brown. Drain dry. Salt as desired (antioxidant salt recommended).

Protein Crackers

"Promine"-D	40 g
Wheat Flour (General Purpose)	360 g
Salt	20 g
Ammonium Bicarbonate	15 g
Hydrogenated Vegetable Shortening	100 g
Water	240 g

Mix dry ingredients, add water, and knead. Roll out on floured board to about 1/16-inch thick. Puncture with fork and cut into squares. Bake at 550°F for about 4 minutes.

If puffy crackers are desired, do not puncture dough.

Cottonseed Flour for Dusting

Graham Crackers: A mixture of 50% cottonseed flour and 50% white wheat flour— well blended—used in dusting at the brake roll eliminates any streaking and gives an exterior finish comparable to the interior of the cracker.

Acid or Acid-Type Cracker (Butter, Sprayed, Etc.): The use of the same blend of dusting flour as used for grahams on the bottom of the form at the last brake, where this type of product is baked on wire pans with or without skeleton, irons, pans, or bands, will give a finish to the bottom of the cracker comparable to that on the top.

To Retard Mold in Baked Goods

Dissolve 1 lb sorbic acid in 20 lb warm (not hot) propylene glycol. Add 2 lb of this solution to every 100 lb of water used in bakery pie filling, cream filling, marshmallow, etc. If the bakery product (marshmallow syrup, pie filling, etc.) will be boiled, the sorbic acid crystals may be added directly to the product to be boiled and stored. Add 1.5 oz sorbic acid crystals to each 100 lb of water used in the filling or marshmallow syrup before cooking. The sorbic acid will dissolve during the cooking or boiling.

Sometimes coconut on the outside of marshmallow becomes moldy, especially if the packages are exposed to sunlight. To concentrate the sorbic acid on the coconut used for topping, 100 lb of coconut may be mixed (preferably in a candy tumbler) with 1 lb of powdered sorbic acid. The total amount of sorbic acid in the entire finished item should not be over 1 part in 1000 (0.1%). For instance, if topping coconut constitutes 10% of the finished item, 1 lb sorbic acid to 100 lb of coconut gives exactly 1 part in 1000. Less than this may be sufficient to give the needed protection.

White Icing Base

Shortening Mix

Dry Mix:

Icing Sugar	44 lb
Dextrose	10 lb
Salt	8 oz
"Starbake-100,"	
Powdered	5 lb, 8 oz
Cellulose Gum,	
CMC-7HSXP	1 lb
Sodium	
Propionate	1 oz

Blend Into:

Creamed, high-ratio	
shortening	15 lb

To Prepare Syrup:

Gelatin, Type	
A, 200 bloom	1 lb
Warm Water	13 lb
Dissolve and add:	
Corn Syrup	10 lb
Total Short-	
ening-	
syrup	100 lb 1 oz

Heat to 150°F. Add 2/3 of syrup to shortening mix, blend in well. Then mix in balance of syrup. Flavor to taste with vanilla.

Buttercream Icing

White Icing Base	6 lb
Water	16 lb

Heat to boiling and while hot, slowly blend into:

Icing Sugar,*	
sifted	100 lb

Flavor with:

Salt	1 oz
Vanilla	to taste
Color	desired

---

* Dextrose may be used for part of sugar.

Blend in:

High-Ratio
shortening   16 lb

and *whip* to desired consistency.

Flat Icing (Boiled)

White Icing Base	6-7 lb
Granulated Sugar	20 lb
Water	20 lb

Heat to boiling and while hot slowly blend into:

Icing Sugar, sifted   100 lb

Apply to goods at 120-130°F.

Instant Fluffy Frosting
Mix for Retail
Distribution

Formula No. 1
(One Package)

Dried Egg Whites	6.60 g
Salt	0.33
Cream of Tartar	1.10
Calcium Tartrate	0.33
Corn Syrup Solids	15.00

Cellulose Gum,

CMC-7HSXP	0.53
"Starvis"	5.00
Powdered Sugar	195.00
Powdered Vanilla, Imitation	0.80
	224.69
Water	½ cup

Procedure Using Electric Mixer (If a high-speed commercial mixer is used, reduce the mixing time.)

Pour one-half cup cold water into small mixing bowl and add contents of bag. Mix at low speed until powder is thoroughly wetted out. Beat at highest speed until stiff peaks form when beater is raised (4-6 minutes). Guide mix to center of bowl occasionally. Rotate bowl during mixing. Fold in color and additional flavor, if desired. This mix has a mild vanilla flavor.

No. 2
(Two Package)

Package A:

Dried Egg Whites	6.60 g
Salt	0.33
Cream of Tartar	0.70
Calcium Tartrate	0.17

Corn Syrup
Solids   10.00
Sugar, Superfine
Granulated   40.00

Package B:

"Starvis"   5.00
Powdered Sugar   159.30
Powdered Va-
nilla, Imitation   0.80
Cellulose Gum,
CMC-7HSXP   0.59
                     223.49
Water   ½ cup

Procedure Using Electric Mixer (If a high-speed commercial mixer is used, reduce mixing time.)

A. Pour ½ cup cold water into small mixing bowl. Add contents of Bag A. Mix at low speed until powder is thoroughly wetted out. Beat at highest speed until straight, stiff peaks form when beater is raised (2-3 minutes). Guide mix to center of bowl occasionally. Rotate bowl.

B. Turn mixer to lowest speed and slowly pour in contents of Bag B in a steady stream. Scrape bowl with rubber scraper. Then, beat at highest speed until stiff peaks form (2-4 minutes). Rotate bowl during mixing. Fold in color and additional flavor, if desired. This mix has a mild vanilla flavor.

Chocolate Icing

a Granulated Sugar   5 lb
  Glucose   8 oz
  "Sweetex"   1.5 lb
  Salt   2.5 oz
  "Cinacoa"   1.5 lb
  Water   2 lb
b 6X Sugar   15 lb
  "Sweetex"   3 ½ lb
  Evaporated Milk   2 ¼ oz
  Flavor   To suit

Boil *a* to 238°F and add immediately to creamed mixture of *b*.

Cream light. After adding the syrup, continue mixing until the icing has reached the proper consistency for use (100 to 110°F). If separation occurs, rewarm, add milk and cream again. Use warm.

Hard-Drying Royal Icing

Meringue Powder   4 oz
4X or 6X Sugar   1 lb

| Cold Water | 1 pt |
| 4X or 6X Sugar | 4 lb |

First mix the meringue powder and the 1 lb of sugar together in a container, add the cold water, and thoroughly dissolve in a machine kettle. Add the 4 lb of sugar and mix. Add just enough liquid blue to make a brilliant white icing. Place the kettle in the machine, attach the spatula, and beat at low or second speed. When ready, keep covered with a damp cloth. Keep a surplus of dissolved meringue powder on hand in the refrigerator for use as required. No acids are needed with the meringue powder.

———

Meringue

"Promine"-R	15.00
Sodium	
Bicarbonate	1.25
Trisodium Phosphate·12H$_2$O	1.25
Boiling Water	200.00
Citric Acid	0.63
Premium	
"Albusoy"	1.85
CMC-7HSP	2.50
Corn Syrup Solids	75.00
Sucrose	225.00

Dry-mix the "Promine"-R, bicarbonate, and phosphate. Sift the remaining dry ingredients together. Pour the boiling water into a heated mixing bowl; with constant stirring add the protein mixture. Whip at high speed until stiff. In a steady slow stream add remaining ingredients. Whip for 2 to 3 minutes more. With pastry tube squeeze onto cookie tray which has been powdered with starch. Dry in oven at 170°F for 3 or more hours until completely dry. Serve with all-vegetable whipped toppings, soy chocolate pudding, or chocolate frozen dessert. Extra powdered sugar may be added as desired to alter properties of the meringue.

———

Meringue Stabilizer

Formula No. 1
(Hot Process)

Cellulose Gum,	
CMC-7HSP	15
Agar Agar	7-15
"Starbake"-100,	
Powdered	15
Dextrose	55-63

Procedure

1. Egg Whites   1 qt
   Whip to soft peak.
2. Sugar   2.5 lb
   Water   1.5 qt
   Stabilizer   2-2.5 oz

Heat to boiling. Slowly stream into egg whites. Whip to desired consistency at medium speed.

No. 2
(Cold Process)

Cellulose Gum,
   CMC-7HSXP   13
Tapioca Flour   18
Dextrose   69

Procedure

1. Egg whites   1 qt
   Sugar   2 lb
   Whip medium stiff.
2. Stabilizer   4 oz
   Sugar   1 lb
   Tap Water   1 qt

Mix. Break up lumps. Add to egg whites and sugar while whipping at medium speed. Whip to desired consistency.

Marshmallow Cream or Topping

Formula No. 1
(Vertical Beater)

Water   3 lb
Egg Albumin   1 lb

Soak the albumin in the water, stirring occasionally until thoroughly dissolved; then place in a beater.

Salt   2 oz
Powdered
   Tartaric Acid   0.25 oz
"Sweetose"   5 lb

Add these to the first mix and whip until light and stiff.

"Sweetose"   45 lb

Meanwhile heat this to 230°F and add gradually while beating at medium speed.

Vanilla   To suit

Continue beating until the desired consistency is obtained. Just before finishing add the vanilla and mix in well. Pack into containers at once while warm.

Use a high-grade egg albumin, either crystalline or powdered, and dissolve thoroughly before using. Keep the machine and ingredients free from fats at all times.

Add hot "Sweetose" immediately after removing from the heat to the beaten meringue in a steady stream. Adjust the flow of hot syrup so that the temperature of the mix will be about 165°F after all the syrup has been incorporated. An 80-qt bowl is required for the given batch.

No. 2
(Vertical Beater)

Water	3 lb
Egg Albumin	1 lb

Soak the albumin in the water, stirring occasionally until thoroughly dissolved; then place in a beater.

Salt	2 oz
Powdered	
Tartaric Acid	0.25 oz
"Sweetose"	5 lb

Add these to the first mix and whip until light and stiff.

Water	8 lb
Confectioners'	
Starch	1 lb
"Sweetose"	10 lb

Suspend the starch in the water, add the "Sweetose," and bring to a boil while stirring; continue to cook until clear.

"Sweetose"	35 lb

Add this and cook to 230°F.

Vanilla	To suit

Remove from the heat and add the hot syrup gradually to the mass while beating at medium speed. Continue to beat until the desired consistency is obtained. Just before finishing, add the vanilla and mix in well. Pack into containers at once while still warm.

Use a high-grade egg albumin, either crystalline or powdered, and dissolve thoroughly before using. Keep the machine and ingredients free from fats at all times. Add the hot "Sweetose," immediately after removing from the heat, to the beaten meringue in a steady stream. Adjust the flow of hot syrup so that the temperature of the mix will be about 165°F after all of the syrup has been incorporated.

An 80-qt bowl is required for the given batch.

No. 3
(Horizontal Beater)

"Sweetose"   200 lb

Heat at 170 to 180°F; then place in a beater.

Water	12 lb
Egg Albumin	4 lb
Salt	8 oz
Powdered Tartaric Acid	1 oz

Soak the albumin in the water, stirring occasionally until thoroughly dissolved. Then add the salt and acid and add this mixture gradually to the first mix while beating.

Vanilla	To suit

Continue to beat until the desired texture is obtained. Just before finishing, add and stir in well. Pack into containers at once while warm. Approximate yield is 216 lb.

Use a high-grade egg albumin, either crystalline or powdered, and dissolve thoroughly before using. Keep the ingredients and machine free from fats at all times. From one eighth to one quarter of the egg albumin may be substituted by gelatin soaked in cold water (part of the 12 lb listed in the formula) and added with the albumin solution while beating.

Adjust the temperature to which the "Sweetose" is heated, so that after the albumin solution is incorporated the temperature of the mix will be about 165°F.

No. 4
(Semiliquid Marshmallow)

To prepare a batch of approximately 54 lb, soak 1 lb powdered egg albumin in 3 lb of water. Occasionally stir until the albumin is completely dissolved and then introduce the solution into the beater with an 80-qt bowl.

Now add 2 oz salt, ¼ oz powdered tartaric acid, and 5 lb corn syrup. Whip the mixture until light and stiff.

Meanwhile, heat 45 lb corn syrup to 230°F. Then gradually add the hot syrup to the mixer, while beating at medium speed. The flow of hot syrup should be adjusted so that the temperature of mix will be about 165°F after all of the syrup has been introduced. Continue beating until the desired consistency is obtained. Add vanilla just before finishing.

Marshmallow Icing

"Promine"-R	6.00
Sodium Bicarbonate	0.50
Trisodium Phosphate·12H$_2$O	0.50
Boiling Water	105.00
Sucrose	150.00
Premium "Albusoy"	0.75
Citric Acid	0.25
"Seakem" 402	0.30
CMC-7HSP (Cellulose gum)	0.50
Vanilla	

Thoroughly mix together the "Promine," bicarbonate, and phosphate. Sift the remaining ingredients together. Pour the boiling water into a heated mixing bowl; with constant stirring add the protein mixture. Whip at high speed until stiff. Add remaining ingredients in a slow steady stream. Add vanilla or other flavor and whip for 2 more minutes. Use immediately on cake (makes about 1 quart of icing).

This can be prepared as a two-package dry mix, with the protein mixture in one package and the other ingredients in the second one.

Whip Topping

Hydrogenated Cottonseed-Oil Shortening	18.0
"Aldo" 40	1.0
Butter	2.0
Coconut Slab Oil (M.P. 118°F)	2.0
Nonfat Milk Solids	9.0
Sugar	12.0
Lecithin	0.5
Water	55.5

The sugar and milk solids are added to the water and heated, with agitation, until they are dissolved. The other ingredients are then added and the mix is heated, with constant agitation, at 150°F for pasteurization. Vanillin extract is added as the mix cools. It is then homogenized and cooled quickly. Finally the batch is refrigerated and aged for at least 48 hours before whipping.

This formula produces a good whip topping with a specific gravity of about 0.5 after whipping.

All-Vegetable
Whipped Topping

Formula No. 1
(White)

Hydrogenated Vegetable Oil*	61.00
Mono- and Diglyceride Emulsifier	1.25
"Promine"-D	9.00
Sucrose	25.00
Hot Water	153.00
Vanilla	

* Color	2R; 20Y max
Free fatty acids	0.5 max
Wiley melting point	86–90°F
A.O.M	150 hr plus

Mix together hydrogenated vegetable oil, emulsifier, "Promine," vanilla, and sucrose to form a dry mix, which will be plastic. When ready to use, place entire mixture into mixer for malted milk or Waring Blender, add the hot water, and mix for 3 to 5 minutes at high speed. Fill into dispenser, add more flavoring, if desired, and cool thoroughly. Charge dispenser and keep under refrigeration. Serve. (Makes about one-half pint of liquid)

For mechanical whipping, fat should be increased to contain between 28 and 40%, based on final weight, including water. The protein content might have to be reduced to compensate for high viscosity.

No. 2
(Chocolate)

Hydrogenated Vegetable Oil*	50.0
Mono-and Diglyceride Emulsifier	0.6
"Centrophil" SM	0.6
"Promine"-D	8.0
Sucrose	37.5
Cocoa	6.5
Sodium Bicarbonate	0.1
Boiling Water	147.0
Vanilla	

Mix together all ingredients except water. When ready to use, put mixture into blender or malted milk mixer, add boiling water and mix at high speed up to 5 minutes. Fill into dispenser, cool and charge. Store in refrigerator until ready to use. (Makes ½ pint of liquid)

Fudge Topping

Formula No. 1
(Cold)

"Sweetose"	40 lb

Sweetened Condensed Whole
Milk 10 lb
Cocoa Powder 4 lb
Salt 4 oz
Bicarbonate of Soda 1 oz
Water 8 lb
Vanilla To suit

Combine the ingredients in the order listed, mixing together thoroughly to insure a smooth blend. Stir while heating to 200°F and hold at this temperature for 15 minutes. Then turn off the heat, add the flavor, and mix well. Homogenize at once to insure a smooth product; then fill and seal in containers at 180°F. The approximate yield will be 60 lb.

No. 2
(Hot)

"Sweetose" 40 lb
Sweetened Condensed Whole
Milk 10 lb
Cocoa Powder 4 lb
Salt 4 oz
Bicarbonate of Soda 1 oz
Water 4 lb
Vanilla To suit

Proceed as indicated under Formula No. 1.

Select the cocoa for flavor and color. A blend of natural (or breakfast) and dutched cocoa is recommended. Be sure to stir the mixture while heating to 200°F, and then stir occasionally while holding at that temperature. The approximate yield will be 56 lb.

Cake-Pan Grease
(For Applying with Brush Machine or by Hand)

Formula No. 1

Shortening 75
Wheat Flour 75
"Proflo" 10

No. 2
(For Standard Spraying Machine)

Shortening 75
Wheat Flour 14
"Proflo" 6

No. 3
(For Y.M. DeVilbiss Machine for Sponge Cake)

Oil 33
Shortening 30
Flour 34
"Proflo" 6-10

No. 4
(For Y.M. DeVilbiss
Machine for Other Cakes)

Oil	45
Shortening	28
"Proflo"	10
White Flour	15

No. 5
(For Dukay Pan
Grease Machine)

Shortening	30
Soft Flour	15
"Proflo"	15
Salad Oil	30

The air pressure on the compressor tank should be 140 lb.

All these formulas give best results when creamed properly at 80 to 82°F and then applied in a thin film to clean pans.

Bakers' Pan Grease

Bread Flour	10
"Proflo"	10
Vegetable Shortening	30

Cream this very light and, while mixing at low speed, gradually add:

Vegetable Oil	50

Fruit Whip
Dessert Powder

U.S. Patent 2,588,307

Skim-Milk Powder	10.00
Gelatin Powder	9.50
Sugar Powder	50.00
Dehydrated-Fruit Powder	30.00
Irish-Moss-Extract Powder	0.51

Alginate Dessert Powder
(Jelly)

U.S. Patent 2,918,375

Formula No. 1

Sugar	100
Sodium Alginate	4
Sodium Hexameta-phosphate	3
Tricalcium Phosphate	1
Fumaric Acid	2

Color and Flavor to suit.

No. 2

British Patent 828,350

Sodium Alginate, Dried	46.0
Sodium Carbonate, Dried	20.0
Calcium Phosphate	6.0
Sugar Powdered	800.0

Citric Acid,
Anhydrous 50.0
Strawberry Flavor,
Powd. 1.6
Red Food Color,
Powd. 0.4

93 g of above is dissolved in 50 cc of water.

All-Vegetable Chocolate Frozen Dessert

Cocoa Syrup 20.0
Dextrose 40.0
Hydrogenated
Vegetable Oil* 50.0
"Promine"-D 22.5
Gelatin 2.0
Sucrose 50.0
Water 320.0
Vanilla Extract 5.0

---

* Color	2R; 20Y max
Free fatty acids	0.5 max
Wiley melting point	86–90°F
A.O.M.	150 hrs plus

Combine all ingredients, except the vanilla extract. Pasteurize at 160°F for 30 minutes. Homogenize and cool. After the mixture has been thoroughly chilled, add vanilla and freeze.

All-Vegetable Non-Starch Chocolate Pudding

Chocolate Liquor 14.0
Sucrose 72.0
Salt 3.2
"Promine"-D 60.0
Lactose 15.0
Hot Water 355.0
Vanilla Extract 5.0

Melt chocolate liquor. Sift dry ingredients together. Now combine all ingredients in blender or by mechanical stirring. Fill into pudding dishes and cool. Serve with all-vegetable whipped topping. (Makes about 4 portions)

Dry Pudding Mix to Be Cooked

6X Sugar 44.75
"Cerelose" 34.00
"Starbake"-100,
Powd. 19.00
Salt 1.00
Vanilla Flavor 1.00
Cellulose Gum,
CMC-7HSXP 0.25

Add 4 oz mix to 2 cups milk and heat to boiling. Pour into molds and cool.

Instant Pudding Mix

U.S. Patent 2,801,924

Cane Sugar	80.0 g
Flavored Corn Sugar	2.0 g
Colored Corn Sugar	2.6 g
Table Salt	1.2 g
Pre-cooked Starch	16.8 g
Tetrasodium, Pyrophosphate, (Anhydr)	1.8 g
Disodium Ortho-phosphate (Anhydr)	2.4 g
Calcium Acetate	0.6 g

One pint of milk is added to this mixture.

Instant Pudding Mix

Formula No. 1

6X Sugar	79.7
Pregelatinized Potato Starch	12.0
Vanilla Flavor	1.8
Salt	2.2
Tetrasodium Pyrophosphate	2.5
Calcium Acetate	1.5
Cellulose Gum, CMC-7HSXP	0.3

Add 4 oz mix to 2 cups milk and beat slowly with an egg beater for about 1 minute. Pour into molds and allow it to set.

No. 2

Cane Sugar (Finely Ground)	65.85
Pregelatinized Potato Starch	26.60
Vanilla Flavor	1.80
Salt	2.00
Sodium Acid Pyrophosphate	1.00
Dibasic Sodium Phosphate	2.00
Cellulose Gum, CMC-7HSXP	0.75

Add 4 oz mix to 2 cups milk and beat slowly with an egg beater for about 1 minute. Pour into molds and allow it to set.

Imitation Fruit Flavor Oils

No. 1
Imitation Raspberry

beta Ionone	10.00
Isobutyl Acetate	23.49
Anisic Aldehyde	1.00
Phenyl Ethyl Alcohol	3.00
Phenyl Ethyl Iso-Butyrate	4.00

Ethylmethyl *p*-		Ethyl Caprate	4.0	
Tolyl Glycidate	35.00	Terpenyl Butyrate	0.5	
Phenylethyl		Ethyl Caproate	2.0	
Anthranilate	7.00	Benzaldehyde	8.0	
Vanillin	3.00	*p*-Tolyl Aldehyde	3.0	
Hexyl Butyrate	0.50	*p*-Methylbenzyl		
Isoamyl Acetate	4.00	Acetate	11.0	
Ethyl Butyrate	5.00	Ethylmethyl *p*-Tolyl		
Rose Oil	4.00	Glycidate	16.0	
Diallyl Sulfide	0.01	Vanillin	7.0	
		Heliotropin	5.0	

No. 2
Imitation Strawberry

Ethylmethylphenyl	
Glycidate	35.0
Amyl Aldehyde	0.5
Bornyl Acetate	0.5
Ethyl Caproate	2.0
Vanillin	3.0
Beta-Ionone	9.0
Ethylmethyl *p*-Tolyl	
Glycidate	12.0
Isobutyl Acetate	30.0
Ethyl Butyrate	8.0

No. 3
Imitation Cherry

Cinnamyl	
Anthranilate	3.0
Isoamyl Acetate	12.0
Isobutyl Acetate	12.0
Cinnamic Aldehyde	
DimethylAcetal	3.0
Benzyl Alcohol	5.5
Geranyl Butyrate	2.0

No. 4
Imitation Peach

Linalyl Formate	10.0
gamma-Undeca-	
lactone	13.0
Linalyl Butyrate	4.0
Heliotropin	14.5
Geranyl Valerate	15.0
Furfural	1.5
Alpha-Methylfuryl	
Acrolein	6.0
Methylcyclopen-	
tenolone Valerate	10.0
Benzaldehyde	5.0
Isoamyl Formate	15.0
Isobutyl Butyrate	6.0

No. 5
Imitation Apple

Geranyl Valerate	10.0
Geranyl Butyrate	8.0
Geranyl Propionate	8.0
Linalyl Formate	10.0

Isoamyl Valerate	15.0
Vanillin	8.0
Allyl Caprylate	6.0
Geranyl Aldehyde (Citral)	5.0
Acetaldehyde	6.5
Methylcyclopen-tenolone Valerate	8.0
Alpha-Methylfuryl Acrolein	2.0
Isoamyl Butyrate	13.5

No. 6
Imitation Coffee

alpha-Furfuryl Mercaptan	10
Ethyl Vanillin	3
Solvent	87

Non-evaporating Tobacco Flavor

U.S. Patent 2,969,795

Silicone Fluid (20 centipoise)	10
Glycerol	10
Apple Syrup	10
Sorbitol	5
Molasses	5
Maple Syrup	5
Rum	5
Brandy	2
Essential Oils	1
Vanilla Extract	1
Licorice	1

Deer's Tongue	1
Cinnamon	1
Tonka Beans	1
Ginger	0.5
Glucose, Sucrose, Ethanol, or Water	To suit

Beverage Powders

The following formulas represent typical formulations for beverages in which the sugar is added by the consumer. The package would contain 1 ounce of powder and make 2 quarts of beverage.

Raspberry

Imitation Raspberry (Dry Flavor) ("Flav-O-Lok" 3X)	1.4
Citric Acid, Anhydrous	25.0
Dextrose	171.6
Certified Red Ponceaux 3R	2.0

The citric acid content of the above formula applies generally to red fruit flavors.

Lemon-Lime

Lemon-Lime (Dry Flavor) ("Flav--O-Lok" 3X)	1.40

Citric Acid,
Anhydrous 77.70
Powdered Cloudifier 2.75
Dextrose 116.15
Certified Green
Color 2.00

The higher citric acid content in this product is usual with the citrus formulations.

Emulsion
(Citrus Beverages)

Orange Oil or
Orange Oil
blended with
Lemon Oil and
Tangerine Oil 8 fl oz
Brominated Vegetable Oil 6 fl oz
32° Bé Sugar
Syrup 8 fl oz
Gum Arabic 6 oz
Sunset Yellow
Color 3 oz
Water 104 fl oz

These ingredients will make one gallon of concentrate; use two ounces to a Gallon of Syrup.

Place in a graduate of suitable size 8 ounces of orange oil or a blend of orange and other oils with 6½ ounces of brominated oil. Stir this mixture until it is well blended.

The object of using brominated oil of about 1.250 specific gravity, is to increase the specific gravity of the citrus-oil mixture, which is about 0.880.

The specific gravity of the bottled beverage is approximately 1.028 and therefore the specific gravity of the brominated citrus oil mixture should be adjusted to this point.

The brominated oil should be added to the citrus oil in small portions and thoroughly mixed and checked with the hydrometer after each addition until the desired specific gravity has been obtained.

The oils must be at room temperature or 70°F when taking the specific gravity. The oils may be brought down to 70°F by placing the container in a water bath or refrigerator. To raise the temperature, the oil containers should be immersed in warm water.

The specific gravity of the oil mixture can be quickly determined by a special hydrometer with a single reading.

Place the adjusted oils in a can or some other suitable vessel of about 1-gallon capacity. Sprinkle the gum arabic on top of the oil and stir with a stick until all the gum has been mixed with the oil and there are no lumps. This is very important, as otherwise you will run into trouble later on. Now, while stirring, mix the syrup with the oil and gum. The gum will congeal at this point. Keep stirring for a few minutes; then add the water while stirring. This will make a total of 1 quart. Then add, while stirring, 1 qt of water. By this time, the oil, gum, and water should form a smooth emulsion. To the remaining water (2 qt), in a separate vessel, add 3 oz of Sunset Yellow color and when it is all dissolved, add this solution to the oil-gum mixture. Stir for about 10 minutes and then pass through the homogenizer twice.

Use a high-speed mixer to blend this premix properly. Mixing by hand will not keep the oils together. If no high-speed mixer is available, the emulsion should be stirred while in the homogenizer bowl.

Citrus flavors should be made fresh. Do not make more than two to three weeks' supply at any time. (All other flavors may be made as much as five or six months in advance.) Be sure to add 5 gr benzoate of soda to each gallon if the flavor is not to be used within 3 weeks, and store in a cool place. Freezing will spoil the emulsion.

NOTES: It was found that a blending of 6 oz orange oil, 0.5 oz lemon oil, and 1.5 oz tangerine oil gives a very fine fruity flavor.

When purchasing brominated oil, find out whether it is heavy or light. Also ask the supplier for instructions on how to use their product.

Basic Flavor Compound

Basic Flavor Compounds of this type may be used in many ways.

1) They may be added to true fruit flavors in small ratio in order to fortify them. In such cases the finished flavor must be designated as an Imitation Flavor.

2) They may be used to make Imitation Fruit Syrups by adding 2 oz of the Basic Compound to three qt propylene glycol, one qt water and suitable amount of Certi-

fied Food Color. Two ounces of this finished flavor is then added to one gallon of simple sugar syrup.

3) They may be used to make finished flavors for hard candy by adding from 12 to 16 oz of the Basic Compound to propylene glycol to make one gallon of finished flavor, and then using the finished flavor at the rate of 1.5 to 2 oz for 100 lb of hard candy.

No. 1

Imitation Raspberry

Isobutyl Acetate	18.750
Amyl Acetate	9.375
Ethyl Acetate	18.750
Vanillin	7.030
Ionone Alpha	3.515
Ionone Beta	4.687
Clove Oil	1.170
Cinnamon Oil	0.586
Citral	0.586
Benzyl Acetate	2.344
Anise Oil	0.586
Gamma Nonyl	
Lactone	0.586
Palatone	0.390
Isoamyl Formate	4.687
Benzyl Benzoate	26.958

No. 2

Imitation Strawberry

Amyl Butyrate	17.750
Ethyl Benzoate	3.125
Ethyl Butyrate	24.000
Benzyl Acetate	3.125
Ethyl Methyl	
Phenyl Glycidate	9.375
Vanillin	3.125
Diacetyl	0.780
Ethyl Anthranilate	0.780
Methyl Heptine	
Carbonate	0.390
Palatone	0.780
Ethyl Acetate	22.708
Isoamyl Acetate	6.250
Ethyl Oenanthate	6.250
Ionone Alpha	1.562

No. 3

Imitation Wild Strawberry

Ethyl Valerate	0.750
Ethyl Butyrate	3.750
Ethyl Oenanthate	0.375
Benzyl Acetate	1.125
Oil Wintergreen	0.375
Aldehyde C-16	
(so-called)	0.750
Ethyl Benzoate	0.375
Methyl Heptine	
Carbonate	0.093

Benzo Dihydro		Allyl Caproate	2.000
Pyrone	0.093	Isoamyl Valerate	0.625
Isoamyl Acetate	1.500	Isoamyl Acetate	1.000
Ethyl Acetate	6.816	Ethyl Pelargonate	0.625
		Terpeneless Orange	
No. 4		Oil	0.093
		Terpeneless Lemon	
Imitation Pineapple		Oil	0.093
Ethyl Butyrate	4.750	Butyl Butyrate	0.187
Isoamyl Butyrate	2.000	Ethyl Valerate	1.000
Ethyl Acetate	2.750	Ethyl Propionate	0.087

Orange Syrup

Formula No. 1

	25% Sweetener Solids "Sweetose"	50% Sweetener Solids "Sweetose"
"Sweetose"	30 lb	60 lb
Granulated Sugar	75 lb	50 lb
Water	51 lb	46 lb
Dry Citric Acid	15 oz	15 oz
Dry Orange I Color	0.75 oz	0.75 oz
Terpeneless Orange Flavor	9 oz	9 oz

No. 2
(Concentrated Syrup; Ratio 1:3)

"Sweetose"	30 lb	60 lb
Granulated Sugar	75 lb	50 lb
Water	51 lb	46 lb
Dry Citric Acid	3 lb, 12 oz	3 lb, 12 oz
Dry Orange I Color	3 oz	3 oz
Terpeneless Orange Flavor	36 oz	36 oz

Bring the "Sweetose," sugar, and water to a boil and add the color and flavor at once. Seal in containers while hot. The flavor and color may be varied to suit. The approximate yield of each batch will be 150 lb or 15 gal.

Cherry-Flavored Syrup

Formula No. 1

	25% Sweetener Solids "Sweetose"	50% Sweetener Solids "Sweetose"
"Sweetose"	30 lb	60 lb
Granulated Sugar	75 lb	50 lb
Water	51 lb	46 lb
Dry Citric Acid	15 oz	15 oz
Dry Amaranth Red Color	1.625 oz	1.625 oz
Cherry Concentrate	0.75 oz	0.75 oz

No. 2
(Using Unsweetened Cherry Juice)

	25% Sweetener Solids "Sweetose"	50% Sweetener Solids "Sweetose"
"Sweetose"	30 lb	60 lb
Granulated Sugar	70 lb	45 lb
Cherry Juice (19% Solids)	30 lb	30 lb
Water	27 lb	22 lb
Dry Citric Acid	15 oz	15 oz
Dry Amaranth Red Color	1.085 oz	1.085 oz
Cherry Concentrate	0.75 oz	0.75 oz

No. 3
(Concentrated Syrup, Ratio 1:3)

"Sweetose"	30 lb	60 lb
Granulated Sugar	75 lb	50 lb
Water	51 lb	46 lb
Dry Citric Acid	3 lb, 12 oz	3 lb, 12 oz
Dry Amaranth Red Color	6 oz	6 oz
Cherry Concentrate	3 oz	3 oz

Instructions as for Orange Syrup.

Lemon Syrup

	25% Sweetener Solids "Sweetose"	50% Sweetener Solids "Sweetose"
"Sweetose"	30 lb	60 lb
Granulated Sugar	75 lb	50 lb
Water	51 lb	46 lb
Dry Citric Acid	15 oz	15 oz
Terpeneless Lemon Flavor	11 oz	11 oz
Dry Tartrazine Color	0.1 oz	0.1 oz

Instructions as for Orange Syrup.

Pineapple-Flavored Syrup

Formula No. 1

	25% Sweetener Solids "Sweetose"	50% Sweetener Solids "Sweetose"
"Sweetose"	30 lb	60 lb
Granulated Sugar	75 lb	50 lb
Water	51 lb	46 lb
Dry Citric Acid	15 oz	15 oz
Pineapple Flavor	2 oz	2 oz
Tartrazine Solution	7 cc	7 cc

No. 2
(Using Unsweetened Pineapple Juice)

"Sweetose"	30 lb	60 lb
Granulated Sugar	70 lb	45 lb
Clarified Pineapple Juice	30 lb	30 lb
Water	26 lb	21 lb
Dry Citric Acid	15 oz	15 oz
Pineapple Flavor	2 oz	2 oz
Tartrazine Solution	7 cc	7 cc
(5 g in 100 cc water)		

Instructions as for Orange Syrup.

Strawberry Flavored Syrup
Formula No. 1

	25% Sweetener Solids "Sweetose"	50% Sweetener Solids "Sweetose"
"Sweetose"	30 lb 15 oz	60 lb
Granulated Sugar	75 lb	50 lb
Water	51 lb	46 lb
Dry Citric Acid	15 oz	15 oz
Dry Amaranth Color	2.5 oz	2.5 oz
Strawberry Flavor	To suit	To suit

No. 2
(With Juice from 5 x 1 Pack Frozen Strawberries)

"Sweetose"	30 lb	60 lb
Granulated Sugar	70 lb	45 lb
Strawberry Juice (22% Solids)	30 lb	30 lb
Water	27 lb	22 lb
Dry Citric Acid	15 oz	15 oz
Dry Amaranth Color	1.75 oz	1.75 oz
Strawberry Flavor	To suit	To suit

Instructions as for Orange Syrup.

No. 3

	%	Procedure
1. Water	26.10	Place in tank
Strawberry Puree*	40.00	
2. Cellulose Gum, CMC-7HSP	0.75	Premix, add, dissolve
Sugar	5.00	
3. Sugar	29.00	Add, heat to 190°F
Salt	0.05	
Sodium benzoate	0.10	
4. 50% Citric Acid	to pH 3.5	Add, package

* 3–1 pack (3 parts fruit, 1 part sugar); 40% soluble solids (refractometer).

Root-Beer Syrup

	25% Sweetener Solids "Sweetose"	50% Sweetener Solids "Sweetose"
"Sweetose"	30 lb	60 lb
Granulated Sugar	75 lb	50 lb
Water	51 lb	46 lb
Root-Beer Concentrate	30 oz	30 oz
Caramel Color	30 oz	30 oz
Foam Producer	To suit	To suit

Instructions as for Orange Syrup

Vanilla-Flavored Syrup

	25% Sweetener Solids "Sweetose"	50% Sweetener Solids "Sweetose"
"Sweetose"	30 lb	60 lb
Granulated Sugar	75 lb	50 lb
Water	51 lb	46 lb

	25% Sweetener Solids "Sweetose"	50% Sweetener Solids "Sweetose"
Vanillin (Dissolved in 6 oz 190 Pure Grain Alcohol)	1.63 oz	1.63 oz
Pure Vanilla Concentrate	2.25 oz	2.25 oz
Caramel Color	To suit	To suit

Instructions as for Orange Syrup.

Rasberry-Flavored Syrup

	25% Sweetener Solids "Sweetose"	50% Sweetener Solids "Sweetose"
"Sweetose"	30 lb	60 lb
Granulated Sugar	75 lb	50 lb
Water	51 lb	46 lb
Dry Citric Acid	15 oz	15 oz
Dry Amaranth Red Color	1.625 oz	1.625 oz
Raspberry Flavor	0.75 oz	0.75 oz

Instructions as for Orange Syrup.

Sugarless True-Fruit Raspberry Syrup

Water (180-190°F)	40 gal	Cold Water	To make 100 gal
Sodium Benzoate	0.5 lb	True-Fruit Raspberry Flavor	3.75 gal
Pectin, N.F. No. 444	6 lb	Citric Acid (50% Solution)	150 fl oz
Calcium "Sucaryl"	12 lb	Color "Carbo-Lok"	4 oz

Add the benzoate and pectin to the water with vigorous agitation. Then stir in the "Sucaryl." Add the flavor, citric acid, and color with agitation and sift in the "Carbo-Lok" with stirring.

Chocolate

	Formula No. 1 %	No. 2
Water	39.05	39.3
Sugar	35.00	35.0
Cocoa	10.00	10.0
Corn Syrup Solids	12.00	12.0
Nonfat Milk Solids	3.00	3.0
Cellulose Gum, CMC-7HP	0.75	0.5
Salt	0.10	0.1
Sodium Benzoate	0.10	0.1
Setting temperature	40°F	5°F

Premix gum with 5 to 10 times its weight of sugar and add to the water. Disperse. Add cocoa and heat to 180°F. Maintain for 10 minutes. Add balance of ingredients, heat to 180°F, and maintain for 10 minutes. Homogenize.

Cocoa Syrup

Cocoa	180
Sucrose	600
Glucose	180
Glycerol	50
Sodium Chloride	2
Vanillin	0.5
Sodium Benzoate	1.6
Distilled	
Water To make 1000	

Low-Calorie Apricot Jam

Crushed Pitted	
Fresh Apricots	25 lb
Water	20 lb
Glycerol	3 lb
Low-Methoxyl	
Pectin	6 oz
Citric Acid	2.66 oz
"Sucaryl"	
Sodium	2.75 oz

Pour the glycerol into the kettle and stir in the pectin to a smooth slurry. Add the water and stir until no lumps remain. Heat and add the fruit. Stir and boil for about 5 minutes. Add the citric acid and "Sucaryl." Boil about 5 minutes more. Pour into containers while hot, and seal immediately.

Orange Marmalade

Orange Shreds   20 lb

Water	3 gal
Corn Syrup	10 gal
Cane Sugar Syrup	10 gal
(7.5 lb/gal)	
Orange Juice	3 gal
Citric Acid (50%)	
Solution	20 oz
Pectin Solution	2 gal
(1 lb/2 gal water)	

Cook all but the citric acid solution to 220°F. Then add the citric acid solution just before filling containers.

Crushed Peach Topping

Formula No. 1
(With Straight Fruit)

	25% Sweetener Solids "Sweetose"	50% Sweetener Solids "Sweetose"
Peaches	100 lb	100 lb
Granulated Sugar	75 lb	50 lb
"Sweetose"	30 lb	60 lb

No. 2
(With Frozen Fruit)

Frozen Peaches (4 x 1 Pack)	125 lb	125 lb
Granulated Sugar	50 lb	25 lb
"Sweetose"	30 lb	60 lb

No. 3
(With Frozen Fruit)

Frozen Peaches (5 x 1 Pack)	120 lb	120 lb
Granulated Sugar	55 lb	30 lb
"Sweetose"	30 lb	60 lb

The approximate yield of each batch will be 160 lb at 68% soluble solids.

Place all ingredients in a kettle and heat rapidly to 222°F. Then fill into containers at once and seal while hot. Cool rapidly to prevent darkening of the fruit. To reduce floating of the fruit, cool the batch to about 150°F before filling; after filling and sealing, pasteurize at 190°F and then cool rapidly.

Tutti-Frutti Topping

Crushed Pineapple Topping	65
Crushed Strawberry Topping	15
Maraschino Cherry Pieces	10
Crushed Peach Topping	10

Drain the cherries well before adding. Blend all fruits together easily but well. Various proportions of the different fruits may be used.

Crushed Pineapple Topping

	25% Sweetener Solids "Sweetose"	50% Sweetener Solids "Sweetose"
Crushed Pineapple	100 lb	100 lb
Granulated Sugar	75 lb	50 lb
"Sweetose"	30 lb	60 lb

The approximate yield will be 160 lb at 68% soluble solids.

Use finely crushed, canned, unsweetened pineapple, packed in juice. Place all ingredients in a kettle and heat rapidly to 222°F. Then fill into containers at once and seal while hot. Cool rapidly to prevent darkening of the fruit.

Nesselrode Topping

Crushed Pineapple Topping	75
Crushed Strawberry Topping	10
Maraschino Cherry Pieces	10
Chopped Pecan or Other Nuts	5

Drain the cherries well before adding. Blend together easily but well. Various proportions may be used.

Crushed Raspberry Topping

Formula No. 1

	25% Sweetener Solids "Sweetose"	50% Sweetener Solids "Sweetose"
Fresh Red Raspberries	100 lb	100 lb
Granulated Sugar	75 lb	50 lb
"Sweetose"	30 lb	60 lb

No. 2
(With Frozen Fruit)

Frozen Red Raspberries (2 x 1) Pack	150 lb	150 lb
Granulated Sugar	25 lb	—
"Sweetose"	30 lb	60 lb

No. 3

Frozen Red Raspberries (3 x 1 Pack)	133 lb	133 lb
Granulated Sugar	42 lb	17 lb
"Sweetose"	30 lb	60 lb

No. 4

Frozen Red Raspberries (4 x 1 Pack)	125 lb	125 lb
Granulated Sugar	50 lb	25 lb
"Sweetose"	30 lb	60 lb

No. 5

	25% Sweetener Solids "Sweetose"	50% Sweetener Solids "Sweetose"
Frozen Red Raspberries (5 x 1 Pack)	120 lb	120 lb
Granulated Sugar	55 lb	30 lb
"Sweetose"	30 lb	60 lb

The approximate yield of each batch will be 160 lb at 68% soluble solids.

Place all ingredients in a kettle and heat rapidly to 222°F. Then fill into containers at once and seal while hot. Cool rapidly to preserve the natural color of the fruit.

Crushed Strawberry Topping
Formula No. 1

	25% Sweetener Solids "Sweetose"	50% Sweetener Solids "Sweetose"
Fresh Strawberries	100 lb	100 lb
Granulated Sugar	75 lb	50 lb
"Sweetose"	30 lb	60 lb

No. 2

Frozen Strawberries (3 x 1 Pack)	133 lb	133 lb
Granulated Sugar	42 lb	17 lb
"Sweetose"	30 lb	60 lb

No. 3

Frozen Strawberries (4 x 1 Pack)	125 lb	125 lb
Granulated Sugar	50 lb	25 lb
"Sweetose"	30 lb	60 lb

No. 4

	25% Sweetener Solids "Sweetose"	50% Sweetener Solids "Sweetose"
Frozen Strawberries (5 x 1 Pack)	120 lb	120 lb
Granulated Sugar	55 lb	30 lb
"Sweetose"	30 lb	60 lb

The approximate yield of each batch will be 160 lb at 68% soluble solids.

Mix the fresh berries with part of the sugar, let this stand, and drain off the juice; for frozen berries, defrost and drain off the juice. Add the "Sweetose" and sugar to the drained juice and heat to 228°F. Then add the drained berries and cook the batch as quickly as possible at 222°F. Cool the batch quickly to about 150°F to prevent floating of berries. Then fill and seal in glass containers; pasteurize at 190°F; and cool rapidly.

No. 5

	25% Sweetener Solids "Sweetose"	50% Sweetener Solids "Sweetose"
Fresh Strawberries*	112 lb	112 lb
Pectin Solution (1 qt)	32 fl oz	32 fl oz
Granulated Sugar*	75 lb	50 lb
"Sweetose"	30 lb	60 lb
Water (1.5 gal)	12 lb	12 lb
Certified Color Solution	8 fl oz	8 fl oz
Standard Citric Acid Solution	8 fl oz	8 fl oz
Benzoate of Soda USP	1.33 oz	1.33 oz

* The fruit and sugar are adjusted in each case to give 112 lb fruit and 100 lb sweetener solids.

No. 6

	25% Sweetener Solids "Sweetose"	50% Sweetener Solids "Sweetose"
Frozen Strawberries*	150 lb	150 lb
Granulated Sugar*	37 lb	12 lb

No. 7

Frozen Strawberries* (4 x 1 Pack)	140 lb	140 lb
Granulated Sugar*	47 lb	22 lb

No. 8

Frozen Strawberries* (5 x 1 Pack)	135 lb	135 lb
Granulated Sugar*	53 lb	28 lb

* The fruit and sugar are adjusted in each case to give 112 lb fruit and 100 lb sweetener solids.

The approximate yield of each batch will be 172 lb at 60% soluble solids.

Place the berries in a kettle and heat to boiling, while stirring. Add the pectin solution, sugar, and "Sweetose" and heat rapidly to 220°F or approximately 65% soluble solids. Then add the water, stirring well, to reduce the soluble solids back to 60%. This will cause the sweetener to penetrate the fruit and reduce the floating of the berries to a minimum.

Mix in well the color, acid, benzoate of soda, and flavor (if desired), and pack into glass containers.

The pectin solution is prepared by stirring 8 oz of 150-grade rapid-set pectin and 2 lb granulated sugar into 16 lb (2 gal) boiling water.

The red-color solution consists of 1½ oz FD & C Red No. 1 dissolved in 32 fl oz of hot water.

Butterscotch Topping

Formula No. 1

"Sweetose"	18 lb
Sweetened Condensed Whole Milk	2 lb, 8 oz
Bicarbonate of Soda	0.5 oz
Water	4 oz

Dissolve the soda in the water, place in a kettle with the "Sweetose" and milk, and cook while stirring at about 245°F, or until the desired color has developed.

Brown Sugar	2 lb
Salt	2.5 oz
Water (½ gal)	4 lb

Dissolve the sugar and salt in the water, add at once to the hot batch, and mix until smooth; then cook the mixture at 222°F.

Butter	1 lb
"Sta–Sol" Lecithin	0.25 oz
Flavor	To suit

Turn off the heat and mix in well with preceding. Homogenize and fill into glass jars. Seal the containers at 180 to 190°F. The approximate yield will be 26 lb.

Adjust the cooking of the "Sweetose" and milk to obtain the desired color. Use a medium-yellow or golden-brown sugar (No. 8 or No. 10 preferred).

No. 2

Brown Sugar	50 lb
"Sweetose"	50 lb
Water	33 lb
Salt	1 lb

Place in a kettle and cook rapidly at 228°F.

Vanilla or Flavor Blend	To suit

Turn off the heat, and mix in well. Then fill into containers and seal while hot.

The approximate yield will be 118 lb.

Use a medium-yellow or golden-brown sugar (No. 8 or No. 10 preferred). The product may be used as it is, or 2 oz of melted creamery butter may be added to each quart of the topping, stirring the mixture well before each serving.

For a rum-buttered topping, include some rum flavor in the blend.

Buttered Pecan Topping

Heat the pecans in an oven at about 350°F until thoroughly roasted. Then add sufficient creamery butter to coat the nuts thoroughly and allow to cool. Place a liberal quantity of cooked, buttered pecans (amount variable, depending on the cost desired) in containers and fill with butterscotch syrup. Seal while hot or pasteurize at 190°F.

If the pack is to be used for buttered-pecan ice cream, the nuts should be broken or chopped into pieces.

Banana Salad Topping

Crushed Pineapple Topping	90
Treated Sliced Bananas	10

Select firm bananas and slice in uniform sections. Drop the slices immediately in a 5% solution of citric acid and allow them to stand for about 1 hour; then drain. Fold the bananas, easily but well, into hot (at least 200°F) pineapple topping. Fill into containers, enameled cans preferred, and seal while hot (at least 190°F). Cool rapidly to prevent darkening of the fruit.

Caramel Topping

"Sweetose"	20 lb
Sweetened Condensed Whole Milk	5 lb
Bicarbonate of Soda	0.75 oz
Water	4 oz

Dissolve the soda in the water, place in a kettle with the "Sweetose" and milk, and heat with stirring, to about 245°F, or until the desired color has been developed.

Water	4 lb
Salt	2 oz

Add, stir until smooth, and heat to 224°F.

Vanilla Flavor Blend	To suit

Turn off the heat, and mix in well. Homogenize and fill into glass jars. Seal the containers at 180 to 190°F.

The approximate yield will be 26 lb.

Adjust the cooking of the "Sweetose" and milk to obtain the desired color. Use a medium-yellow or golden-brown sugar (No. 8 or No. 10 preferred).

Orange-Pineapple Topping

Pineapple Topping	50 lb
Orange Concentrate	32 fl oz
Orange Color	To suit

Blend all together well while the pineapple topping is still hot. In place of the orange concentrate, 4 oz of orange oil may be used. The amounts of flavor and color may be varied to suit.

Walnuts in Imitation Maple-Flavored Syrup (Prepared Sundae Dressing)

Formula No. 1

"Sweetose"	48 lb
Granulated Sugar	17 lb, 4 oz
Water	7 lb, 12 oz
Maple Flavor Caramel Color	To suit

No. 2

"Sweetose"	60 lb
Granulated Sugar	7 lb, 4 oz
Water	5 lb, 12 oz
Maple Flavor	To suit
Caramel Color	To suit

Bring the "Sweetose," sugar, and water to a boil and then add the flavor and color.

Place the desired quantity of walnut pieces in each container; then fill with hot syrup and seal at once. A half-gallon container will require about 2 lb of nut meats and a No. 3 can, about 1 pound.

To Prevent Mold in Maple Syrup

Add 0.02% sodium propyl p-hydroxybenzoate.

Chop-Suey Topping

Dried Figs	40 lb
Water (Variable)	40 lb

Soak the figs in the water for about 12 hours; drain and grind the figs to a fine pulp; then mix the pulp with the drained-off water.

Pitted Dates	10 lb	

Grind fine and add to the figs.

"Sweetose"	30 lb
Granulated Sugar	10 lb
Water (Variable)	10 lb

Add and heat to 218°F.

Walnut Pieces	2 lb

Add to mixture and bottle or can while hot.

The approximate yield will be 130 lb.

This topping will require reduction with an equal volume of simple syrup when used at the soda fountain.

Walnut Bleaching

Walnuts are bleached to improve their appearance. As the nuts are harvested, the shells are badly stained, but bleaching produces a light tan color of the shells without disturbing the meat.

Bleaching consists of treating the nuts with a sodium hypochlorite solution of 2%, draining this solution, and then treating the nuts in a 10 to 15 volume hydrogen peroxide solution. Approximately 12 to 15 lb of 130-volume hydrogen peroxide is employed for a ton of walnuts.

Preserving Unpopped Corn Kernels

U.S. Patent 2,518,247

Vegetable Oil	5 lb
Artificial Butter Color	1.5 oz
Artificial Butter Flavor	33 oz
Salt	100 lb

The flavor and moisture content are preserved by coating the corn with this mixture.

Proper Moisture for Corn to Pop

Experimenters in home economics have found that corn kernels must contain just a certain amount of moisture to pop satisfactorily. For 1 lb of popcorn, dissolve as much salt as possible at 68°F in 1 oz of water. Soak up this salt solution with strips of paper toweling, scatter these through the corn, and seal in a jar for 10 days to 2 weeks. The salt gives off or absorbs water as needed, thus imparting the proper amount of moisture to the corn for satisfactory popping.

Coating for Nuts

"Promine"-D	40
Water	450
Trisodium Phos-	
phate·12H$_2$O	3
Glycerol	10

Add phosphate to water and then very carefully mix with the "Promine" to prevent lumping and foaming. Heat to 140°F and stir until completely dissolved. Add the glycerol. Dip nuts into liquid and let them dry under a constant stream of dry warm air. Coating, when completely dry, will prevent leakage of oil from nut kernel. One can use a second coating if so desired.

Coating for Rice
and Nuts

Zein G200	100
Stearic Acid	17
Oleic Acid	8
Aq. Ethanol, 90%	240

The stearic acid is dissolved in the ethanol at 35-40°C, as it is not soluble in 90% aque- ous ethanol at room temperatures. The propylene glycol is added and the solution cooled to 30°C. The Zein G200 is added slowly with good agitation and stirred until dissolved. If desired the formulation may be further diluted with 90% aqueous ethanol.

Sugared Peanuts

A gum-arabic mucilage is prepared by dissolving 8 lb of powdered gum in 1 gallon of cold water, stirring to obtain a good solution.

Eighty pounds or a convenient quantity of selected blanched peanuts are placed in a standard revolving pan, which is set in motion, and the gum mucilage is poured over the nuts so that the mass is well coated. To prevent sticking, a few pounds of powdered sugar are then sprinkled over the moist nuts as they revolve. The mass is allowed to revolve until thoroughly dry, taken from the pan, and sifted.

The gum-coated nuts are returned to the pan, which is set in motion. A syrup consisting of 5 lb of sugar and 1 lb of corn syrup is prepared by dissolving in water and tested by a hydrometer to read 34° Bé. This syrup is slowly poured on the revolving nuts and liberal charges of icing sugar are added, the operator running his hand through the mass to ensure that all pieces are covered. The mass is allowed to run until dry, then the process is repeated until the required size is attained. The goods are then placed in wire or starch trays to dry thoroughly in a warm dry room and are then packed.

Salting Peanuts in Shells

Prepare a supersaturated salt solution and pump this brine into a steam-jacketed kettle, into which a wire-mesh basket containing peanuts in shells is lowered. Secure the cover on the kettle and heat the brine just short of the boiling point. Continue heating so that the low vapor pressure created within the kettle will force the brine into the permeable peanut shell.

Now turn off the steam, open the kettle, and remove the basket of nuts. After draining the brine from the peanuts, transfer them into a rotary dryer. Thus driving off the moisture leaves salt crystals inside the shells.

Finally, put the peanuts in a revolving oven, heated to about 800°F, timing this operation according to the roast desired.

Milk-Cream Fudge
(Short, Coarse Texture)

Granulated Sugar	60
Corn Syrup	10
Invert Sugar	10
Sweetened Condensed Whole Milk	20
Cream (26% Butter Fat)	10
Water	4
Coconut Oil	3
Vanilla, Maple, or Other Flavor	To suit

Heat all the ingredients in a standard caramel-cooking kettle, equipped with agitators, to 242°F. Cool the batch slowly to 165°F; then add half a pound of salt and flavor and beat the batch until it has grained.

Add nut meats, coconut, etc., and spread the batch directly into boxes previously lined with waxed paper, or spread to form a sheet which may be later scored or cut.

The rather dark color of the fudge is the result of slow cooling.

Fondant Fudge
(Smooth, Plastic)

Granulated Sugar	70
Corn Syrup	10
Invert Sugar	10
Evaporated	
Unsweetened Milk	20
Dairy Butter	5
Coconut Oil	
(M.P. 76°F)	2

Heat all the ingredients to 243°F, then pour the batch into a grain-free fondant beater, permit it to cool to approximately 130°F, and then add half a pound of powdered salt and flavor as desired.

Beat the batch until it becomes plastic and shows evidence of graining. At this stage, it may be taken from the beater and filled into boxes or trays lined with waxed paper to form a heap or block.

This fudge may also be extruded on a rolled cream center machine, or it may be stored in covered containers and later remelted by heating at 145 to 155°F, then poured into trays lined with waxed paper, or it may be deposited into impressions in starch.

Fudge of this type is frequently referred to as *sea-shore fudge*, which is formed into heaps or transferred into trays or boxes lined with waxed paper while the fudge is in a semigrained or plastic condition. The rather smooth texture of this fudge is the result of cooling the cooked batch before beating.

Fondant for Fudge

Granulated Sugar	100
Corn Syrup	20
Invert Sugar	15
Water	25

Heat the ingredients to 244°F. Cool to 130 to 125°F; then beat into fondant.

The percentage of corn syrup and invert sugar recommended in this recipe will produce fondant by any process on any type of fondant-processing equipment.

Frappé

Formula No. 1

Corn Syrup	50
Invert Sugar	50
Egg Albumin	1

Dissolved in

Water	2

Heat the corn syrup to 245°F, add the invert sugar, and mix until melted. Then place the resulting syrup into a beater, start the beater, add the dissolved albumin, and beat until light.

No. 2

Invert Sugar	50
Spray-Process Finely Powdered Albumin	1
Corn Syrup	50

Place the invert sugar cold into an upright beater, add the spray-process albumin, and mix thoroughly. Meanwhile, heat the corn syrup at approximately 200 to 220°F, add this to the invert sugar-albumin mixture and beat until light.

Milk Fudge
(Semi-short, Smooth)

Corn Syrup	45
Coconut Oil (M.P. 76-96°F)	15
Sweetened Condensed Whole Milk	60
Granulated Sugar	60
Salt	1

Heat all together to 246°F, add 50 lb of basic fondant at once, mixing until the fondant is melted; then add flavor and 20 lb of frappé, and mix thoroughly. Spread the batch on heavy oiled waxed paper and allow it to cool to approximately 170°F.

The relatively high percentage of corn syrup used in the recipe, plus the corn syrup and invert sugar content of the frappé, will yield a fudge of semichewing semi-short consistency.

Chocolate-Milk Fudge
(Plastic, Moderately Light)

Corn Syrup	30
Sweetened Condensed Nonfat Milk	45
Liquor Chocolate	7
Cocoa Powder	5

Coconut Oil		Coconut Oil	
(M.P. 76-96°F)	3	(M.P. 96°F)	8
Granulated Sugar	60	Salt	½

Place all the ingredients into a caramel-cooking kettle, equipped with agitators, and heat to 246°F. Cool to 230°F, then add 45 lb of basic fondant, 30 lb Frappé No. 1, flavor, nuts, etc., and spread the batch at approximately 165°F.

Whipped Milk Fudge
(Nougat Type)

a	Invert Sugar	20
	Vegetable or Soy Albumin	2

Beat together cold until light.

b	Granulated Sugar	70
	Corn Syrup	30
	Water	15

Heat to 260°F and add this cooked syrup to a, mixing well. Then add 40 lb basic fondant, mixing until the batch becomes plastic and graining is visible.

c	Spray-Process Powdered Nonfat Milk	16

Melt the coconut oil and add the powdered milk and salt, mixing well before adding this to the batch together with flavor and 20 lb of diced candied fruits.

This nougat-type of fudge may be deposited in the form of peaks or kisses on a special depositor designed to handle plastic candy batches; or, the batch may be filled into boxes lined with waxed paper to form blocks or slabs.

This fudge is frequently referred to as "divinity fudge."

Chocolate Fudge

a	Liquor Chocolate	10
	Coconut Oil (M.P. 96°F)	5
	Enzyme-Treated Powdered Whole Milk	15
	Salt	1
	Powdered Sugar	100

Melt the liquor chocolate and coconut oil, add the powdered milk and salt, and mix well.

Then add the powdered sugar and mix thoroughly.

b Water	10
Corn Syrup	15
Invert Sugar	15
Flavor	As desired

Heat the water, corn syrup, and invert sugar to approximately 200°F; add this to *a*, mixing and heating at 165 to 175°F. Then add the flavor and deposit into impressions in starch.

This fudge may also be deposited on waxed paper or rubber mats to form wafers.

Vanilla-Milk Cast Fudge

Corn Syrup	15
Invert Sugar	10
Water	8
Coconut Oil (M.P. 76 to 96°F)	10
Salt	1
Flavor	q.s.

Heat the ingredients to approximately 200°F, then add 15 lb nonfat powdered milk and 100 lb powdered sugar. Mix and heat to approximately 175°F, add vanilla, butterscotch, maple, or butter flavor, and cast into impressions in starch.

By using different percentages of crystal controllers, such as corn syrup and invert sugar, crystallization of the sugar dissolved in the syrup portion of the fudge can be retarded, thus aiding in the retention and distribution of the available moisture. The ratio and composition of the syrup phase to the sugar crystals developed will determine the physical characteristics of the fudge.

Licorice Paste

Wheat Flour	28
Corn Syrup	7
Gelatin	10
Raw Sugar	14
Licorice Juice	5
Caramel (Burnt Sugar)	14

Soak the gelatin until it is pliable. Dissolve the raw sugar in 2.25 gal of water in a steam pan, then add and dissolve the juice from the broken licorice. Strain this solution to eliminate foreign bodies and lumps.

Place the strained solution in a steel stirring cooker, add

the corn syrup, and dissolve the soaked gelatin in the batch.

Start the cooking and add the flour slowly to allow gradual gelatinization without lumping. Continue the cooking until a ball of medium texture is obtained in a cooled test piece. Pass the cooked batch through the licorice refiner to eliminate lumps and then transfer it to the extruding machine.

Place the extruded sheets or rods in iron trays in a well-ventilated room.

Nougat-Type Candy
(High Protein)

"Promine"-D	30
Sucrose	150
Water	140
Corn Syrup Solids	50
CMC-7HSP	2
Calcium Carbonate	1

Blend dry ingredients, add water, and whip all in Hobart mixer at speed 3 with wire beater. Form kisses, or place in cups, and bake at 400°F for 15 minutes.

Turkish Candy

Granulated Sugar	55 lb
Invert Sugar	
Syrup	45 lb
Water	110 lb
Thin Boiling	
Starch	11 lb
Cream of Tartar	3 oz

Put 50 lb of the water into a candy kettle. Add the granulated sugar and invert syrup, boil. Mix the starch and cream of tartar with the remaining 60 lbs of water. Add this to the boiling sugar and cook all to a heavy sheet. Cast to the desired thickness. This batch should finish to a weight of 128 pounds.

Marzipan Candies

Basic Almond	
Paste	15 lb
Short or Standard	
Casting Fondant	10 lb
Finely Powdered	
Sugar	15 lb
"Convertit"	$\frac{1}{2}$ lb
Salt	1 oz
Color and	
Flavor	As desired

Place all ingredients into a

mixing machine, or mix and knead by hand, to obtain a paste or dough sufficiently firm to retain its shape when the marzipan is later formed.

The basic almond paste, fondant, sugar and invertase may be mixed together to form a master batch, then later divided into sections to which color and flavor may be added to each portion or section.

Bonbon Centers

The marzipan may be rolled into sheets approximately three-eighths to one-half inch high, and then cut into various shapes with a small hand lozenge or cookie cutter. Powdered sugar may be used when forming the sheets.

After the marzipan sheet has been cut into smaller pieces, place them on trays lined with waxed paper and permit them to dry on the surface before dipping in fondant creme or chocolate.

Crystallized Marzipan

Portions of the marzipan batch may be colored and flavored with vanilla, orange, lemon, Roman Punch, pistachio, or rum flavor, adding colors to harmonize with the flavors.

When rolling or cutting the marzipan, use powdered sugar as a dusting medium. The rolled marzipan candies are then permitted to dry on the surface before they are immersed in a cool crystallizing syrup, cooking to approximately 33.5 Bé (at boiling point).

The crystallizing syrup is made by heating 72 pounds of sugar with 28 pounds of water to the boiling point. The batch is then tested while boiling to obtain a Bé 33.5.

The surface of the resulting crystallizing syrup is then sprinkled lightly with cool water and the syrup is allowed to cool to approximately 85-75°F, at which temperature it is then ladled over the marzipan candies, which have been placed into wire crystallizing baskets or crystallizing pans.

Metal screens are placed over the candies in the crystallizing pans to keep them immersed in the syrup. The candies are permitted to remain in the syrup from 7 to 12 hours, after which the syrup is carefully drained from the candies and they are then turned out on heavy waxed paper to dry.

Chocolate Cream

Sweet Butter	20
High-Ratio Shortening	20
Chocolate Liquor	12
Milk Powder	14
Granulated Sugar	20
Water	72

Bavarian Cream

Make 2 qt of custard cream. Cool to 50°F. Place 1 oz of gelatin in 2 oz of cold water and dissolve in a double boiler. Mix the dissolved gelatin into the cooled custard cream.

Whip 2 qt of homogenized cream. Add vanilla and rum flavor to taste. Then fold the whipped cream and custard cream together with a wire whip.

All kinds of fruit jellies, bananas, or other fillings may be folded into the cream. This makes a very good pie filling which can be decorated with whipped cream on top.

Coffee Cream
(All-Vegetable: Kosher)

"Promine"-D	2.5
Hydrogenated Vegetable Oil*	12.5
Dextrose	5.0
"Centrophil" SP	0.1
"Viscarin" B	0.025
Disodium Phosphate	0.25
Citric Acid (To pH 7.0)	0.075
Sufficient water to bring to 100 ml	

---

* Color	2R; 20Y max
Free Fatty Acids	0.5 max
Wiley Melting Point	86–90°F
A.O.M.	150 hr plus

Melt the hydrogenated vegetable oil and mix with the "Promine," Dextrose, and "Viscarin" B.

To the water, add the disodium phosphate and citric acid. Heat the water to 160°F, add the dry mix. Pasteurize at 160°F for 30 minutes. Following pasteurization, immediately homogenize at 2000 pounds pressure. Cool the product.

Depending on the type of water used, the citric acid may be varied to get the desired *p*H. It appears that the final *p*H of the product should be about 6.8; however, more information may be necessary from tests on different types of waters.

Ice-Cream Mix

Formula No. 1
14% Butterfat

Sweet Butter	15 lb
Skim- or Whole-Milk Powder	14 lb
Sugar	13 lb
Powdered Egg Yolk	1½ lb
Gelatin	6 oz
Water	30 qt

Dissolve the milk powder in the water and heat them to 110°F. Mix together thoroughly the sugar, dry egg yolk, and gelatin powder and add this mixture to the milk, while stirring continuously. Heat to 145°F, stir vigorously, pour the mass into a homogenizer bowl, and homogenize. Cool immediately by placing the can containing the finished product in a water bath. Let it stand for 24 hours in order to age; then freeze.

No. 2

19% Cream	60
Skim-Milk Powder	4
Whole Milk	20
Sugar	15
Powdered Egg Yolk	1.5
Gelatin	0.5

Modified Whipping Cream

Formula No. 1

Sweet Cream, Sweet Butter	1.5 lb
High-Ratio Shortening*	1.5 lb
Spray-Process Skim Milk or Whole Milk Powder	14 oz
Granulated Sugar	10 oz
Water	72 oz

This will make 4.5 qt of cream.

* For these four recipes, use only "M.F.B." "Quickblend," "Sweetex," "Covo Super-mix," or "Bakerite" 400 high-ratio shortening.

Dissolve the milk powder in

the cold water. Add the fat to the milk-water mixture in small pieces. Heat to 145°F. If the fat is not dissolved completely, remove the bowl from the heat and stir until dissolved. Check the temperature and if lower than 145°F, apply heat until this temperature is reached. Do not apply heat directly to the mixture. Make a double boiler by placing a smaller bowl inside of a larger bowl that contains water. When ready, pour the mixture through a strainer into the bowl of the homogenizer. Place an empty clean container under the homogenizer for a receiver. Start the machine and pump through the mixture. Cool the finished cream as quickly as possible by placing the container with the cream in a water bath and stirring every 15 minutes until the temperature reaches 50°F. Then place in the refrigerator. Or, place the cream in a refrigerator immediately, and stir it every 15 minutes.

No. 2

High-Ratio	
Shortening	3 lb

Spray-Process

Whole or Skim	
Milk Powder	14 oz
Granulated Sugar	10 oz
Water	72 oz

This will make 4.5 qt; 1 qt of light cream added to each 3 qt of this mixture will make a very fine cream.

No. 3

Sweet Butter	1.5 lb
High-Ratio	
Shortening	1.5 lb
Fresh Milk	2.5 qt
Milk Powder	3 oz
Granulated Sugar	10 oz

Use no water in this recipe. The milk takes the place of the water. This will make 1 gallon.

No. 4

19% Cream	16 qt
Skim-Milk	
Powder	16 oz
High-Ratio	
Shortening	12 lb
Sugar	4.5 lb

Add 1 qt, more or less, of 19% cream to each 2 qt of this mixture when whipping.

For whipping, the cream must be aged 48 hours. In

whipping homogenized cream, the cream should be very cold, not above 35°F (a thermometer should be used for checking). The mixer bowl and wire whip should be chilled.

Start the machine at second speed and operate the machine at this speed until the cream has expanded to double the original amount. Then finish at high speed until the desired stiffness is obtained. If the finished whipped cream is too heavy, add fresh milk or light cream until the desired consistency is obtained.

Homogenized cream requires slightly longer whipping time than cream that has not been homogenized.

For best results, put not more than 5 qt in a 20-qt bowl at one time. If larger quantities are desired, use the same proportions. Beat for 17 minutes at second speed and 1 minute at high speed.

It is important to use clean utensils and to strain the mixture before pouring into the bowl of the machine.

Reconstituted Milk

Dry Skim Milk   10.5 lb

Unsalted Butter   4.5 lb
Water   86 lb

This formula makes 100 lb or 50 qt of milk.

Put cold water in a steam-jacketed kettle, add powdered milk, and beat with a wire hand whip until the milk powder is dissolved. (Milk powder will not dissolve in hot water.) When the milk powder is dissolved, add the butter in small pieces. Heat the mixture to 145°F and then shut off the steam. If all the butter has not dissolved, stir until it has and then check the temperature. If it is less than 145°F, apply heat.

In taking the temperature of liquids, be sure to stir the mixture thoroughly from the bottom up, as the temperature at the bottom may be 5 to 10° higher than at the top.

If a steam kettle is not available, make a satisfactory double boiler by placing a smaller vessel inside of a large vessel filled with water.

If direct heat is used, the flame should be very low and the mixture should be stirred constantly in order to prevent scorching.

Cool immediately by placing the cans of milk into a water bath until the temperature is reduced to 50°F, stirring the milk occasionally to aid the cooling.

Short Cuts for Large Quantities of Milk

Concentrate the mixture of milk powder and butter with a little water. Add the rest of the water after the mixture has been homogenized. The following amounts are needed to make 400 lb of milk (3.5% butter-fat).

42 lb skim-milk powder, or four times the amount used for 100 lb of milk.

18 lb sweet butter, or four times the amount used for 100 lb of milk.

86 lb of cold water, or same quantity as used for 100 lb of milk.

Put together, heat to 145°F, and pass through the homogenizer. To each 15 qt of the homogenized mixture, add 25 qt cold water while the mixture is still hot. Then place it in the refrigerator for cooling.

Reconstituted Milk

The following represent standard formulations for reconstituting milk and cream, using milk and powder, unsalted butter or other fats, and water.

3.5% Reconstituted Milk

Dry Skim Milk	10.5
Unsalted Butter	4.5
Water	86

10% Cream

Unsalted Butter	10
Dry Skim Milk	9
Water	79

15% Cream

Unsalted Butter	18
Dry Skim Milk	8
Water	74

20% Cream

Unsalted Butter	23.25
Dry Skim Milk	7.5
Water	59

25% Cream

Unsalted Butter	30
Dry Skim Milk	7
Water	63

40% Cream

Unsalted Butter	47.0
Dry Skim Milk	5.6
Water	47.4

The percentages of butter fat and milk solids are the same as are contained in fresh milk and cream. If richer milk is desired, add more butter. Be sure to use only spray-process milk powder.

Heat the batch to 145°F and pass it through a homogenizer. Place an empty and thoroughly clean container beneath the homogenizer for a receiver.

To make a chocolate-milk drink, add 3 lb of chocolate liquor or cocoa powder and 7 lb sugar to the formulation for 3.5% milk.

A good grade of shortening or other fat may be substituted for the butter with very good results.

Reconstituted Whole-Milk Powder

If whole-milk powder is to be used for beverages, to each quart of cold water, add 4 oz of whole-milk powder, stir un-til dissolved, and pass through a homogenizer cold.

Soy Milk
(All-Vegetable)

a	Hydrogenated Vegetable Oil*	3.5
	"Promine"-D	3.5
	Lactose	5.0
	"Centrophil" SM	0.1

* Color	2R; 20Y max
Free Fatty Acids	0.5 max
Wiley Melting Point	86–90°F
A.O.M.	150 hr plus

Mix all ingredients and add 50 grams of hot water. Pasteurize at 160°F for 30 minutes.

b	Calcium Hydroxide	0.18
	Calcium Chloride	0.10
	Dipotassium Phosphate	0.40
	Monopotassium Phosphate	0.10

Add mixture b to 35 grams of water then add to mixture a. Homogenize and adjust pH to 6.5 to 6.8 with lactic acid.

This product may be flavored to taste.

Chocolate Drink
All-Vegetable,
High Protein

Formula No. 1
(Liquid)

"Promine"-D	25
Sucrose	25
Cocoa Syrup	100
Vanilla Extract	58

Blend "Promine"-D and sugar then combine all ingredients. Pasteurize at 160°F for 30 minutes.

No. 2
(Powder)

"Promine"-D	5.00
Cocoa Powder	2.50
Corn Syrup Solids	5.00
Sucrose	15.00
Salt	0.10
"Vescarin"-B	0.01
Vanilla Extract	0.50

For preparation, mix one part of the powder to 3 parts of water.

High-Stability Edible
Fat Spread

Hydrogenated Acetylated Peanut Oil	77.30
Hydrogenated Cottonseed Oil	20.00
Salt	2.20
Soybean Lecithin	0.50
Propyl Gallate	0.01
Butyric Acid	3.00 ppm
Diacetyl	3.00 ppm
Vitamine A	1.00 ppm
Butter Color	To suit

The peanut oil and cottonseed oil are mixed with the salt at or below 145°F. Dry nitrogen is passed through the mixture for a period of time sufficient to remove any dissolved oxygen. The influx of nitrogen is stopped and, while the mixture is under a blanket of nitrogen, the remaining ingredients, i.e., lecithin, gallate, flavor, color, and vitamin A, are added and thoroughly blended by continuous agitation.

While the blended product is under an atmosphere of nitrogen and at 130 to 140°F, it is passed through a "Votator," into which dry nitrogen is introduced so as to increase the volume of the product 11 to 12%. The product leaves the "Votator" at 40 to 45°F, is filled into containers, and sealed under an atmosphere of nitrogen.

To Retard Rancidity in Shortening

Melt 10 lb shortening (not over 200°F, well below the smoke point.) Add 3.9 oz of butylated hydroxyanisole and stir to dissolve. Use 1 lb of this mixture for each 100 lb of shortening in the mix. This mixture may become firm on cooling or stay liquid, depending on the shortening used. When creaming the shortening and sugar, or in sandwich fillings, add the mixture of shortening and butylated hydroxyanisole at the beginning of the mixing. The stock mixture may be prepared in large quantities as it keeps almost indefinitely.

Coloring for Margarine

Lactoflavin	3 g
Carotene	1½ g

This is sufficient for 1 ton of margarine.

All-Vegetable "Cheese" Spread

Water	250.0
Hydrogenated Vegetable Oil*	165.0
"Promine-D"	40.0
Salt	2.5
Calcium Lactate·5H₂O or 50% Lactic Acid	4.0

* Color	2R; 20Y max
Free fatty acids	0.5 max
Wiley melting point	86–90°F
A.O.M.	150 hr plus

At pasteurizing temperatures mix hydrogenated vegetable oil, salt, and "Promine-D" together, add the water and stir thoroughly. Pasteurize at 160°F for 30 minutes, add the lactate or the lactic acid and homogenize. Add desired flavor before or after cooling. Store under refrigeration. (Makes approximately 1 pound)

Wax Coating for Cheese

Formula No. 1

French Patent 1,188,140

Glyceryl Monostearate	91.2
Lecithin	8.0
Water	0.6
Food Coloring	0.2

No. 2

Microcrystalline Wax (M.P. 155°F)	22

Paraffin Wax   49
Petrolatum   29

Melt and mix thoroughly.

No. 3
(For Gouda Cheese)

Paraffin Wax   85-88
Petrolatum   5-10
Ceresin Wax   1-2.5
Oil-Soluble Carmine   2-3

Blue Cheese Flavor

U.S. Patent 3,034,902

a Acetone   3.00
   2-Pentanone   3.00
   2-Heptanone   1.50
   2-Octanone   1.00
   2-Nonanone   1.00
   2-Undecanone   0.50
   Ethanol   1.00
   Acetaldehyde   0.04
   Cotton Seed Oil,
      Stabilized   80.00
b Butyric Acid   16.8
   Caproic Acid   24.1
   Caprylic Acid   5.7
   Capric Acid   15.5
   Cotton Seed Oil,
      Stabilized   27.9

Use 0.018-0.9 of *a* and 0.22-
0.45 of *b* to 100 of product to
be flavored.

To Prevent Adhesion of Press Cloth to Rindless Cheese

Soak the pressing cloth in
2% solution of Sodium tetra-
phosphate in soft water before
applying it to the cheese.

Base for Cordials

Sugar   17 oz
Water   8 oz
Ethanol 96%   40 oz
Water   To make 1 gal

This mixture can be varied
depending upon the proof desired, as well as the sweetness
required for the various types.
For example, some formu-
lations for anisette contain
twice the amount of sugar
given above, whereas a black-
berry-flavored cordial may
contain only 12 to 14% of
sugar.

Cherry Brandy (Kirsch)

Mash 3 parts ripe sour
cherries and 1 part sweet
black cherries. Place in a
wooden or earthenware container and add enough 90%
ethanol to cover. Let it stand

four to six weeks with occasional agitation, then put it through a filter press. Add 250 to 300 g of sugar to the liter and enough water to give a final content of 35-40% alcohol.

Grenadine

Mash 2 kg of black sweet cherries and 1 kg of dark sour cherries. Heat in porcelain to thick consistency. Add 1 kg mashed raspberries and 3 kilos 90% ethanol. Let the mixture stand for four weeks, then add 4 kg of sugar dissolved in 2.5 to 3 kg of water.

Advokaat (Egg Liqueur)

Mix the yolks of 15 eggs, 300 g of sugar, and 400 g of boiled milk. Slowly stir the mixture into 1 liter of brandy. Warm the mixture on a steam bath till it is thick. Fill while warm into dry bottles.

Citrus Fruit Color

U.S. Patent 2,943,943

Dipentene	20 cc
Triethanolamine	5 cc

Oleic Acid	5 cc
Bixin	2 g

Heat to 220°F and add

Water	100 cc

Synthetic Cinnamon Flavor

Imitation Cassia Oil	1	
"Cinacoa" 40		100

Low-cost Cinnamon

"Cinacoa" 40	10
Cinnamon	10

Low-Cost Cocoa Powder

"Cinacoa" 65	24
"Proflo"	6
Cocoa Powder	70

To Increase Strength of Mustard Seed Flavor

Add a small amount of citric acid or sodium carbonate to the mustard seed before crushing it to increase the amount of flavoring oil.

Green-Tea Sticks

Japanese Patent 3200 (1953)

Powdered Green Tea	100
Starch	5
Water	65

Heat and mix for 2 minutes and mold into sticks of 2 mm diameter. Dry at 70 to 80°C for 15 minutes and at 150°C for 20 minutes.

Moist Shredded Coconut

U.S. Patent 2,631,104

Moist Shredded Coconut	91
Sorbitol	8
Propylene Glycol	1

To Preserve Grated Horseradish

Add 0.15 to 0.20% potassium *m*-bisulfite to grated horseradish in vinegar.

It maintains pungency, aroma and odor, color, and crispness.

To Control Decay of Fruits and Vegetables

U.S. Patent 2,604,409

Wash with a 1 to 5% solution of thioacetamide at 60 to 100°F.

To Prevent Darkening of Bananas during Dehydration

An addition of 0.05 to 0.10% aqueous solution of allyl thiourea to bananas controls darkening during dehydration.

To Clarify Whiskey and Other Beverages

A water solution of polyvinylpyrrolidone (0.005%) added to a liquor with stirring, followed by filtration with "Celite" or other filter aid, gives a clear product which does not form a haze on chilling to 2-3°C.

A similar amount and treatment works well with wine.

For grape juice, 0.2% PVP is used, followed by 24 hours storage at 4°C.

For beer, 0.01-0.02% PVP is added to the hot wort in the brewing kettle.

Root Beer Foam Stabilizer

U.S. Patent 2,942,978

A 1% solution of polyethylene oxide of viscosity

about 3000 cp is added to root beer to give a concentration of 160 ppm.

To Stabilize Beer-Foam

U.S. Patent 2,712,500

An addition of 0.005 to 0.050% of hydroxypropyl methylcellulose increases the life of foam on beer.

To Improve Flavor and Foam of Beer

U.S. Patent 3,026,204

Add monosodium glutamate in an amount equivalent to 200 ppm of glutamic acid.

Fining Agent for Effervescent Wines

French Patent 1,171,754

Bentonite	150
Tannin	1
Water, Distilled	1000

Use 1 cc to one liter of wine.

Fish Cakes

Sliced Potatoes	200 lb
Shredded Salt Cod	130 lb
Milk Powder	10 lb
Hydrogenated Shortening	5.5 lb
Toasted Onion Powder	8 oz
White Pepper	2 oz
Garlic Powder	0.25 oz
Citric Acid	To taste

Soak the shredded cod in running water about 2 hours. Boil the potatoes and drained cod in a kettle until cooked (about 30 to 40 minutes). Grind the fish and potatoes, then add the remaining ingredients, and mix well. Can the product while it is still hot.

Fish Balls
(Norwegian Style)

Cooked Ground Haddock or Cod	60 lb
Milk (Skim-Milk Powder May Be Used)	5 gal
Fish Broth	5 gal
Potato Flour	1.75 lb
Wheat Flour	0.75 lb
Cracker Crumbs	0.5 lb
Salt	1 lb
White Pepper	0.25 oz
Nutmeg and Ginger	To taste

Steam the fish and then grind it. Combine the remaining ingredients in a mixer, add the fish, and beat well until blended. Form into balls and set in hot water for about 12 minutes. Put into cans and fill with fish broth; then seal.

All-Vegetable "Frankfurter"

Formula No. 1

"Promine"-D	210
Water	610
Hydrogenated Vegetable Oil	120
Hydrolyzed Vegetable Protein	26
Seasonings and Flavorings	10
Lecithin	4
Food Coloring	

Mix "Promine"-D, hydrolyzed vegetable protein, and seasonings together in a Hobart-type mixer (or equivalent) using a flat beater at 2nd speed. Melt hydrogenated vegetable oil and lecithin together, add this to the dry mixture, and blend for 15 to 20 minutes, to make mixture plastic. Add the cold water with the food coloring and continue blending until the emulsion is smooth. To remove any excess of air from the mixture either use a vacuum mixer or draw a vacuum on the finished emulsion. This procedure prevents the possible bursting or shriveling of the casings during cooking. Stuff the mass into casings, tie-off, and cook them in steam at 212°F for 30 minutes. Chill in cold water for about 10 minutes and refrigerate.

No. 2

"Promine"-R	205
Water	600
Trisodium Phosphate ($12H_2O$)	26
Hydrogenated Vegetable Oil	120
Hydrolyzed Vegetable Protein	26
Seasonings and Flavorings	10
Lecithin	4
Food Coloring	

Add "Promine"-R to 500 parts of water and mix thoroughly in the Hobart-type mixer (or equivalent) using a flat beater. Dissolve the trisodium phosphate in the remaining 100 parts of water, add this to the "Promine"-R mixture, and mix for about 5 minutes. Melt the hydrogenated vegetable oil and lecithin together, adding this to the mixture with the rest of the ingredients, and mix to a smooth emulsion. Finish as described for No. 1.

Protein Dumplings

Water	45.0
Trisodium Phosphate·12 H_2O	0.5
"Promine"-D	32.0
Salt	2.0
Fat	3.0
Monosodium Glutamate	0.4

Dissolve trisodium phosphate in water, then mix it with the "Promine," add the other ingredients, and knead well. Form into small dumplings, boil in bouillon and serve.

Alternately, grate cheese and add it to the dough; then boil in slightly salted water and serve with tomato sauce. These dumplings may also be fried in deep fat until light brown and served as a snack.

Meat Tenderizer

Formula No. 1

Australian Patent 141,755

Papain	3
Salt	450
Sodium Pyrosulfate	20-25

Soak the meat in 30 parts of this mixture and 600 parts of water below 60°C for 5 to 60 minutes.

No. 2

Spanish Patent 211,836

Salt	80
Dextrose	10
Pepsin	10

Sprinkle on the meat before cooking and keep it in the refrigerator up to 24 hours.

No. 3

	Powder	Liquid
Salt	40-80	10-20
Sugar	10-30	4-8

	Powder	Liquid
Propyleneglycol	—	10-20
Water	—	60-70
Monosodium Glutamate	10-20	5-10
Hydrolyzed Vegetable Protein Enzyme	—	5-10
Spices	To suit	

No. 4

U.S. Patent 3,033,691

Protease	1-15
Propylene Glycol	5-20
Sodium Chloride	17
Dextrose	1-44
Water	To suit

Soup Seasoning

Powdered Thyme	0.5
Powdered Sweet Marjoram	0.25
Powdered Rosemary	0.25
Powdered Savory	0.125
Powdered Celery	0.25
Powdered Onion	0.75
Powdered Pepper-cream	0.125

Beef-Soup Base

Hydrolyzed Plant Protein	18.00
Salt	15.00
Beef Extract	12.00

Cream Gravy

Fat	14 oz
Sifted Flour	12 oz
Nonfat Milk Solids	1.75 lb
Soup Base	7 oz
Water	3.5 qt

Melt the fat and flour and blend well. Add the milk solids and soup base to the water and beat with even whipping until well blended; then heat until hot. Add gradually to the fat and flour mixture; cook until thick, stirring constantly; and season with salt and pepper to taste.

Beef Pot Pie Gravy

Vegetable Shortening	4.00
Cellulose Gum, CMC-7HSP	0.50
Beef Broth	90.80
Flour (All-Purpose)	4.00

Salt	0.40
Paprika	0.25
White Pepper	0.05

Coloring: 6 ml of combination of 0.1 g each of caramel brown shade and burnt sugar shade, dissolved in 25 ml of water.

Piquant Sauce

Powdered Whole Egg	1.5 lb
Powdered Egg Yolk	1 lb
Corn Oil	3 gal
Salt	2 lb
Vinegar (100-grain)	1 qt
Powdered Mustard	6 oz
Sugar	5 lb
"Spiceolate" Sauce Dressing	2 oz
Cornstarch	5 lb
Water	5 gal

Dissolve the salt and sugar in about four gal of the water and when boiling, add the cornstarch that has been previously mixed with one gal of cold water. Cook until transparent, watching that it does not stick or burn; then cool. Place the oil in the beater, add the "Spiceolate," then the powdered egg, and beat until perfectly smooth. Then slowly add the vinegar, and finally the cornstarch. Beat until perfectly smooth and continue for ten minutes more.

Beefsteak Sauce

Malt Vinegar (40-grain)	50 gal
Powdered Dried Mushrooms	5 lb
Finely Ground Dry Walnuts	5 lb
Powdered Dry Onions	5 lb
Powdered Dry Garlic	5 lb
Apple Pulp	20 gal
Sugar	20 lb
or	
Molasses	30 lb
Salt	20 lb
Imitation Maple No. 600 D & O	0.5 gal
"Spiceolate" Beefsteak Sauce	0.5 lb
Soy Sauce	25 gal

Cook slowly for two hours, replacing with water any evaporation. Bottle hot at 180°F. This will make 100 gallons.

Sauce Dressing for Salads

Salad Oil	10 gal
Cider Vinegar (50-grain)	5 gal
Salt	2 lb
Paprika	0.5 lb
Tomato Juice	1 gal
Pectin No. 100	6 oz
"Spiceolate" Sauce Dressing	2 oz

This formula will make 16 gallons. The less oil used in this formula, the less flavor is required.

Mix the pectin with the paprika, the "Spiceolate" and salt with the oil. Then add, all at one time, the vinegar and tomato juice. Agitate continually for 10 or 15 minutes until the sugar and salt are dissolved and the product is smooth.

If garlic is desirable in this dressing, add 0.5 oz "Spiceolate" garlic, to the batch.

Sea-Food Cocktail Sauce

Water	25 gal
Sodium Benzoate	24 oz
Paprika	50 lb
Dehydrated Onion	5 lb

Distilled Vinegar (100-grain)	12 gal

Mix to a smooth paste, then heat and add:

Heavy Tomato Puree (Sp.Gr. 1.06)	50 gal

Heat to 210°F and add:

Salt	30 lb
Sugar	100 lb
"Spiceolate" Cocktail Sauce	4 oz

Heat to 210°F and bottle at not less than 180°F. Cap with spot crowns.

Barbecue Sauce

Water	25 gal
Sodium Benzoate	24 oz
Paprika	50 lb
Powdered Dehydrated Garlic	2 lb
Powdered Dehydrated Onion	5 lb
Distilled White Vinegar (100 grain)	10 gal

Mix these, to remove all lumps, until a heavy paste is secured. Then add:

Heavy Tomato
Puree (Sp.Gr.
1.06)   50 gal

Mix and heat to 180°F. Continue mixing and add:

Salt   30 lb
Sugar   100 lb

Dissolve and add:

"Spiceolate"
Barbecue Sauce
D & O   8 oz

Heat mixture to 210°F and bottle and cap at not less than 180°F to prevent formation of mold.

Tomato Sauce for Beans

Tomato Puree
(Sp.Gr. 1.04)   125 gal
Salt   75 lb
Monosodium
Glutamate   20 lb
Sugar   275 lb
Corn Syrup   125 lb
Chopped Onions   25 lb
Cider Vinegar
(45 grain)   4 gal
"Spiceolate"
Tomato Sauce   4-8 oz
Maple Flavor   1 qt
Bicarbonate of
Soda   6 oz

This will make 600 gallons. Cover the coils with water, add the onions and puree, and bring to a boil. Mix the corn syrup, vinegar, salt, sugar with about 70 gal of water, and run in this syrup into the preceding. Add the soda and water to make 600 gal, and bring to a boil. Run through a finisher and keep agitated at points where the sauce is held. Add the flavors just before running through a finisher.

Tomato Sauce

Vegetable Oil   4 oz
Chopped Onions   12 oz
Chopped Fine   1 small
Garlic   clove
Soup Base   7 oz

Worcestershire-Type
Sauce

Apple Vinegar   60 gal
Soy Sauce   25 gal
Black Strap
Molasses   5 gal
Tomato Puree   5 gal
Maple Flavor
No. 600 D & O   32 oz
"Colloidex" Lime
Flavor No. 1   1 gal

Dehyd. Powd.
Onions   5 lb
Dehyd. Powd.
Garlic   5 lb
"Spiceolate"
Worcestershire
Sauce D & O   0.5 lb
Salt   40 lb

This will make 100 gallons.

Sauce for Pork and Beans

Salt   80 lb
Sugar   190 lb
Corn Syrup   250 lb
Monosodium
Glutamate   20 lb
Ground Onions   35 lb
Cider Vinegar
(45 grain)   4 gal
"Spiceolate"
Bean Sauce   2-4 oz
Maple Flavor   0.5 gal
Hot Water   To make
                    600 gal

Catsup

Tomato Pulp
(Sp.Gr. 1.022)   120 gal
Sugar   70 lb
Salt   13 lb
Vinegar (100
grain)   5 gal

Ground Onions   13 lb
Catsup "Spice-
olate" No. 1   3 oz
Paprika   1 lb
Cayenne Pepper   1.5 oz
Cook to a finish of 55 gal

Salad Dressing

a Starch Paste:

Vinegar   15 lb
Water   30 lb
Sugar   10 lb
Tapioca Flour   1.5 lb
Corn Starch   3.5 lb

Cook mixture at 185 to 195°F for 10 minutes. Allow it to cool thoroughly.

b Emulsion Base:

Corn Oil   26 lb
Whole Egg Yolk   2.5 lb
Water   1 lb
Salt   0.5 lb
Mustard   0.25 lb

Mix the egg yolk, water, salt, and mustard. While beating, slowly add the corn oil to form a stiff emulsion.

Blend the 60 lb of cold starch paste with the 30 lb of Emulsion Base to form the finished salad dressing.

Mustard Sauce

Vinegar (90 grain)	55 gal
Cornstarch	60 lb
Mustard Flour	60 lb
Paprika	10 lb
Powdered	
Turmeric	3 lb
Sugar	85 lb
Salt	35 lb
Water	25 gal

Mix the dry ingredients, add to the vinegar and water slowly to avoid lumps, cook to the desired consistency, and then add ½ lb of "Spiceolate" mustard sauce. Mix well and fill hot into wide-mouth jars.

Sweet Liquor for Pickles

White Vinegar	
(100 grain)	85 gal
Granulated	
Sugar	2350 lb
Water	115 gal
"Spiceolate"	
Sweet Pickle	4 lb

Heat the vinegar and sugar to boiling for 15 minutes to invert the sugar; add the water and continue heating. Mix the "Spiceolate" with an equal quantity of water and add this to the syrup. Mix thoroughly for several minutes and strain or filter to obtain a bright product.

Pickle-Flavor Emulsion

Oil of Dill Weed	25.00
Oleoresin Capsicum	1.00
Gum Arabic	1.25
Gum Tragacanth	1.00
Benzoate of Soda	0.10
Water To make	100.00

Mix the oil of dill weed and oleoresin thoroughly with agitation. Disperse the gums into the oil phase with vigorous, high-speed agitation. Slowly add the aqueous phase with continuous agitation. Homogenize in a Manton-Gaulin type of pressure homogenizer.

Ham-Pickling Salt

Formula No. 1

Sugar	40
Salt	132.5
Sodium Nitrate	1
Sodium Nitrite	1.5
Polyglycol 400	
Monostearate	1

Dissolve in warm water.

No. 2

U.S. Patent 2,596,067

Salt	18.00
Sugar	1.00
Sodium Nitrite	0.10
Sodium Nitrate	0.10
Disodium	
Phosphate	4.75
Water	76.05

The ham is injected with this and then stored for 15 to 20 days in this solution. The cured ham is washed, boned, smoked, and cooked at 155°F.

Bacon-Pickling Solution (For Hypodermic Injection)

Water	71.9
Salt	16.8
Sugar	11.1
Sodium Nitrite	0.2

To Preserve Dried Yeast

U.S. Patent 2,523,483

Dried granulated yeast is coated with at least 10% of shortening warmed slightly to make it fluid.

Moldproofing Foods

Cheese is protected against mold by incorporating 2.5 g of sorbic acid for 1000 sq inches of waxed cellophane wrapping.

Bloating of pickle brine is controlled by 0.05 to 0.10% sorbic acid.

Molding of chocolate syrup is prevented by the addition of 0.1% sorbic acid.

Improving Powdered Eggs

Japanese Patent 172,804

Powdered Eggs	100
Sodium Citrate	1

The addition of 48 cc water to 11 g of this mixture gives, on heating, the same coagulation as fresh eggs.

Appetite Duller

U.S. Patent 2,714,083

Sodium Carboxy-methylcellulose	4
Tartaric Acid	1

Form into tablets. If a tablet is allowed to dissolve slowly in the mouth, it covers the tongue and lining of the mouth temporarily and thus lowers keenness of taste and appreciation of food.

To Reduce Dust in Handling Grain

U.S. Patent 2,585,026

Explosion hazards of (wheat) grain dusts are overcome by spraying while agitating in a screw conveyor with:

Refined Mineral Oil	4.00
Propylene Glycol Monolaurate	0.08
Water	95.92

Use 1% of this emulsion on weight of grain.

CHAPTER IX

INKS AND CARBON PAPER

White Ink for Photographic Negatives

1. Glycerol   20.0
2. CMC (Low Viscosity)   2.0
3. Titanium Dioxide   10.0
4. "Victawet" solution (5%)   1.0
5. "Marasperse" C   0.3
6. Phenol   0.1
7. Water   66.6

Dissolve 2 in 1 and 7, then add 5 and 4. With good mixing add 3; mix until uniform. Pass through colloid mill.

Ink for Correcting Blueprints (Non-erasable)

U.S. Patent 2,931,724

Potassium Oxalate   12.50
Gum Arabic   3.75
Sodium Benzoate   0.10

Tartrazine   0.10
Lead Chromate   6.00
Lead Sulfate   1.80
Barium Sulfate   3.60
Water   72.15

Opaque X-Ray Ink

Spanish Patent 240,867

Zinc Bromide   16.69
Litharge   36.66
Pigment   3.33
Benzol   21.66
Varnish   16.66
Acetone   5.00

Dissolve the last three items and then mix in the first three in the order shown.

Fluorescent Marking Ink (Removable)

U.S. Patent 2,950,256

Anthracene   400 g
Naphthacene   4 g

Ethyl-
hexanediol   10,000 cc
Xylene   4,000 cc

Marks made by this ink can be removed by heating at 300°F for less than two seconds.

Indelible Ball Point Ink

German Patent 1,064,663

Zapon Echt Orange   4.0
Hydroxyethylated
  Fatty Alcohol   58.0
Methylpyrrolidone   27.0
Sodium Chloride   1.8
Water   2.7

The paste from these ingredients is formed into an emulsion with water and good mixing. The viscosity depends on the amount of water added.

To Revive Ball Point Pens

Heat the metal tip gently with a match, or soak tip in alcohol or cleaning fluid.

Marking Ink

This is a dye-based ink for use on metal cans and coated cellophane surfaces.

PVP K-90 (20%
  solution)   7
"Igepal" CO-710   5-10
Dye in Isopropanol q.s.

Heat the dye and isopropanol to 180°F. Add PVP and "Igepal" CO-710.

Permanent Marking Ink for Metal

U.S. Patent 2,962,398

Zirconium
  Oxychloride   20
Copper Chloride   100
Hydrochloric Acid   300
Water   700

After applying, neutralize with ammonia or other alkali.

Ink Eradicator

Spanish Patent 252,659

Citric Acid   2-20
Sodium Hypo-
  chlorite Solution
  (2-20 g free
  chlorine)   1000

Printing Ink for Aluminum

U.S. Patent 3,014,822

Zein	80.0
Rosin	20.0
Methanol	150.0
Water	50.0
Phosphoric Acid	16.1
Methyl Violet	5.3
Victoria Blue (Conc.)	7.5

Magnetic Ink, High-Speed Scanning

British Patent 857,820

Pentaerythritol Alkyd Varnish	37
Linseed Oil	10
Lecithin	1
Carbon Black	1
Iron Blue	1
Iron Oxide	50

Smear-Resistant Ink

Formula No. 1

Super Spectra Carbon Black	3.71
"Norlig" Solids	4.30
"Cellosize" HEC WP-09	4.38
Glyoxal (Dry basis)	2.19
Water	85.42

Similar formulations yield colored inks with good wet-rub resistance. Here are two examples:

	No. 2 (Red)	No. 3 (Green)
Toluidine Red	3.72	—
"Monastral" Green	—	4.65
"Cellosize" HEC WP-09	9.50	9.50
Glyoxal (Dry basis)	4.28	4.25
Water	82.50	81.60

Gravure Ink

Formula No. 1

Chrome Yellow, Light	50.0
"Amberol" 801	7.5
Linseed Alkyd	5.0
Chlorinated Rubber (10 centipoises)	8.0
Toluene	19.5
Tolusol	10.0

No. 2

Chrome Yellow, Light	25.0
"Amberol" 750	13.0
Dibutylphthalate	9.0

SS Nitrocellulose
(¼ sec.)   15.5
Ethyl Alcohol   37.5

For both: In mills requiring a batch of low viscosity such as the Kady, all the pigment plus enough solvent to produce a slurry and just enough vehicle solids to initiate flow is put into the mill for the dispersion phase. After the dispersion is complete the remaining ingredients are added.

Moisture-Set Printing Ink

Zein G200	100
Resin*	250
Diethyleneglycol	750
Pigment	375-750

* Suitable resins are "Amberol" 750, "Arochem" 462, "VBR" 757, etc.

The resin and Zein G200 are dissolved in the glycol at either room or elevated temperature. When solution is complete the desired type and quantity of pigment is introduced into the finished solution and ground on a roll-mill in the conventional manner.

Water-Base Printing Ink

Zein G210	100
Resin*	100
Water	450-500
Aq. Ammonia 28%	15-20
Solvent**	150-100

* Shellac, "Amberol" 750, "Arochem" 462, and "VBR" 757. Others are also applicable.
** Proprietary alcohol, isopropyl alcohol, n-butyl alcohol, or the glycol ethers may be used.

Dissolve the resin in the solvent. Slurry the Zein G210 in the water (30°C, or cooler) and add the ammonia, stirring constantly. When the Zein is dispersed, add the resin solution slowly with constant agitation. For a finished ink, the desired pigment is ground in this vehicle with a ball or colloid mill.

Flexographic Printing Ink

Zein G200	100
"Arochem" 462 Resin*	100
Proprietary Alcohol	320
Water	20
Pigment	115-200
"Tamol" 731 (Dry Basis)	5

* "Amberol" 750 or "VBR" 757 may be used.

First dissolve the resin in the alcohol. When solution is complete add the Zein G200 and stir until dissolved. Stir in the "Tamol" 731, add the pigment and grind in a ball or colloid mill.

Electrically Conducting Printing Ink

U.S. Patent 3,043,784

Carbon Black	4-15%
Calcium	
Lignosulfonate	5-20
Water	40-55

Mineral Oil	15-30%
of total above.	

Decalcomania (Flexible Ink)

French Patent 1,179,355

Smoked Sheet	
Rubber	100.0
Sulfur	2.5
Tetramethylthiuram	
Disulfide	0.1
Styrene	15.0
Benzene	382.0
Color	To suit

Heat-Transfer Inks

These special inks operate on the principle of decalcomania except that they are heat-activated. The ink which is applied as a relatively heavy coating on a conventional paper is transferred by means of a typewriter or addressograph type (flat bed) plate impression to a special master paper. This latter, when brought into contact with the copy paper and impressed on the reverse or uncoated side by means of a heated bar, causes a small amount of the ink to be released onto the copy paper. This process may be repeated several times, depending upon the specific properties of the ink and the amount of heat and pressure used in making the transfer. These latter conditions are closely controlled by the automatic equipment. The requirements of the ink are that the intitial transfer be clean and sharp and that the secondary multiple transfers be designed for specific temperatures and stroke pressure. Both the coating (ink) and the primary or master impression must be stable on standing.

The following formulations, although not completely applicable in all respects (commercially operable on available automatic equipment), will illustrate the general requirements:

Component	Formula No. 1	No. 2
Carnauba Wax	10	16
Beeswax	14	11
"Oxiwax"	10	4
Montan Wax, Crude	7	10
Gilsonite Super Selects	7	6
Petrolatum	7	6
Nigrosine SSB	14	16
Nigrosine Base	10	—
Oleic Acid	4	5
"Baker's 15" Oil	7	16
Molacco Black	10	10

Grinding Procedure:

These inks are best prepared on a roller mill. The waxes, oils,
and petrolatum are melted together in a heated vessel equipped
with a slow-speed stirrer. When mixed thoroughly, the gilsonite
is added and the heating and stirring are continued until a
reasonable degree of dispersion is effected. The pigments and
dyes are then added and the stirring is continued until the
mass is reasonably uniform. The mix is then transferred to a
hot roller mill and milled until uniform.

A desirably strong coating is obtained by using a 14 rod and 9#
Kraft paper. Depending upon the effect desired, the coating
should be taken off the coating roll onto either hot or cold rolls.

Heat-Transfer Inks

Component	No. 3	No. 4	No. 5	No. 6	No. 7	No. 8	No. 9	No. 10	No. 11	No. 12	No. 13	No. 14
"AA" Castor Oil	13.7	90.5	56.3	60.0	27.4	49.3	85.5	77.0	86.0	40.2	58.0	36.3
#3 Castor Oil	—	—	—	—	—	—	—	—	—	12.8	—	27.4
Lard Oil	—	—	—	—	54.5	—	—	—	—	—	—	—
Oleic Acid	43.3	—	—	—	—	2.9	—	—	—	—	—	—
Carnea Oil	13.7	—	—	—	—	—	—	—	—	—	—	—
Milori Blue	11.3	—	—	—	—	—	—	—	—	—	—	—
Blue Lake 4921	6.7	—	—	—	—	—	—	—	—	—	—	—
Pulp Toner	—	9.5	—	—	—	—	—	—	—	—	—	—
Purple Lake 8400	—	—	—	6.4	—	—	—	—	—	—	—	—
Red Duplex 20-4520	—	—	—	—	18.1	—	—	—	—	—	—	—
Brown Lake A-276	—	—	—	—	—	29.0	—	—	—	—	—	—
Peerless Carbon Black	—	—	—	—	—	1.4	—	—	—	—	—	—
Green Lake	—	—	—	—	—	14.5	—	—	—	—	—	—
Purple 9673	—	—	—	—	—	—	14.5	—	—	—	—	—
Watchung Red	—	—	—	—	—	—	—	19.3	—	—	—	—
Red Lake 198	—	—	—	—	—	—	—	3.7	—	—	—	—
Green 2038	—	—	—	—	—	—	—	—	14.0	—	—	—
Victoria Blue B Base	11.3	—	—	—	—	—	—	—	—	—	—	—
Crystal Violet APN	—	—	43.7	—	—	—	—	—	—	47.0	42.0	—
Bismarck Brown	—	—	—	33.6	—	2.9	—	—	—	—	—	—
Fuchsine	—	—	—	—	—	—	—	—	—	—	—	36.3

These formulations are utilized as follows:

	Type of Grind or Dispersion		Type of Grind or Dispersion
	Ball or roller mill	No. 9	Roller Mill
No. 4	Roller Mill	No. 10	Morehouse mill
No. 5	Roller Mill	No. 11	Roller Mill
No. 6	Roller Mill	No. 12	Roller Mill
No. 7	Colloid Mill	No. 13	Roller Mill
No. 8	Roller Mill	No. 14	Roller Mill

3. One color blue
4. Two color purple
5. Purple hecto
6. Puple copy
7. Two color red (bichrome)
8. Brown copy
9. Two color purple (bichrome)
10. Two color red (bichrome)
11. Two color green (bichrome)
12. Semi-hecto purple
13. Full hecto purple
14. Red hecto

398

Stencil Duplicating Inks

Stencil duplicating inks make impressions through cut stencils superimposed upon a porous blanket, by means of a slow flow of the stencil ink. Originally, these inks were entirely non-aqueous in composition. Water inks are now available, but since they have certain serious limitations, discussion will be limited to the oil-based types. Several commercial formulations are as follows:

Component	Formula								
	No. 1	No. 2	No. 3	No. 4	No. 5	No. 6	No. 7	No. 8	No. 9
Vehicle*	95.4	75.5	83.1	64.7	71.4	82.3	81.0	—	—
"AA" Castor Oil	—	—	—	—	—	—	—	80.5	83.4
London Rosin Oil	—	—	—	—	—	—	—	13.4	10.2
Super Spectra Black	0.2	—	—	—	—	—	—	—	—
Peerless Carbon Black	3.2	—	—	—	—	—	—	4.5	—
Molacco Black	—	—	—	0.5	—	—	—	—	—
Milori Blue	1.2	4.2	—	—	—	—	—	0.8	—
Ultramarine	—	15.8	—	—	—	—	—	0.8	6.4
Gloss White	—	4.5	10.0	31.7	7.2	11.1	13.0	—	—
Purple Toner 6025	—	—	6.9	—	—	—	—	—	—
Brown 2605	—	—	—	3.1	—	—	—	—	—
Chrome Yellow 1085	—	—	—	—	14.2	—	—	—	—
Zinc Yellow 1425	—	—	—	—	7.2	—	—	—	—
Watchung Red	—	—	—	—	—	6.6	—	—	—
Green 2133	—	—	—	—	—	—	6.0	—	—
Color of Ink	Black	Blue	Purple	Brown	Yellow	Red	Green	Black	Blue

* Vehicle:	
"AA" Castor Oil	39.1
London Rosin Oil	60.9

Prepare vehicle by charging proper amounts into vessel and stirring till uniform with slow agitation. The inks are prepared by mixing the vehicle and pigments in a ball mill and cold grinding for 4–6 hours.

No. 10

U.S. Patent 2,980,543

Petroleum Sulfonate	20
Mineral Oil	17
Sorbitol Monooleate	8
Triethylene Glycol	3
Nigrosine WSJ	5
Water	47

No. 11

Water-Soluble (Aerosol)

PVP K-30	10.0
Alcohol-Soluble Dye	0.5
Ethanol	14.5
Propellent 12/11	
(30/70)	75.0

Duplicating Stencils

Duplicating stencils are semi-plastic mats used for the preparation of copy by means of a duplicating machine. The stencil consists generally of a coating of highly plasticized nitrocellulose lacquer deposited on one side of a specially prepared paper commonly referred to as *yoshino tissue*. This paper is characterized by a fibrous structure with a relatively open weave. It is highly absorbent and has greater tear strength in one direction. The combination of special lacquer and paper weave permits the surface of the stencil to be typed (cut) with a stylus so that the impression acts as a screen opening for the ink when the stencil is held against an inked pad. Another specific component of duplicating stencils is a commercial chlorinated hydrocarbon known as "Halowax," which, when allowed to recrystallize slowly from an appropriate solvent, produces a fibrous or needle-like structure. This, superimposed on the paper as part of the lacquer, reinforces the fibrosity of the sheet.

The following formulations are commercial lacquers used in the manufacture of stencils:

Formula No. 1

a	U.S.P. Collodion	56.7
	(13A) Alcohol	14.2
	Triacetin	1.8
	Benzyl Alcohol	2.2
	"AA" Castor Oil	13.3
b	Prussian Blue	1.3
	"AA" Castor Oil	2.5
c	"Halowax"	2.7
	"AA" Castor Oil	2.7
	Stearic Acid	2.6

Procedure:

The grouped components are prepared separately and added to the batch as mixtures. (b) The Prussian blue is ground in the "AA" Castor Oil on a roller mill till uniform. (c) The "Halowax" is dispersed in the "AA" Castor Oil with heat and stirring, the stearic acid is added, after which the entire mix is allowed to cool to room temperature.

(a) The collodion is weighed into a closed container fitted with an explosion-proof, slow-speed agitator (equipment should be suitably grounded). The triacetin is added, followed by the benzyl alcohol. All additions should be followed by adequate stirring to ensure uniformity. Next follows the b mixture, the c mixture, the rest of the "AA" Castor Oil, and finally, the 13A alcohol. After a final stirring, the solution should be allowed to stand. To avoid bubbles, it is drawn for use from the bottom of the vessel.

No. 2

"Hyvis" Washed Film Scrap	3.4
Ethyl Acetate	14.3
Acetone	0.8
13A Alcohol	26.8
White Oleic Acid	10.5
"AA" Castor Oil	2.6
"Halowax"	1.8
"Fastolux" Blue	1.5
"Ocenol"	5.8
Butyl Stearate	5.5
Orthophenyl Phenol	0.9
Methanol, Anhydrous	26.1

Procedure:

The "Halowax" is dispersed in the "AA" Castor Oil with heat and stirring. It is then allowed to cool to room temperature. The film scrap is next cut in a closed container fitted with a slow-speed, explosion-proof mixer. The acetone, ethyl acetate, and 13A alcohol are filled into the mixer and the scrap added. When the film has been completely dissolved, the "Fastolux" blue paste is added, followed by "Halowax"/"AA" Castor Oil mixture, the "Ocenol," and the orthophenyl phenol. The mixture should be thoroughly stirred after each addition. Then follows butyl stearate, oleic acid, and the mixture is then brought to final dilution with the anhydrous methanol. The entire mixture is thoroughly stirred and allowed to stand to remove bubbles. The mixture is drawn from the bottom through a strainer before use.

Mimeograph Ink

	Formula No. 1	No. 2
Lampblack	6.4	—
Violet Toner	0.6	—
Aluminum Hydrate	2.2	1.00
Long Oil Varnish	0.6	—
Castor Oil	78.5	40.00
Lanolin	11.7	3.00
Lithographic Varnish	11.7	20.00
Carbon Black	—	2.50
Victoria Blue in Oleic Acid	—	0.25
Benzene	—	8.00
Milori Blue in Varnish (1:1)	—	1.50

Pigment Dispersion:

"Aroclor"4465 is a useful resin in rotogravure and other printing inks. It is an important ingredient in the following mimeograph ink suitable for use on bond paper:

"Aroclor" 4465	40
Lubricating Oil (SUV 1200 @ 100°F)	35
Paraffin Oil (SUV 76 @ 100°F)	20
Carbon Black	4
Oil-Soluble Dye	1

Grinding Procedures:

All these carbon inks may be adequately processed by ball milling at 180-200°F for four to eight hours. Compositions containing Milori Blue require the longest periods. Optimum results are obtained by first reducing the vehicle to a liquid state and then adding dyes or pigments. Mills should be frequently vented to prevent emulsification, and temperatures should be carefully controlled to avoid decomposition.

General Coating Information:

Blue Pencil—This type of carbon is generally coated on either 9# or 10# Kraft. On occasion a 10# Sulfite may be used. The coating rod will vary from 5-9 depending upon use intended.

Blue Record—A variety of tissues and rods are used, depending upon the weight and finish desired.

One-Time—These low-cost carbons are generally applied on 9-10# Kraft with either a 0 or 3 rod.

Soluble Blue—These are generally applied with a 6 or higher rod on a better grade of paper. A 9# fifty percent rag or even 15# parchment may be used. Certain of these carbons have a waxed back.

Specific Recommendations:

Formula 7—Use a 10# Sulfite or a 10# Kraft with a 6 or 7 rod. Take off on chilled rolls for an intense sharp write, or on warm rolls for a less sharp write of improved cleanliness. This formula may be sharpened by increasing the content of higher-melting paraffin and reducing the paraffin of lower melting point.

Formula 8—Use a 10# Sulfite with a 7 rod. Take off on warm rolls and chill before rewinding. The write may be sharpened and hardened by adding small amounts of Sunwax 5512.

Formula 9—Use a 6 rod on 10# Kraft or 10# Sulfite. Take off on a warm roll before chilling and rewinding. Take-off on a cold roll directly after coating will provide a more intense write of higher gloss.

Formula 10—A 6 or 8 rod should be used on 10# Kraft. Take-off on cold rolls will provide maximum intensity. Warm rolls will produce a sharper and cleaner write. This formula may be hardened by the addition of small amounts of "Sunwax 5512" or "Oxiwax."

Pressure-Sensitive Carbon Paper		Alkali Blue	2.1
		Toluene	15.0
U.S. Patent 2,984,582		Ethyl Acetate	52.0
"Vinylite" VYHH	10.0	This composition is applied	
Lanolin	10.0	to a suitable paper base.	
Mineral Oil	8.3		

Carbon Paper

	Formula No. 1	No. 2	No. 3	No. 4	No. 5	No. 6	No. 7	No. 8	No. 9	No. 10
Palm Wax	25.5	—	26.7	31.2	19.3	—	—	—	—	—
Ceresine Wax	5.9	—	6.3	8.0	—	—	—	—	—	—
Ouricury Wax	—	43.3	—	—	—	35.6	—	—	—	—
Montan Wax	—	—	—	—	13.4	—	11.5	—	—	—
"Sunwax" 5512	—	—	—	—	—	—	7.4	44.8	8.0	—
"Oxiwax"	—	—	—	—	—	—	4.6	—	32.0	—
Paraffin Wax 124°F	—	—	—	—	—	—	9.0	—	—	—
Carnauba Wax	—	—	—	—	—	—	0.9	—	—	—
Petrolatum	43.1	—	43.5	34.2	42.3	—	—	—	—	—
Carnea Oil	—	—	—	—	—	—	—	—	13.3	—
"AA" Castor Oil	—	—	—	—	—	—	—	42.8	20.7	21.5
#3 Castor Oil	1.9	46.4	2.2	2.2	1.9	56.3	36.7	5.7	—	21.7
Milori Blue	19.6	—	17.7	22.4	23.1	—	—	—	—	5.5
Toning Blue	1.9	—	2.4	—	—	—	8.2	—	—	—
Victoria Blue BOC	2.1	10.3	1.2	—	—	8.1	4.0	6.7	6.0	5.1
Victoria Blue BGO	—	—	—	2.0	—	—	—	—	—	—
Pulp Toner	—	—	—	—	—	—	0.7	—	—	—
Methyl Violet	—	—	—	—	—	—	0.7	—	—	—
China Clay	—	—	—	—	—	—	16.1	—	20.0	—

These formulations are utilized as follows:

1. Low cost, general use, blue pencil
2. Oil-soluble, blue pencil, normal write
3. One-time blue pencil
4. Typewriter blue carbon ink
5. Low cost, blue pencil (substitute for oil-soluble type)
6. Conventional high intensity, oil-soluble, blue pencil
7. Hard blue pencil
8. Soft, oil-soluble, blue pencil, non-vegetable wax type
9. Medium hard, blue pencil carbon
10. Medium hard, blue pencil carbon (combined type)

Thermosensitive Colored Carbon Paper

U.S. Patent 2,936,247

Carnauba Wax (Fatty gray)	12.2
Ceresin Wax	11.9
Mineral Oil	30.7
Strong Red Lake	27.7
Graphite	17.5

Black Typewriter Carbon Paper Ink

Formula No. 1

"Petrolite" WB-5 Wax	30
Methyl Violet Base	2
Victoria Blue Base	2
Alkali Blue, CP-812	2
Peerless Carbon Black	5
Black Toner 8100	25
Mineral Oil (300 SUS)	41

No. 2

One-Time Black, Medium

"Petrolite" WB-2 Wax	12.0
Paraffin Wax (156°F)	25.0
Raven 15 Carbon Black	10.0
ASP-100 Clay	18.0
Methyl Violet Base	1.0
Nigrosine	1.0
Mineral Oil (300 SUS)	16.5
Petrolatum	16.5

No. 3

One-Time Black, Soft

"Petrolite" WB-2 Wax	10
Paraffin Wax (156°F)	20
Raven 15 Carbon Black	17
Methyl Violet Base	1
Nigrosine	1
Mineral Oil (100 SUS)	51

Heat-Sensitive Copy Sheets

U.S. Patent 2,940,866

Paper is coated with

Sodium Polyacrylate (5% solution)	20.0
Benzamide	20.0
Crystal Violet Carbinol Base	0.4
Tetramethyl-ammonium Hydroxide (10%)	1.5
Caustic Soda (30%)	0.5

No developing or fixing is needed.

Spirit and Hectograph Carbon Inks

Component	Formula No. 1	No. 2	No. 3	No. 4	No. 5	No. 6	No. 7	No. 8	No. 9	No. 10
Carnauba Wax	12.8	12.9	15.1	14.7	10.3	13.5	8.2	7.0	12.8	20.7
"Sunwax" 5512	4.4	9.0	11.6	—	—	4.8	11.3	—	—	—
"Oxiwax"	—	—	—	—	—	—	7.1	9.0	—	—
"Telura" Oil	13.5	8.5	3.8	21.4	1.9	13.5	10.0	—	17.8	13.0
"Carnea" Oil	—	—	—	—	—	—	—	10.0	—	—
#3 Castor Oil	1.1	1.2	3.7	2.8	7.0	1.5	—	7.0	1.1	10.3
"Baker's 15" Oil	4.4	4.5	4.0	3.8	7.0	4.5	—	—	4.7	—
Anhyd. Lanolin	6.0	6.3	7.7	7.2	7.0	6.0	6.4	—	5.9	8.2
Petrolatum	—	—	—	—	—	—	—	10.0	—	—
Lecithin	0.7	0.4	—	0.4	0.4	0.5	—	—	0.4	—
Crystal Violet APN	39.5	—	—	39.5	49.0	39.5	57.0	—	—	—
Crystal Violet, Hidalco	—	—	—	—	—	—	—	57.0	39.9	—
Crystal Violet, HL-2	10.0	—	—	—	—	—	—	—	—	—
Material 83	—	55.2	42.5	—	6.2	16.0	—	—	15.8	—
Methyl Violet Conc.	1.8	2.0	11.6	10.2	1.5	3.2	—	—	1.7	—
Ethyl Violet	5.8	—	—	—	—	—	—	—	—	—
Fuchsin	—	—	—	—	—	—	—	—	—	35.6
Rhodamine 5 GDN	—	—	—	—	—	—	—	—	—	12.2

These formulations are utilized as follows:

1. Sharp-writing, purple, fast take-off
2. Sharp-writing, purple, medium cost
3. Very sharp, purple, low cost
4. Long wear, medium sharp, purple
5. Top grade purple, general use
6. Electromatic
7. Low viscosity, general use
8. High grade purple, special wax base
9. Extremely brilliant purple
10. General purpose, medium cost, red spirit, on orange side

Spirit and Hectograph Carbon Inks

Component	Formula No. 11	No. 12	No. 13	No. 14	No. 15	No. 16	No. 17	No. 18	No. 19	No. 20
Carnauba Wax	21.2	—	13.1	12.7	26.2	12.8	11.3	13.2	5.8	12.5
"Sunwax" 5512	—	—	7.4	—	—	7.4	—	8.6	—	—
"Oxiwax"	—	15.6	—	—	—	—	—	—	—	—
Ouricury	—	—	—	—	—	—	—	—	4.2	—
Telura Oil	12.8	9.5	—	17.6	36.6	10.9	15.6	8.6	17.2	17.3
Carnea Oil	—	—	9.2	—	—	—	—	—	—	—
Peanut Oil	—	—	—	—	—	—	—	—	1.5	—
"AA" Castor Oil	10.2	—	—	—	—	—	—	—	0.8	—
#3 Castor Oil	—	6.6	1.5	1.2	2.2	1.2	1.1	1.4	—	1.1
"Baker's 15" Oil	—	—	4.7	4.8	8.8	4.8	5.5	4.3	6.7	7.1
Anhydrous Lanolin	8.1	1.0	6.4	5.7	12.7	6.0	15.6	5.9	7.8	5.6
Petrolatum	—	9.0	—	—	—	—	—	—	—	—
Lecithin	0.9	—	0.6	0.5	0.7	0.5	0.1	0.1	—	0.4
Crystal Violet, APN	—	—	—	—	12.8	39.7	35.0	56.5	56.0	—
Crystal Violet, Hidalco	—	58.3	57.1	—	—	—	—	—	—	—
Crystal Violet, HL-2	—	—	—	56.0	—	14.9	—	—	—	56.0
Material 83	—	—	—	—	—	—	14.3	—	—	—
Methyl Violet, Conc.	—	—	—	1.8	—	1.8	1.5	1.4	—	—
Fuchsin	46.8	—	—	—	—	—	—	—	—	—

11. Brilliant red spirit, long running, medium cost, on bluer side
12. Brilliant purple, long running
13. General use, medium cost, purple
14. Very hard purple, sharp writing, for printing and fine detail
15. Hectograph, short run, purple
16. Hard purple, electromatic and printing
17. General quality, medium cost, purple
18. Medium cost, extremely hard, purple
19. Medium cost, extremely brilliant write
20. Sharp writing, long run, non-beading, electromatic and printing purple

As basic dyestuffs are organic in nature, they are heat-sensitive. Therefore, avoid excessive heating. As they are also water-soluble and the vehicles are hydrophobic, incorporation of water in the ink invariably gives rise to an unwanted emulsion. Furthermore, such moisture as is in the ink may also produce excessive viscosity or even possible decomposition. Ground (finished) inks should, therefore, be stored under *moisture-free* conditions, or if this is not possible, should be reground (warm) immediately before use.

Spirit inks may be processed either in a ball mill or on a roller mill. Procedures vary from plant to plant. Some companies premix their inks in a slow-speed mixer and follow with a roller mill grind of sufficient length to ensure complete dispersion. Others premix, ball mill for a few hours and then finish with several mill passes.

Plant experience has proved that optimum reproducible results with specific formulations are usually obtained when the dye is added after the vehicle is reduced to a liquid in a kettle or Day Mixer. The mixed, unground ink is then dispersed on a heated roller mill seven to nine times. This procedure eliminates the possibility of mechanically entraining water in the ink. An alternate procedure involves premixing the ink by melting the vehicle in a hot ball mill, adding the dyes, grinding two to four hours with frequent venting, followed by three to five passes on a heated roller mill. Proper dispersing (grinding) procedure in a ball mill must not involve temperatures in excess of 190°F. Frequent venting is necessary to remove moisture.

General Considerations on Coating with Spirit Inks

Spirit carbons are generally coated by means of a wire-wound rod although slot-coating has yielded excellent results. The size of the rod can vary widely depending on the size of the wire. Coating fountains should be operated at the lowest possible temperatures consistent with the melting point of the ink and the optimum viscosity for coating. Excessive quantities of liquid ink should not be stored in heated kettles but should be made up immediately before use. All inks should be strained through a fine mesh strainer before addition to the coating fountain.

Coating speed of spirit machines should be as slow as possible. As the amount of spirit ink deposited on a unit area is considerably greater than in the case of either record or pencil, greater care must be exercised to ensure uniformity and smoothness. Ideal coating speed for most spirit inks is about 40 ft/minute. Temperature of the machine rolls is also important. Most inks will yield satisfactory results providing the first roll (after the coating roll) is just warm enough to prevent the ink from setting. Generally this roll will reach such a temperature of its own accord when the machine has been in operation for a short time. The coating should then be allowed to cool more rapidly. Proper heat balance is necessary to obtain excellent finish and writing properties. Excessive heat imparts a bronzy cast to the sheet and forces the vehicle into the paper with consequent shortening of shelf life. The "write" will also be excessively sharp and shorter than usual. Excessive cooling, on the other hand, gives rise to a glossy sheet with very rapid take-off, with possibly a tendency to offset, block, or to give a much broader "write."

A frequently recurrent problem is the appearance of a bead on the clean edge. This results in rolls with an offset on the back of the carbon and if the bead is sufficiently pronounced may cause cracking of the roll. If the spirit carbon has to be recoated, then the problem becomes extremely serious. This problem has been extensively studied and has been found to be due, for the most part, to improper or uncontrolled grinding. Actually, it does not always appear immediately but becomes apparent after the specific ink and machine have been in operation for some time. Controlled studies on particle size in ink samples removed from the coating fountain after specific operation intervals show that the bead is generally the result of a selective separation of larger ink particles from the ink as a whole. Such a condition arises either as the result of faulty grinding operation or possibly to an aging deterioration of the ink due to oxidation or excessive heating. Formula 20 is a special ink which was formulated to eliminate beading. It is unique in that it can be used to completion on successive days without regrinding and without formation of bead.

Recommended Paper Requirements for Spirit Carbon

Paper requirements for the production of quality spirit hectograph carbons are just as important as for record carbons and possibly even more, as there are variables present in the duplicating process which are not encountered in record use. Some excellent production papers follow:

Paper	Starch	Mullins	Weight	Ground-wood
Cantine No. 90 C2S	—	15-16	60#	None
Newton Falls A-43	—	14-16	37	”
St. Regis A-40	—	16-18	47	”
Crocker Burbank A-32	—	15-16	48	”
Crocker Burbank A-76 ClS	—	15-16	58	”

Specific examples of production stating requirements for paper and rod with specific inks are as follows:

Ink No.	Paper	Rod No.	Use
2	40# Book	12-14	General use, sharp printing
2	57# ClS	12-14	Very sharp printing
2	A-76 or A-32	14	Sharp writing, sharp printing
4	A-76	10	General use
4	A-32	7-14	General use
4	40# Book	9-14	General use
5	A-32	14	Sharp writing, general use
6	A-32	12	Electromatic, very sharp printing

Ink No.	Paper	Rod No.	Use
16	A-76	9-14	Sharp writing, general use
16	57# ClS	9-14	Sharp writing, general use
17	57ClS	9-14	Sharp writing, sharp printing
20	Milo	9-14	General use, sharp printing

Fairly light-weight sulfite papers may be employed for regular "write," long-running, and fairly brilliant spirit carbons. Copies requiring sharp "writes" and exceptional brilliance are best made with coated one-side paper. It should also be noted that papers free from starch (preferably casein based) should be used. Starch coated, or papers containing starch in the coating, generally produce short-lived sheets with questionable writing and printing properties. Some sheets tend to shorter shelf life. Master paper should also be starch-free. The best grades are two-side coated and of a minimum 60# weight.

Hectograph carbons should be prepared, using uncoated paper, as the "write" obtained is generally broad. Inks required for hectograph use (as opposed to spirit) are generally softer and of higher dye content. Normally, hard spirit formulations may be converted to hectograph use (gelatin copy) by increasing the rate of take-off and softening the composition. The use of petrolatum and/or lanolin at the expense of some of the hard wax will generally prove advantageous here. Dyes of maximum strength should be employed.

As most spirit carbons are top-coated, some comments about this phase of operation may be of interest. Top-coats are generally inert as regards dye solubility. They are also of lower melting point than the base coat (spirit ink), most of them ranging from 145-155°F. They are best applied with a 0 or 3 rod. The carbon is usually aged before top-coating. A period of 2-5 days is generally best although, that is coating and re-coating, continuous operations have been successful. In this case the base formula is specially compounded and carefully applied so as to have a certain measure of dryness and the spirit coating is applied so as to be fairly dry. The coating is then chilled and the top-coat applied immediately.

To Regenerate
Typewriter Ribbons

French Patent 1,271,300

Immerse used ribbons for less than a minute in

Ethylene Dichloride   70
Diethyl Phthalate   30

Drain and dry in air.

Chapter X

LUBRICANTS

Metal-Drawing Compound

Formula No. 1

Polyethylene Glycol 4000	10
Tallow Soap	20
Borax	70

Add 1 pound of this mixture to 1 gallon of water.

No. 2

Graphite	30
Calcium Chloride (Pure) (Saturated water solution)	70

Stable Transparent Lubricating Grease

An excellent water-resistant grease can be formulated by combining one part aluminum stearate with four parts "Oronite" Polybutene No. 32 in the following manner:

Place one-half of the Polybutene in a kettle and heat to a temperature of 150°F. Introduce the soap and stir the slurry until the soap and the oil are thoroughly mixed. Heat the mixture to 180°F and while agitating add the rest of the Polybutene. Continue heating to 300°F until the mass becomes entirely fluid and hold for a few minutes. This solution can then be cooled and homogenized in the conventional manner.

Watch Lubricant

U.S.S.R. Patent 136,503

Bone Oil	75-80
Dioctyl Sebacate	20-5
Stearic Acid	0.02-0.05

Antioxidant 0.2-0.5
Phenyl Mer-
curic Oleate 0.025-0.05

Zinc Stearate Dispersion
(Lubricant)

U.S. Patent 2,965,589

1. Water Deionized 98
2. Polyglycol
   Monostearate 1-10
3. Zinc Stearate 1-10

Heat 1 and 2 to 70-80°C and mix, then add 3 with mixing.

Rubber Air Bag
Lubricant

	%
"Carbowax" 400	10-30
Mica	10-20
Water	60-80
Soap (if desired)	About 0.1

Lubricant for
Cold-Forming Plastics

German Patent 1,041,682

Sodium Salt of Sul-
fated Olive Oil 200
Water To make 1000

Adjust to pH 7.4

Oil for Wool Weaving

Spindle Oil 76
Wool-Wax 5
Polyglycol Oleate 10
Olein 6
Ammonium Hydroxide 3

This washes out of wool with soap or synthetic detergents.

Non-Staining Dry Lubricant for Weaving Loom

French Patent 1,152,822

Talcum Powder
(250 Mesh) 200
Mica Powder
(200 Mesh) 725
Zinc Stearate (Low
density) 75

Soluble Cutting Oil

Formula No. 1

"Atlas" G-3300 2.5
"Atlas" G-1086 2.5
Petroleum Sulfonate 12.5
Paraffinic Base Oil 82.5

No. 2

"Atlas" G-3335 2.0
Petroleum Sulfonate 15.0
Naphthenic Base Oil 83.0

No. 3

Neutral Mineral Oil	6
"Petromix" #9	1

Steel-Wire Cable Lubricant

German Patent 1,037,713

Petrolatum	25-35
Gasoline	25-30
Diacetone Alcohol	0-5

Gas Turbine Threading Lubricant

Glycerol	60
Copper Powdered	25
Graphite	15

Molybdenum Disulfide for Lubricants

Polish Patent 41,459

Ammonium Molybdate	1
Sulfur	2

Heat at 360-380°C until fumes cease.

High Temperature Solid Lubricant

1. Molybdenum Sulfide	70
2. Graphite	7
3. "K" Silicate of Soda	23

Moisten 1 with water to form a smooth paste and then work in 3 and 2.

Lubricating Oil Additive

"Paratone" N	365.00
45° NW Spermoil	1.75
Mineral Oil 340-355	1.75

This product, when added to the crankcase, mixes with the motor oil in the crankcase and travels through the pistons, valves, bearings, and piston rings. It gives added protection to these parts of the engine and prolongs its life.

The product should be added to the crankcase after every change of oil.

What it does for the engine:

Decreases motor noise
Stops burning Oil
Prolongs motor life
Increases gasoline mileage
Frees sticky valves
Aids compression

CHAPTER XI

METALS

Hot-Tinning of Cast Iron

The method depends for its success on careful and complete grit-blasting of every portion of the surface to be tinned. The operations are:

1. Blast thoroughly with No. 70 angular steel grit, i.e., grit that passes a 70-mesh screen.
2. Degrease by trichloro-ethylene vapor method. If a very smooth surface is vital, water-barrelling may be included at this stage.
3. Dip in aqueous flux of following composition:

Zinc chloride	24 lb
Sodium chloride	6 lb
Ammonium chloride	3 lb

Hydrochloric acid 2 pt
Water 10 gal (100 lb)

All surfaces of the piece must be wetted by the flux. The delay between operations 1 and 3 should be kept as short as possible.

4. The piece is then lowered slowly into the first tinning pot, the surface of which is covered by a layer of flux at least one inch thick. The flux contains:

Zinc Chloride	8
Sodium Chloride	2
Ammonium Chloride	1

and must be kept boiling freely by frequent additions of water. With large articles, which must necessarily be immersed rather slowly, there may be some risk of the aqueous flux drying on the upper parts before immersion is complete. This will inevitably lead to imperfect tinning and must be prevented by spraying with water or with the aqueous flux from a fine rose or any simple spray gun. The temperature of the tin should be 300°-320°C.

5. Although the piece should accept the tin within one minute, it is advisable to allow it to remain in the tin pot for at least five minutes. This length of time allows flux residues to escape and greatly minimizes risk of "sweating" in the finished coating.

6. After removal from the first tin pot, subsequent tinning and other manipulations are as normally practiced.

Although the direct chloride process has been used mostly for protective tin coatings, some information is available on results to be expected when the method is used as preliminary to bonding babbitt alloys to cast iron. The figures for adhesion should be as good as, or better than, those usual for the fused chloride process; about 2 tons/sq in. is a reasonable expectation. This figure is admittedly not so good as that obtainable by graphite oxidation, but the latter processes are more complicated.

Plating on Uncommon Metals

In preparing the surface of a metal for electroplating, the properties of the metal must be carefully considered in determining the procedure to be used. Because of its brittleness, beryllium must be machined very carefully. A centerless ground beryllium rod which had a surface that appeared to be entirely satisfactory was subjected to chemical polishing. After only one-fourth mil had been removed, machining cracks were exposed.

Satisfactory electroplating on a surface thus damaged by machining would be impossible. Uranium, thorium, zirconium, and titanium, must be ground very slowly and with adequate lubricant or coating agent to prevent the development of heat-checked surfaces. Chemical polishing of zirconium has been successfully used. But chemical polishing of beryllium and electropolishing of thorium and uranium have been unsatisfactory as finishing operations before plating because they tend to open up subsurface defects.

Activation of the surface to be plated is an important step. The following procedures are recommended for the following base metals:

Aluminum: Zinc coating by immersion in concentrated sodium hydroxide solution containing sodium zincate, used at 70 to 80°F.

Magnesium: Zinc coating by immersion in a solution of sodium pyrophosphate, zinc sulfate, potassium fluoride, and potassium carbonate, operated at pH 10.0 to 10.4 and 180°F.

Beryllium: Same as for magnesium, but at pH 7 to 8. Alternatively, beryllium may be activated by anodic treatment in a phosphoric-hydrochloric acid solution, followed by treatment in nitric acid. When activated by the second method, beryllium may be directly plated with copper, iron, lead, magnesium, nickel, rhodium, silver, tin, or zinc.

Zirconium: Immersion in a solution of ammonium fluoride and hydrofluoric acid. Nickel or iron may be deposited over the hydride layer thus formed, the composite heated at 400°F to decompose the hydride, and finally heated at 1300 to 1500°F to form a diffusion bond. Or, a zinc coating may be applied to zirconium by immersion in molten zinc chloride at a temperature above the melting point of zinc.

It is noted that where an intermediate layer, such as an immersion zinc coating, is used, galvanic effects and alloying may act to limit the service of the parts.

The choice of plating baths is governed largely by the chemical activity of the base metal. In the case of zirconium, which is quite passive in most solutions, the electroplating characteristics and the properties of the deposit may be the determining factors in the choice.

Nickel-Plating Baths

Formula No. 1

	To liter
Nickel Sulfate	100 g
Ammonium Chloride	30 g
Sodium Sulfate	100 g
Boric Acid	15 g
"XXXD"*	20 cc
pH	5-6
Temperature	100°F
Plating Rate	80 min/mil

No. 2

	To liter
Nickel Sulfate	145 g
Ammonium Chloride	15 g
Magnesium Sulfate	75 g
Boric Acid	15 g
"XXXD"*	20 cc
pH	5-6
Temperature	100°F
Plating Rate	80 min/mil

No. 3

	To liter
Nickel Sulfate	300 g

Nickel Chloride	50 g
Boric Acid	30 g
Hydrogen Peroxide	As needed
pH	2
Temperature	140°F
Plating Rate	30 min/mil

No. 4

	To liter
Nickel Sulfide	300 g
Nickel Chloride	50 g
Boric Acid	30 g
Sodium Formate	15 g
"XXXD"*	20 cc
pH	4
Temperature	140°F
Plating Rate	30 min/mil

---

* Harshaw Chemical Co.

Nickel-plating formulas are recommended for plating on aluminum, beryllium, thorium, uranium, and zinc. Their high pH makes them particularly desirable for plating on these active metals. Formula 2 is said to give a more ductile deposit and is not limited in thickness. The nickel deposits from these baths, however, are somewhat porous. If an impervious coat on beryllium is required, formula 2 as a strike, followed by plating in formula

4 is recommended. Formula 3 is recommended for plating on zirconium.

Black Nickel Plating

Nickel Sulfate 7H$_2$O   50
Zinc Sulfate 7H$_2$O   25
Pot. Sulfocyanate   32
Ammonium Sulfate   15
Water   To make 1000
pH4.5-5.5
Current density 0.1-0.15 amp/sq dm at 20°C.

After plating treat with potassium dichromate (5% solution) at 60°C for 30 minutes.

Non-Electric Gold Plating

Gold Chloride   0.6
Potassium Cyanide   10.0
Disodium Phosphate   6.0
Sodium Sulfite   3.0
Sodium Hydroxide   1.0
Water   1000.0

Immerse cleaned pieces in boiling solution of this mixture.

Zinc Brush Plating Solution

U.S. Patent 3,017,334

Zinc Sulfate   150 g
Zinc Acetate   200 g
Ammonium
   Chloride   30 g
Boric Acid   30 g
Mannitol   50 g
Hydrochloric
   Acid   100 cc
Distilled Water   1000 cc

Bright Dip for Brass

Sulfuric Acid   2 gal
Nitric Acid   1 gal
Water   1 qt
Hydrochloric
   Acid   0.5 fl oz

Nickel-Silver Bright Dip

Sulfuric Acid   1 pt
Sodium
   Dichromate   0.5 lb
Water   5 gal

The water should be added very slowly to the acid while stirring.

Nonpoisonous Copper Strike

Copper Sulfate   5.00
Sodium "Tetrine"
   Liquid Conc.   37.00
"Duponol" MF   0.07
Sulfuric
   Acid   To give pH 4-5

Immerse for one to five minutes at 40 to 60°C.

Iron-Plating Bath

Formula No. 1

To liter

Ferrous Ammonium Sulfate	350 g
"Duponol" ME	1 g
pH	4.0-5.5
Temperature	140°F
Plating Rate	20 min/mil

No. 2

To liter

Ferrous Chloride	300 g
Calcium Chloride	335 g
pH	0.8-1.5
Temperature	190°F
Plating Rate	20 min/mil

No. 3

To liter

Ferrous Sulfate	300 g
Ferrous Chloride	42 g
Ammonium Sulfate	15 g
Sodium Formate	15 g
Boric Acid	15 g
"Duponol" ME	1 g
pH	4.1-4.3
Temperature	140°F
Plating Rate	30 min/mil

Although formula 1 is the best-known of the three, it does not give consistently uniform deposits of pit-free iron. For plating on beryllium, formula 3 is recommended, and either 2 or 3 may be used on zirconium. The addition of the wetting agent in formula 3 results in a weaker diffusion bond.

Zinc-Cadmium Plating

The following phenomenon is probably not universally known. If you use zinc anodes in a cadmium barrel plating bath instead of the cadmium anodes, a bright bluish zinc-cadmium plate is obtained which has very much the appearance of chromium plating and which gives a better protection against corrosion than a coating of either zinc or cadmium alone. Articles plated in this way and kept stored for several months retain their brightness, whereas zinc or cadmium plating grows dull or unsightly much earlier.

As the bath becomes gradually depleted of cadmium more cadmium salts must be added periodically to the bath.

Plating Steel Hardware

Hardware parts of formed steel are finished in various shades of copper plate, or oxidized silver plate, and coated with clear lacquer. Polishing precedes plating and is accomplished generally in three operations:

1. Polish on a 180 emery setup wheel, using an 8-inch wheel turning at 2,300 rpm.
2. Polish on a 180 emery setup wheel, using a tallow lubricant.
3. Polish on a sewed buff, using Stevens greaseless buffing composition.

After being polished, the steel parts are plated with either copper or silver. The copper-plating process is as follows:

1. Rack and electroclean in a reverse-current cleaner containing 12 oz/gal of "Anodex." Temperature: 190 to 212°F; current density: 50 amp/sq ft.
2. Rinse with cold running-water.
3. Dip in 10% hydrochloric acid.

4. Rinse with cold water.
5. Plate with nickel for 30 minutes at 5 to 15 amp/sq ft in a solution composed of:

Nickel (as Metal)	3 oz/gal
Ammonium Chloride	2 oz/gal
Boric Acid	4 oz/gal
pH	5.8 to 5.9

Use no agitation; add hydrogen peroxide to control pitting.

6. Rinse with cold water.
7. Plate with acid copper for 30 minutes at 10 to 15 amp/sq ft in a solution containing:

Copper Sulfate	27 oz/gal
Free Sulfuric Acid	6.5 oz/gal

8. Rinse with cold water.
9. Rinse with hot water.

The process is completed by buffing and applying clear lacquer, or other similar treatment, as specified.

Steel parts to be silver-plated are polished and nickel-plated in exactly the same manner as described. After being plated with nickel, they are rinsed and, without being reracked, are processed as follows:

1. Silver strike 10 to 15 seconds in a solution containing:

Free Sodium
  Cyanide*   8.0 oz/gal
Silver Metal   0.4 oz/gal

-----
* Highly poisonous.

2. Silver plate 1 hour to 1 ¼ hours at 5 amp/sq ft in a solution containing:

Silver Metal   1.5 oz/gal
Free Sodium
  Cyanide*   5.0 oz/gal

Add carbon disulfide brightener, 2 oz/qt of silver strike solution, as required.

3. Immerse in silver-recovery concentration rinse.
4. Rinse in cold water.
5. Oxidize in a solution containing one-eighth oz/gal of sodium polysulfide, operated at 180 to 190°F.
6. Rinse in cold water.
7. Rinse in hot water and dry.

If a plain silver finish is desired, the process is terminated at step 4. The oxidizing process, steps 5 to 7, can be applied, following the cycle for copper-plating.

Bronze parts are silver-plated by the same process as that described for steel parts.

Hard Tin Plating

An electrodeposit of tin may be hardened by codepositing copper or antimony or both with tin.

An acid tin-alloy bath may be made by adding soluble salts of either of these metals to the following fluoborate tin bath:

	To liter
Tin	40 g
Free Fluoboric Acid	40 g
Cresol Sulfonic Acid	20 g
Gelatin	1 g
β-Naphthol	1 g

Copper carbonate may be used as a source of copper and antimony fluoride, as a source of antimony. One g/liter of copper will result in about 1% of copper in the deposit and 0.5 g/liter of antimony will yield about 1% of antimony in the deposit at a current density of 15 amp/sq ft.

In continuous use, the bath can be maintained by adding the following agents at the following approximate times; plating tests are essential for continuous control.

Gelatin: 0.2 g/liter once a week; β-naphthol in alcohol to saturate the solution (about 0.5 g/liter) every 3 to 4 weeks; cresol sulfonic acid about 2 g/liter every 3 to 4 weeks.

Tin-Nickel Alloy Plating

Stannous Chloride	27
Nickel Chloride	300
Sodium Fluoride	28
Ammonium Bifluoride	35
Temperature	150°F
Cathode Current Density	About 25 amp/sq ft

Tests have shown that this plating of alloy is resistant to a wide variety of corrosive agents and is in the same class as chromium plate. The alloy plate is electrochemically noble to steel, with a potential close to that of copper. On steel, an undercoating of about 0.0005 inch of copper, followed by plating with the tin-nickel alloy provides good resistance to corrosion.

High-Current-Density Silver Plating

U.S. Patent 2,613,179

	g/liter
Silver Cyanide	40
Potassium Cyanide	55
Potassium Nitrate	100
Potassium Selenite	1

Use at 10 amp/sq ft of area of cathode.

Alkaline Iron Plating

Iron	20 g
Caustic Soda	100 g
Triethanolamine	154 g
Ethylenediamine-tetraacetic Acid	132 g
Water	1 liter

Use at 82°C and 20 amp/sq ft.

Plating with Platinum Group

The platinum group comprises six metals, all of which

are inert, stable, resistant to tarnish and corrosion, and generally white in color. Of the six metals in the group, only platinum, palladium, and rhodium have found practical applications in electroplating. A steadily increasing demand for white, reflecting, and tarnish-resistant finishes has stimulated the development of improved plating processes for these metals.

Chromium Plating

The conventional chromic acid bath is used. As chromium does not adhere well when plated directly on beryllium, intermediate layers of zinc and copper are used. To obtain an adherent chromium plating on zirconium, it is necessary to apply it over an intermediate layer of nickel or iron which has been diffusion-bonded to the zirconium.

Bismuth-Plating Bath

	To liter
Bismuth	
Trichloride	100.0 g
Sodium Chloride	18.5 g

Hydrochloric	
Acid	210.0 g
Temperature	70-90°F
Periodic	10.5 sec,
Reverse	cathodic
	5.0 sec,
	anodic
Plating Rate	33 min/mil

Black Oxide Coating for Aluminum

U.S. Patent 2,681,873

Potassium	
Permanganate	27.1
Sulfuric Acid	10.8
Sodium Dichromate	27.1
Hydrochloric Acid	33.6
Sodium Alkylaryl	
Sulfonate	1.5
Water	280.0

Cleaned aluminum is dipped for 30 seconds to 3 minutes in this solution heated at 85 to 190°F.

Yellow-Gold Coating on Aluminum

Japanese Patent 4755 (1953)

Zinc Chromate	4.0
Nitric Acid	3.5
Zinc Fluoride	1.5
Water	1000.0

Immerse aluminum in this solution for 10 minutes.

Bright Etching of Aluminum

Formula No. 1

U.S. Patent 2,673,143

Caustic Soda	6 lb/gal
Sodium Gluconate	1 lb/gal
Sodium Nitrate	1-1.25 lb/gal

Immerse in an aqueous solution of these salts for 1.5 minutes at 160 to 205°F.

No. 2

U.S. Patent 2,678,875

Silicic Acid	5-50 g
85% Phosphoric Acid	80 g
Nitric Acid	5 g
Glacial Acetic Acid	15 g
Water	To make 1000 cc

Treat at 85-110°C for two to five minutes.

No. 3

U.S. Patent 2,613,141

Phosphoric Acid	60-80

Hydrogen Peroxide	1-10
Water	15-35

Agitate the object in this solution at 70 to 110°C from 15 seconds to 5 minutes.

No. 4

German Patent 812,494
(Bright Finish)

Phosphoric Acid	70
Water	15
Nitric Acid	15

Dip into this at 100 to 110°C.

No. 5
(Mat Finish)

Phosphoric Acid	50
Water	30
Nitric Acid	20

No. 6
(Frosted Finish)

Phosphoric Acid	30
Water	50
Nitric Acid	20

No. 7

U.S. Patent 2,637,634

Nitric Acid (42°Bé)	3 gal

Hydrofluoric Acid
(48-52%)   20 gal
Boric Acid   9-1/3 lb
Water   To make 100 gal

Vat Etching of Aluminum Lithographic Plates

Immerse plates in cold 6% solution of caustic potash for 2 minutes. Rinse in cold water. Immerse in 1.5% solution of sodium carbonate at 188°F for 12 minutes. Rinse in cold water. Dip into mixture of 1 part glacial acetic acid and 35 parts of water for a few seconds. Rinse in cold water, while brushing off the sludge, and dry.

White Etch for Aluminum

Aluminum and aluminum alloys may be cleaned and etched white by immersing in sodium hydroxide followed by a dip in nitric acid.

For a 1-minute etching time in sodium hydroxide use:

Sodium   10%
  Hydroxide   Solution
Temperature   130°F

Rinse in water and dip in:

Nitric Acid   1
Water   1

or:

Nitric Acid   1
Sulfuric Acid   1
Water   2

Metal-Etching Solution

U.S. Patent 2,684,892

Ferric Chloride
  Solution
  (35-48°Bé)   1000 cc
Ammonium
  Persulfate   20-50 g
Hydrochloric
  Acid   20-30 cc

Etchant for Magnesium and Zinc Printing Plates

*Japanese Patent 15,562
(1960)*

	%
Turkey Red Oil	0.30
Oleic Acid	0.15
Nitric Acid	To make
(7%)	100.00

Semiconductor Etch

U.S. Patent 2,927,011

	cc
Nitric Acid	30 cc

Hydrofluoric
  Acid (48%)   20
Acetic Acid,
  Glacial   40-80
Aniline   1

To Mark Metal Articles by Electrolytic Etching

British Patent 551,771

On an electrical conducting block, conveniently made of graphite, with the top surface machined or ground flat, a sheet of blotting paper is placed, which is wetted with an electrical conducting solution, such as:

Formula No. 1

Hydrochloric Acid
  (Conc)   7
Nitric Acid   5
Water   88

No. 2

Hydrochloric Acid   50
Water   50

No. 3

Sodium Chloride   10
Water   90

An electrically insulating stencil, such as a waxed stencil paper, on which the required marking or wording has been typed or printed is placed on the absorbent, face down. The flat face of the metallic article to be marked, e.g., of high-speed steel or of "Stellite" is placed on the stencil and electrical contact is made by placing a weighted electrically conducting block (conveniently made of graphite) on top of the article to ensure good contact between the stencil and the article.

The graphite blocks are then connected to a source of electrical direct current supply, the block on top being the anode. Any force of direct current from 3 to 220 volts can be used, but 12 volts is used without danger to the operator.

The time required for the etching varies according to the voltage, the current, and the number of words to be etched. If a force of 12 volts is used, connected through the apparatus, 5 seconds are required to etch six syllables on "Stellite." With a voltage of 220, connected through the apparatus in a series with a 100-watt lamp, 8 seconds are required for the same lettering.

Cylindrical articles can be etched by rolling them over the stencil while under pressure from the upper block or alternatively the absorbent and stencil can be placed on a rotating drum and the articles passed underneath in contact with the stencil.

The stencil can be used many times before it is worn out.

Electropolishing Titanium

Formula No. 1

Acetic Anhydride	191 cc
Perchloric Acid	46 cc
Distilled Water	12 cc
Anode Current Density	200-250 A.F.S.
Voltage	60 volts
Temperature	Room

To prevent overheating of the anode, the time of electrolysis should not exceed 45 to 60 seconds.

No. 2

Sulfuric Acid	80-85
50% Hydrofluoric Acid	10-18
Water	5-8
Bath Voltage	5-6 volts
Anode Current Density	500-1000 A.F.S.

No. 3

90% Phosphoric Acid	57-82
50% Hydrofluoric Acid	11-19
Acetic Anhydride	8-27
Bath Voltage	20-30 volts
Anode Current Density	500-1000 A.F.S.

Electropolishing Solution for Nickel-Chromium and Stainless Steel

"Carbowax" 600   40 gal
Phosphoric Acid
 (85% ortho)   100 gal
Water   15 gal

Chemical Polishing of Nickel

U.S. Patent 2,680,678

a Copper Chloride   30
   Concentrated Hy-
    drochloric Acid   500
   Water   1000
b Glacial Acetic
    Acid   60-70
   Nitric Acid   40-30
   Hydrochloric Acid   0.5

Immerse nickel in *a* for at least 5 minutes at 82°F to remove the film of oxide. A few hours later immerse in *b* for 15 to 30 seconds at room temperature.

Chemical Polishing of Steel Wire

Japanese Patent 5004
(1953)

Phosphoric
 Acid   44.5-76.0

Nitric Acid   5.5-19.0
Sulfuric Acid   7.0-24.0
Chromium
 Trioxide   0.5-4.0
Sodium Nitrate
 or Nitrite   5.5-20.0

Immerse wire for 8 to 15 seconds in this solution at 110°F. Drain, wash with water, and dry.

To Blacken Stainless Steel

British Patent 834,834

Sodium Dichromate 500
Iron Sulfate   500
Sodium Alginate   10
Surfactant   1
Water   To make 1000

Mix and apply to cleaned surface and heat to 450°. This coating can be polished to a high gloss.

Blackening of Steel

Sodium Hydroxide   85
Sodium Nitrite   5
Water   10

Steel will turn a deep black when treated in this solution

for 15 to 30 minutes at 300 to 320°F.

To Color Galvanized Iron

Japanese Patent 6610
(1959)

Diallylphthalate Semipolymer	60
Diallylmaleate Semipolymer	15
Aluminum Powder	25
Cyanine Green	1.5
Benzoyl Peroxide	1

Apply this mixture to metal, heat to 100°C for two minutes at 200°C for 15 minutes and cool.

To Dye Anodized Aluminum

Dye	8 g
Glycerol	32 cc
Water	60 cc
Acetic Acid	A few drops

The hygroscopic effect of glycerol makes for slow drying, which allows the dye to penetrate. After drying, the articles are steam-sealed at 100°C to fix the dye and they can then be buffed, using soft

wheels to increase metallic luster and gloss.

To Blacken Aluminum Die Castings

Sodium Molybdate	29.00
Ammonium Chloride	35.00
Ammonium Acetate	3.00
Ammonium Thiocyanate	32.25
"Maprofix" Powder LK	0.75

Use 12-16 oz/gal of this mixture at 175-195°F. Be sure a non-etch cleaner is used followed by proper neutralization.

To Insulate Molten Zinc Surface

A 3-inch layer of vermiculite insulation, placed on top of a vat that contains molten metal at 787°F, is now being used by a large southern wire rolling mill to protect workers' hands. This granular, micalike material helps to maintain even temperatures and prevent unnecessary loss of heat.

To Preserve
Bronze Statues

Lanolin	40
Paraffin Wax	7
Mineral Spirits	53

Warm these compounds to-
gether in a double boiler and
mix until uniform. If used on
weather-bronze, first clean and
dry it.

Germanium Mirror

U.S. Patent 2,941,875

Place the germanium in a
solution of sodium hypochlorite, 0.01 to 5%, and bubble
carbon dioxide into it to make
pH of 6.5-6.6.

Pyrophoric Alloy
(Flint for Lighters)

Austrian Patent 172,660

Formula No. 1

Thorium	66
Tin	34

No. 2

Thorium	67
Cadmium	33

No. 3

Thorium	52
Bismuth	48

No. 4

U.S. Patent 2,490,570

Lead Ingot	1.10
Zirconium Powder	0.67
Titanium Powder	0.23

Place the metals in a cruci-
ble in a vacuum furnace;
evacuate; admit argon; heat
to 1400°F for 5 minutes, and
pour into a mold.

Pen Nib Alloy

U.S. Patent 2,681,276

Chromium	33-40
Molybdenum	5-10
Cobalt	1-10
Nickel	To make 100

Alloy for
Ball-Point Pens

German Patent 1,008,006

Chromium	86.0
Cobalt	8.0
Tungsten	3.0
Boron	3.0

This mixture gives a very

hard, non-corrosive, durable alloy.

Ductile Bismuth Alloy

U.S. Patent 2,610,913

Bismuth	99.83-99.00
Nickel	0.10-0.34

Hot-Working of Red Brass

U.S. Patent 2,577,426

Copper	50-55
Aluminum	0.5-2
Maganese	0.5-2
Silicon	0.5-2
Lead	0-2
Others	0-2
Zinc	To make 100

For Tinning of Copper Wire

U.S. Patent 2,515,022

Tin	0.25-10
Cadmium	0.35-1.5
Antimony	0.05-0.75
Lead	To make 100.00

Degassing Agent for Aluminum Alloys

Hexachloroethane	25.0
Sodium Chloride	36.5
Potassium Chloride	28.5
Sodium Fluoride	5.0
Sodium Aluminum Fluoride	5.0

Add 0.25 to 0.50% of this mixture to the molten alloy before casting.

Recovery of Zinc from Dross

U.S. Patent 2,701,194

Calcium Chloride	48
Sodium Chloride	21
Barium Chloride	31
Fuse and add	
Calcium Fluoride	3

This flux melts at 850°F and can be heated to 1300°F without fuming.

From 10 to 20% of this flux is stirred into the dross and rabbled until the oxides separate from the metal.

Aluminum Solder and Filler

Zinc	86-90
Aluminum	4-6
Tin	4-6
Copper	2

No flux is needed to solder aluminum to other metals.

To Prevent Firestain of Hardened Silver

The addition of 1% aluminum to silver hardened with copper prevents staining when it is annealed or hard-soldered.

Imitation Gold Alloy

U.S. Patent 2,935,400

Copper	88.86
Tin	0.47
Zinc	7.55
Aluminum	1.95
Gold (Optional)	1.13

Heat metals to 1950°C; stir with a carbon rod and quench in oil or water.

Moldable Permanent Magnet

German Patent 1,020,461

Cobalt	48-56
Vanadium	3-13
Chromium	3-9
Iron	To make 100

Mix metals and cold-work or draw into wires.

Antifriction Zinc Alloy

Aluminum	12.50-12.78
Copper	5.70-5.96
Lead	0.41-0.68
Iron	0.34-0.57
Zinc	To make 100.00

Alloy for Flange-Packings

Japanese Patent 3106 (1953)

Antimony	1-2
Tin	4-6
Calcium	0.01-0.20
Lead	To make 100.00

Precision Die-Casting Alloy

Japanese Patent 807 (1954)

Lead	73.15-5.00
Antimony	11.00-1.50
Tin	10.00-4.50
Copper	0.40-0.90
Bismuth	0.30-0.60

Neutron-Absorbing, Low-Melting Alloy

U.S. Patent 2,680,071

Indium	54-62
Cadmium	8-18
Bismuth	To make 100

Dental Amalgam

U.S. Patent 2,698,231

Silver	60-70
Tin	>25
Copper	<6
Zinc	<2

Mix with mercury before use.

To Solder Stainless Steel

Add 1 teaspoonful of glycerol to each cup of stainless soldering flux. The glycerol overcomes the tendency of the flux to stand up in drops on the smooth surface of the stainless steel.

Solder for Sintered Tungsten Carbide or Cast Iron

Japanese Patent 5611 (1954)

Copper	10-50
Zinc	10-25
Cadmium	7-20
Manganese	1-10
Nickel	0-3
Silver	To make 100

Magnesium-Lithium Alloy Melting Flux

U.S. Patent 2,706,679

Lithium Bromide	75
Lithium Chloride	25

Noncorrosive Soldering Flux

U.S. Patent 2,674,554

Hexadecyltrimethyl-ammonium Bromide	9
Aluminum Stearate	2-3
Xylene	39-40
Rosin	To make 100

Flux-Containing Wire Solder

U.S. Patent 2,713,315

Cadmium	2-5
Tin	95-8

A hollow wire is drawn from an alloy of these two metals. The hollow core is then filled with:

Diethanolamine Fluoride	35
Tin Fluoborate	10
Polyethylene Glycol	15
Cadmium Dust (350 Mesh)	2
Triethanolamine	38

Flux-Free Soldering of Aluminum to Stainless Steel

A flux-free method of soldering aluminum to stainless steel at low temperatures, providing a gas-tight, durable bond has been developed as part of a research project sponsored by the Atomic Energy Commission. With other soldering techniques, occlusion of flux invariably destroyed the joint through electrolytic action.

The stainless steel is cleaned with an abrasive, washed in trichloroethylene, coated with a solution of equal parts of water and concentrated hydrochloric acid, preheated to 450°F, tinned with a solder of 50% lead, 50% tin, and rinsed with hot water to remove any acid that may not have been removed by preheating.

The aluminum is cleaned with abrasive and washed in trichloroethylene. After preheating to about 650°F, it is tinned with aluminum solder, consisting of approximately 1% silver, 0.01% aluminum, 0.01% copper, 1% iron, 1% lead, 0.1% silicon, and the rest tin (aluminum solder of this composition is available from various commercial sources).

The two surfaces are placed in contact and heated to 500-600°F with an oxyacetylene flame; the aluminum solder and lead-tin solder alloy readily form a gas-tight joint.

Fluxless Aluminum Solder

U.S. Patent 2,907,105

Copper	2-5
Aluminum	4-7
Zinc	To make 100

Solder for Titanium

British Patent 836,450

Silver	29-50
Nickel	0.05-2
Aluminum, To make 100	

Arc-Welding Flux

U.S. Patent 2,681,875

Roasted Manganese Ore	42.6
Silica	42.6
Fluorspar	3.3
Silicomanganese	4.4
Anhydrous Sodium Silicate	7.1

Grind all ingredients to a fine powder, mix, and heat at 1900 to 2100°F until sintered. Then crush and grind.

Cast Iron Solder Flux

U.S.S.R. Patent 140,664

Ferrovanadium	58-68
Ferrosilicon	0-10
Marble	0-10
Fluorspar	15-25
Iron Powder	15-25
Caustic Potash	2-10
Ferrotitanium	2-10
Titanium Dioxide	0-10
Sodium Silicate	10-20

Tungsten Hard-Soldering Flux

U.S. Patent 2,658,013

Potassium Tetraborate	11
Boric Acid	47
Water	35
Powdered Iron	7

Flux for Lead Burning

U.S. Patent 2,596,466

Zinc Chloride	60
Stannous Chloride	2
Silver Chloride	1
Ammonium Bromide	3

Ammonium Acid Fluoride	1
Ammonium Hydroxide	23
Water	10

This mixture is heated at 180 to 200°F, cooled at 100 to 120°F, and water is then added to suit.

To Solder Nickel- or Chromium-Plated Articles

As ordinary solders will not adhere to either nickel or chromium, it is necessary to remove the plated coating, best with strong nitric acid, before attempting to solder.

Brazing Solder for Tungsten Carbide to Stainless-Steel Tools

Copper	46.5
Zinc	42.5
Nickel	10.5
Tin	0.1

Aluminum Solder

Formula No. 1

Austrian Patent 169,090

Zinc	46.0

Tin	23.0
Ammonium	
Chloride	6.9
Zinc Chloride	6.9

No. 2

U.S. Patent 2,622,035

Aluminum	1
Graphite	1
Methyl Acetone	2
Acetone	2
Sulfur	4

No. 3

Swiss Patent 248,042

Zinc	65-92
Tin	2-25
Lead	1-3
Copper	2 ½-4 ½
Aluminum	2-8

Apply as wire or rod without flux to a joint heated to about 300°C.

No. 4
(For Foil)

Swiss Patent 249,426

Bismuth	15.4
Germanium	7.7
Tin	57.7
Cadmium	15.4
Silver	3.8

No. 5
(Nontarnishing)

Swiss Patent 259,880

Silver	5-30
Aluminum	70-95

No. 6
(Salt-Spray-Resistant)

U.S. Patent 2,552,935

Tin	69.9-98.9
Zinc	1-30
Cerium	0.1-5
Aluminum	0.1-5

Spray for Wax Molds

U.S. Patent 2,701,902

Sodium Silicate	
Solution	225.0 cc
15% Colloidal	
Silica Solution	30.0 cc
Water	450.0 cc
Soap	3.6 g
Flint Powder	
(200 Mesh)	1400.0 g

Rapid-Setting Foundry Sand

Sand	93
Coal Dust	5
Water	1
Sodium Silicate	
(36-38°Bé)	4.5-5.0

Mix for 5 to 10 minutes.

Foundry Molds and Cores

U.S. Patent 2,686,728

Fortified Binder

Powdered Coal-Tar Pitch	98.5-97.0
Diethylene Glycol	1.5-3.0

Foundry Mold or Core

Binder	3
Water	7
Sharp Silica Sand	97

Mull all ingredients together.

Foundry Mold

Formula No. 1

Sand	45
Burnt Sand	54
Bentonite	5

No. 2

Sand	30
Burnt Sand	60
Bentonite	4-1/2
Sulfite Liquor	1/2

Wash for Molds

Water Glass	2
Water	2
Clay	1

This formula has been used both for washing molds and as a metal stop-off to prevent the wetting of steel with tin during tinning.

Molding Sand

Swiss Patent 266,423

Quartz Sand	100.00
Linseed Oil	3.00
Cobalt Naphthenate	0.10
Lead Naphthenate	0.30
Sodium Perborate	0.09

Foundry Sand Core

U.S. Patent 2,930,709

Ottawa Sand	96.00
Sodium Polyphosphate	1.32
Water	1.12
Zirconium Dioxide	1.17
Kaolin	0.39

Investment-Molding Pattern

U.S. Patent 2,525,984

Stearic Acid	8
Magnesium	1/4-3

Precision-Casting Investment

U.S. Patent 2,479,504

Magnesium Oxide	3
Magnesium Dihydrogen Phosphate	20
Silica	77

Dental-Mold Investment

Japanese Patent 673
(1951)

Powdered Sodium Alginate	20
Gypsum	20
Calcium Carbonate	25
Magnesium Carbonate	35

Wash for Foundry Cores

Zirconium Flour	88.50
Sugar	10.00
"Carbopol" 934	0.75
Diisopropanolamine	0.75

Mull the ingredients together carefully. (Make certain the amine is completely dispersed.) The resulting blend can be added to water, isopropanol, methanol or blends of water, methanol and isopropanol at concentrations up to 50% total solids. The resulting wash has good bonding properties and does not settle. The viscosity of the wash depends on the solvent used but it can be adjusted as desired by altering the amount of "Carbopol" 934 and amine.

The wash can be sprayed, brushed, or dipped as desired. Thicker suspensions give thicker coats. Excellent coatings are obtained on baked cores in this fashion with drying and with alcohol systems when ignited.

Release for Graphite Molds

Water	50.00
"Carbopol" 941	0.25
Sodium Hydroxide (10% solution)	1.40
"Ethomeen" C-25	0.25
Mineral oil	50.00
Graphite	25.00

Carefully disperse the "Carbopol" 941 in the water. After complete dispersion, add the sodium hydroxide and then the "Ethomeen" C-25. Next, disperse the graphite in the mineral oil. It is advisable to deaerate this mixture before the next step because of the large amount of air adsorbed on the graphite particles. Lastly, thoroughly mix in the oil-graphite blend and stir to uniformity. The viscosity of the finished suspension-emulsion is 33,000 cps (Brookfield, 20 rpm) and the pH is 10.2.

As the nature of the graphite has a major effect on the efficiency of the "Carbopol" resins, it is important to carefully consider each graphite separately. For instance, when "No. 4XX Special Graphite" is substituted for the "5XX," flocculation occurs. An excellent, stable suspension-emulsion is prepared using the "No. 4XX Special Graphite" by slightly altering the formulation: the sodium hydroxide and "Ethomeen" C-25 content is reduced to 0.70 and 0.25 respectively, resulting in a pH of 6.4. Because of its wide operating pH range, "Carbopol" 941 is effective in both cases.

Both formulations are designed to be diluted 3 to 1 with water and to remain stable.

Lost Wax Precision Casting Core

French Patent 1,185,891

	Formula No. 1	No. 2
Polystyrene	25	40
Castor Wax	50	40
Rosin	25	—
Coumarone Resin	—	20

This mixture is injected into lead-tin-bismuth molds at 80-85°C with pressure of 75-100 kg/sq cm. Cores may be removed at 50°C. They are well polished, solid, and precisely dimensioned.

To Seal Porous Light Metal Castings

U.S. Patent 2,680,081

Waterglass	50 cc
Chromic Acid	7 g
Potassium Dichromate	5 g
Water To make 100 cc	

The castings are placed in a vacuum and then in the solution at 95 to 100°C for half an hour at 60 to 80 lb/sq inch pressure.

Binder for Investment Dip Coating

Sodium Silicate (40°Bé)	50
Water	150
Hydrochloric Acid	2
"Tergitol" Penetrant EH	½

If a defoamer is needed add:

2-Ethylhexanol	0.1%

If a plasticizer is needed add:

Glycerol	5.0%

For thickening add:

"Cellosize" WP-3	0.5%

Pickling-Acid Inhibitor

Use 1% of a 2% tincture of iodine or 0.05 to 0.10% of iodine crystals.

Inhibited Pickling of Steel Solution

Formula No. 1

U.S. Patent 2,482,104

Phosphoric Acid (25% Solution)	99.8
Arsenic Pentoxide	0.1
4-Methyl-2-Hydroxyquinoline	0.1

No. 2

U.S. Patent 2,586,331

a Aniline Thiocyanate	1940 cc
b 40% Formaldehyde	495 g

Add *a* to *b* and stir at 12°C; filter and wash with water.

No. 3

98% Sulfuric Acid	50 cc
38% Hydrochloric Acid	6 cc
Crystalline Ferrous Sulfate	8 g

Stannous
Chloride 3.25 g
Water 1 liter

The steel objects are placed in this liquid for one night then thoroughly rinsed with clean water or, alternatively, they are first treated with lime water or a solution of soda ash for neutralizing and then rinsed with water, after which they are dried to be painted, etc. This treatment is suitable for derusting or descaling steel products practically without dissolving the steel base.

Rust Preventive

Formula No. 1

| "Alox" 365 | 5 |
| Naphthenic Oil (100 seconds at 100°F) | 95 |

This oily solution has excellent lubricating properties in addition to preventing rust and it can be used as a light machine oil. It leaves a thin, fluid film of oil on the metal surface which is highly protective during indoor storage. It is recommended for the protection of bright-finished steels, tools, firearms, fishing equipment, and metal parts during and after processing.

No. 2

"Alox" 600	12.5
Crude Petrolatum	12.5
Naphthenic Oil (200 seconds at 100°F)	50.0
Stoddard Solvent	25.0

This mixture leaves a semi-solid, viscous, self-healing, protective film on metal surfaces when the solvent evaporates. The solution can be applied by brushing or dipping at slightly elevated temperatures, not above the flash point of the solvent. This mixture is recommended especially for the protection of bar stock and semifinished steel for 6 to 12 months of indoor storage.

No. 3

"Alox" 940	20
Butyl "Cellosolve"	5
Stoddard Solvent	75

This solution can be applied by brushing, spraying, or dipping to all finished or semi-finished metal surfaces. It leaves a very thin, almost invisible, coating which is highly protective after evaporation of the solvent. Although the protective film is actually oily, it is so thin that it appears to be dry. The solution can also be used after the last washing in phosphating processes. It is equally effective after the washing and cleaning of metals in other processes because it inhibits surface moisture which may remain on the parts. It is usually in large tanks into which the metal parts are dipped. The water that accumulates at the bottom of the tank is drawn off periodically.

To Protect Sour-Crude Pipe Lines

Even after separation of water, sour crude oils present a serious problem in corrosion of pipelines. Laboratory and pilot-run tests indicate that "Arquad S" will effectively inhibit corrosion in this type of service. In tests using "Arquad S" at a rate of 8 lb to 1000 barrels of oil, 95% protection resulted.

"Armac C" was equally promising in this application.

Results of tests, using 8-12 pounds of "Armac C" to 1000 barrels of oil indicated about the same reduction in corrosion as with "Arquad S."

To Protect Acid Systems

Plugging of water-injection wells either in secondary recovery of oil or in water-disposal operations is usually due to the swelling action of clay minerals, insoluble corrosion, and/or bacterial deposits. Strong brine is usually injected to prevent clay minerals from swelling and sealing off the porous sands. Adjustment to low pH is an excellent preventive measure against swelling of clays. As acid waters have severe corrosive effects on metal equipment, selective corrosion-inhibitors must be used in such systems. "Arquad S" and S-2C materially assist this preventive action because they

1. Protect the water system from corrosive deposits, by forming protective films on metal surfaces.
2. Kill or inhibit growth of bacteria and algae, thus

preventing the formation of insoluble and acid by-products.

Corrosion rates of S.A.E. 1020 steel in hydrochloric acid solutions are as follows:

Corrosion Rate*

pH	Control (No Inhibitor)	Inhibited (10 ppm "Arquad 2C")
0.5	1700	22
1.0	1320	20
2.0	600	16
3.0	35	12
4.0	16	11
5.0	10	10
6.0	8	8

* Expressed in mg/sq dm/day.

Vapor-Phase Corrosion Inhibitor

U.S. Patent 2,629,619

Use 1 to 4 g cyclohexyl ammonium laurate, oleate, or benzoate for each cubic foot of space.

Solid Phosphating for Metal Protection

British Patent 699,308

Sodium Acid Phosphate	106
96% Sulfuric Acid	72

Mix until uniform and then add:

Sodium Nitrate	8
Sodium Dichromate	1

Antirust Steel Coating

French Patent 971,926

Formula No. 1

Nitric Acid	6.65
Phosphoric Acid	10.00
Zinc	6.75
Water	To make 1000.00

No. 2

Zinc Monophosphate	25
Sodium Nitrate	27
Zinc Phosphate	7
Water	To make 1000

Immerse the steel for 1 to 20 minutes at 70 to 80°C.

Phosphating Bath for Steel

U.S.S.R. Patent 136,996

Barium Nitrate	30-40
Zinc Nitrate	10-20
Zinc Dihydrogen Phosphate	5-10

Treat steel for 5-15 minutes at 85-90°C.

Rustproofing of Tools

Dip into aqueous solution of 5% sodium benzoate. Drain and dry.

Volatile Corrosion Inhibitor

Formula No. 1

Diammonium Phosphate	32-40
Sodium Bicarbonate	5-12
Sodium Nitrite,	To make 100

Dissolve separately in cool water, mix, and filter. Use the solution to impregnate paper.

No. 2

Ethanolamine	100.3
Carbon Dioxide	29.3

The gas is fed in slowly with good mixing. Paper is impregnated with the mixture.

No. 3

Sodium Metasilicate (Anhydrous)	6.0
Ground Caustic	15.0
Soap (Low Titer)	2.0
"Dresinate" X	5.0
"Xynomine" Powder 20	1.0
Sodium Metasilicate (regular)	71.0

Tumble 5-10 minutes using 4-6 oz/gal.

No. 4

"Aerosol OT" (100%)	10
Lanolin	50
Kerosine	40

Before use, dilute 1 part of this mixture with 9 parts of mineral spirits, such as VM & P naphtha. Treat the metal by dipping or wiping.

To Inhibit Corrosion by Pentachlorophenol

U.S. Patent 3,039,843

Add 0.025-0.1% orthophosphoric acid to a 1-5% solution of pentachlorophenol.

CHAPTER XII

PAINT AND VARNISH

Aluminum Varnish

Formula No. 1

70% Piccodiene 2215 in Mineral Spirits	143.0
Z-8 Kettle Bodied Linseed Oil	80.0
Mineral Spirits	104.0
6% Cobalt Naphthenate	0.8
	327.8

Viscosity (Gardner)	B
Non-Vol Solids	55%
Weight/gal	7.8 lb

Pigment with 2 lb/gal of lining paste.

No. 2

"Panarez" 7-60 Resin Solution	129.0
Bodied Linseed Oil, Z-9	188.1
Xylene	79.7
Mineral Spirits	266.8
Cobalt Naphthenate, 6%	0.9
Aluminum Paste, Leafing	109.3

Aluminum Paint Vehicle

Formula No. 1

"Panarez" 6-60 Resin Solution	163.00
Bodied Dehydrated Castor Oil, Z-3	350.20
"Panasol" RX-3 Aromatic Solvent	45.00
Mineral Spirits	190.00
Cobalt Naphthenate, 6%	0.15

No. 2

"NevcoVar"	528

447

"Neville" 2-50-W
Hi-Flash Naphtha 207
Cobalt (6%)
Naphthenate 5

Aqueous Aluminum Paint

U.S. Patent 2,858,230

Aluminum (Flaked) 44.0
Water 75.0
Ammonium Acid
Phosphate 1.7
Polyvinyl Acetate
Emulsion 250.0
pH = 5.8

Bronze Paint

"Panarez" 6-210
Hydrocarbon
Resin 197.0
"Parlon" 20 cps 65.6
"Aroclor" 1254 44.8
"Stabilizer" A-5 5.1
"Panasol" RX-3
Aromatic
Solvent 468.7
Bronze Powder
(MD #650) 194.4
——
975.6

Non-Volatile
Matter, wt % 40.00
Bronze Powder
Content (lb/gal
of vehicle) 2.00

"Panarez" 6-210 in
Resin Mix-
ture % 75.00
Plasticizer Content,
wt % on resin 17.00
Stabilizer Content,
wt % on
"Parlon" 0.75

While stirring the bronze
powder, add slowly solutions
of "Panarez" and "Parlon" in
"Panasol" RX-3. Then add
plasticizer, stabilizer, and remaining solvent.

Crown Cap Varnish
(25 gal oil length)

"Piccodiene" 2215 100.0
Tung Oil 117.0
Alkali-Refined
Linseed Oil 39.0
Z-4 Kettle Bodied
Linseed Oil 40.0
Mineral Spirits 360.0
6% Cobalt
Naphthenate 1.6
6% Iron
Naphthenate 1.6
——
659.2

Heat all resin, tung oil, and
alkali-refined linseed oil to
540°F, hold for 20 minutes,
and check with the Z-4 bodied

oil. Thin with mineral spirits at 400°F and add driers.

Spar Varnish

Formula No. 1

Base

"Epon" 1001	336
Methylisobutyl Ketone	75
"Cellosolve"	75
Xylene	75
"Beetle" 216-8	28

Curing-Agent

Versamid 115	151
Methylisobutyl Ketone	34
"Cellosolve"	34
Xylene	34

No. 2

Polyurethane 1210 M/S	625.00
"Uvinul" D-49 (4% Soln in Toluene)	93.75
Mineral Spirits	30.25
Antioxidant	1.00
Lead Naphthenate (24%)	6.50
Cobalt Naphthenate (6%)	1.30
Manganese Naphthenate (6%)	1.30

No. 3

"NevcoVar"	545
Mineral Spirits	182
Cobalt (6%) Naphthenate	3
Manganese (6%) Naphthenate	2
Lead (24%) Naphthenate	5
	737

No. 4

"Panarez" 6-60 Resin Solution	131.3
"Bakelite" CKR 2400 Resin	78.8
Tung Oil	93.3
Bodied Linseed Oil, Z-7 to Z-8	156.2
Mineral Spirits	268.9
Isopropanol	11.8
Lead Naphthenate, 24%	2.8
Cobalt Naphthenate, 6%	2.3
"Exkin" No. 2 (Anti-Skinning Agent)	1.0
	746.4

Prepare a 40% (by wt) solution of CKR 2400 by dissolving the resin in the isopropanol and 107 parts (by wt) of mineral spirits. Disperse this solution in the tung oil; add the remaining ingredients.

Polyamide-Epoxy Resin Enamels

Composition	Clear lb	gal	wt %	White lb	gal	wt %	Yellow lb	gal	wt %	Blue lb	gal	wt %
Pebble or Ball Mill Base:												
Epoxy Resin (Epoxide Equivalent 500)	600	60.0		600	60.0		600	60.0		600	60.0	
SR-82 or Equivalent Silicone Resin (as 100% N.V.)	12	1.5		12	1.5		12	1.5		12	1.5	
Xylene	300	41.5		300	41.5		300	41.5		300	41.5	
Methylisobutyl Ketone	300	45.0		300	45.0		300	45.0		300	45.0	
"Titanox" RA50				850	24.3		300			850	24.3	
C.P. Medium Chrome Yellow							1295	25.9				
Copper Phthalocyanine Blue										24	1.9	
Total Base	1212	147.0	38.1	2062	172.3	51.3	2507	173.9	56.2	2076	174.2	51.6
"Versamid" 401 Solution:												
(60% non-volatile in a 9:1 ratio of Xylene and "Cellosolve")	1000	129.6	31.7	1000	129.6	25.0	1000	129.6	22.5	1000	129.6	24.8
Reducer:												
Xylene	850	118.2		850	118.2		850	118.2		850	118.2	
Butanol	100	13.0		100	13.0		100	13.0		100	13.0	
Total Reducer	950	131.2	30.2	950	131.2	23.7	950	131.2	21.3	950	131.2	23.6
Sum Total	3162	408.8	100.0	4012	433.1	100.0	4457	434.7	100.0	4036	435.0	100.0

Short-Oil Varnish

"Panarez" 6-210
Hydrocarbon
Resin   169.1
Tung Oil   165.1
Bodied Linseed
Oil, Z-6   33.8
Mineral Spirits   368.0
Lead Naphthenate,
24%   2.4
Cobalt Naph-
thenate, 6%   2.0
                         —————
                         740.4

Non-Volatile
Matter, wt %   50
Tung Oil/Linseed
Oil, Vol   5/1

Charge the kettle with all the resin and tung oil. Heat to 450°F in 45 minutes. Hold for body (approximately 2 hours and 30 minutes). Check with the linseed oil. Cool to 400°F and thin to 50% solids. Blend in the driers.

Red Toy Enamel

Lithol Rubine Toner
(light)   109
Rutile Titanium
Dioxide   12
Blending Varnish
168-112-B   269

Soya Lecithin   4
High Flash Naphtha   54

Three-Roll Mill
(2 passes)

Blending Varnish
168-112-B   338
Dehydrated Castor
Oil Z-3   40
High Flash Naphtha   36
6% Cobalt
Naphthenate   1
6% Zirconium
Octoate   2
                         —————

Red Baking Enamel

"Panarez" 6-60
Resin Solution   162.7
"Duraplex" C-57
Resin   379.5
Toluidine Toner   83.4
"Surfex" Calcium
Carbonate   156.3
Aluminum Stearate
Gel, 10%   10.0
Mineral Spirits   99.2
Iron Tallate, 6%   7.6
                         —————
                         898.7

Non-Volatile
Matter (Vehicle),
wt %   50
"Panarez" Resin in
Vehicle Solids,
wt %   30

Premix toluidine toner, Surfex, aluminum stearate gel, and 215 pounds of alkyd. Grind the paste on a three-roll mill. Add in the following order: (1) remainder of alkyd, (2) "Panarez" 6-60, (3) mineral spirits, (4) drier.

White Baking Enamel

Roller Mill Grind:

Titanium Dioxide	285.8
"Acryloid" AT-52 (50% Solids)	285.8
(Add toners as desired)	

Mix with

"Acryloid" AT-52 (50% Solids)	133.0
Epoxy Resin (50% Solids in "Cellosolve" Acetate)	280.0
Xylene	55.5
"Cellosolve" Acetate	18.5
"Raybo" 3 (Anti-silk agent)	0.5

White Air-Drying Enamel

"Panarez" 6-70 Resin Solution	48.8
"Duraplex" D-65-A Resin	435.6

Titanium Dioxide	142.0
Titanium-Calcium	149.1
Aluminum Stearate Gel, 10%	20.0
Mineral Spirits	174.6
Lead Naphthenate, 24%	3.8
Cobalt Naphthenate, 6%	1.5
"ASA" Anti-Skinning Agent	0.5
	975.9

Non-Volatile Matter, wt %	50
"Panarez" Resin in Vehicle Solids, wt %	10

Premix titanium-calcium, TiO_2, aluminum stearate gel, and 119 pounds of the alkyd. Grind the paste on a three-roll mill. Add the following in order: (1) remainder of alkyd, (2) "Panarez" 6-70, (3) mineral spirits, (4) driers and ASA.

Low-Cost Black Baking Enamel

"Panarez" 7-70 Resin Solution	198.6
Bodied Linseed Oil, Z-8	224.4
Mineral Spirits	301.8

Iron Drier, 6%	7.4
"Neo Spectra"	
(Mark I)	
Carbon Black	18.0
	750.2

Non-Volatile Matter	
(Vehicle), wt %	50
Gallon Length of	
Vehicle	20

Charge the Neo Spectra, linseed oil, and 224.4 pounds of mineral spirits into a high-speed ball mill and grind to 7+ Hegman. Dissolve the "Panarez" 7-70 in mineral spirits and add this solution along with iron drier, with stirring, to the material that has been milled.

Clear Baking Finish for Metals

"Panarez" 7-210	
Hydrocarbon	
Resin	148.7
"Bakelite" CKR	
2103 Resin	37.2
Tung Oil	89.3
Linseed Oil, Z-2	91.4
Mineral Spirits	366.6
Iron Drier, 6%	8.9
	742.1

Non-Volatile Matter,	
wt %	50
Gallon Length of	
Varnish	12.3

Charge the kettle with all the oil and "Bakelite" CKR 2103. Heat to 565°F in 45 minutes and hold that temperature for 10 minutes. Check with the "Panarez" 7-210 and continue heating until the resin dissolves. Cool and thin with mineral spirits and drier.

White Satin Wall Enamel

"Panarez" 6-70	
Resin Solution	123.4
Bodied Dehydrated	
Castor Oil, Z-3	173.2
Soya Lecithin	4.0
Mineral Spirits	222.6
Cobalt Naph-	
thenate, 6%	1.7
Lead Naph-	
thenate, 24%	2.2
"Ti-Cal" R-22	
Pigment	650.0
"Surfex" Calcium	
Carbonate	100.0
Aluminum Stearate	3.0

White Interior Enamel

"Panarez" 6-60 Resin Solution	197.8
Bodied Soya Oil, Z-8	189.9
Mineral Spirits	219.7
Aluminum Stearate Gel, 10%	10.0
Lead Naphthenate, 24%	2.4
Cobalt Naphthenate, 6%	1.9
Titanium Dioxide	200.0
"Surfex" Calcium Carbonate	163.5
	985.2

Non-Volatile Matter (Vehicle), wt %	30
PVC, %	30
Gallon Length of Varnish	20

Premix the pigments, oil, aluminum stearate gel, and 50 pounds of mineral spirits. Grind the resulting paste on a three-roll mill. Add the following with stirring: "Panarez" 6-60, mineral spirits, and remaining ingredients.

Fire-Retardant Enamel

38% Phthalic Anhydride-Soya Alkyd	63.0

"Aroclor" 5460 Solution ("Aroclor" 5460, 53; Hi-Flash Naphtha, 22)	75.0
Whiting	84.0
Colloidal Grinding and Dispersing Agent	2.7
Antimony Oxide	42.0
Rutile Titanium Dioxide	33.0

Grind all these ingredients, and thin with:

38% Phthalic Anhydride-Soya Alkyd	96.0
Xylene	38.0
Cobalt Naphthenate (6% cobalt)	1.0
Manganese Naphthenate (6% manganese)	1.0
Lead Naphthenate (24% lead)	2.3

Siliconized Linseed Oil 40% Silicone, Fusion-Processed

Linseed Oil	56.500
Glycerol	3.500
Calcium Acetate Catalyst (0.1% wt of oil)	0.056
"Dow Corning" Z-6018	40.000

Charge glycerol, linseed oil, and catalyst. Heat to 490°F and hold for 30 minutes. Cool below 390°F and add "Dow Corning" Z-6018. Heat to 455°F and hold for 90 minutes. A small amount of xylene in the last stage is helpful in removing water

---

Silicone-Phenolic Resin
45% Silicone,
Fusion-Processed

"Dow Corning" Z-6018	45
"Bakelite" Resin BR-9400	55

Charge phenolic resin to kettle and heat to 280°F. Hold at this temperature until resin softens. Start agitation and charge the Z-6018. Heat to 280°F and remove any water that forms. Pour into cooling tray. The resin cools into a clear, amber-colored, brittle resin.

Silicone-Alkyd Resin

Formula No. 1

This is a nonoxidizing, short-oil coconut alkyd resin, 20% silicone, solvent-processed.

"Dow Corning" Z-6018	13
"Duraplex" ND-76 (60% NVM)	87

Charge ingredients to the resin kettle and cook at reflux (289°F), removing water azeotropically until compatibility and the desired viscosity are obtained. Total cooking time is about 6 hours. Thin with xylene.

No. 2

This is a nonoxidizing short-oil coconut alkyd resin, 50% silicone, solvent-processed.

"Dow Corning" Z-6018	30
"Duraplex" ND-78 (60% NVM)	50
Xylene	20

Charge ingredients to resin kettle and cook at reflux (290°F), removing water azeotropically until desired viscosity is obtained. Total cooking time is about 5 hours. Thin with xylene.

No. 3

This is an oxidizing, long-oil soya isophthalic alkyd resin, 30% silicone, solvent-processed.

Soya Oil	28.6
Isophthalic Acid	8.7
Pentaerythritol	5.8
Xylene	3.9
High Flash Mineral Spirits	35.3
"Dow Corning" Z-6018	17.7

Heat the oil and isophthalic acid quickly to 580°F, and hold this temperature for 10 minutes. Cool to 390°F and add the pentaerythritol and enough xylene to azeotrope the water formed. Raise the temperature rapidly to 450°F, and hold it there until the water has azeotroped off and the alkyd has the desired viscosity.

Cool the alkyd, dilute with the mineral spirits and remaining xylene, and add the Z-6018. Heat at the reflux temperature of the solvent (320 to 330°F) until a sample of the resin will cure to a clear film at 300°F. Dilute to 50% NVM with additional high flash mineral spirits (about 20 parts).

Silicone-Epoxy Resin
(25, and 50% Silicone, Solvent-Processed)

	Formula No. 1	No. 2
"Dow Corning" Z-6018	15.0	30.0
"Epon" 1001 or equivalent	45.0	30.0
"Cellosolve" Acetate	20.0	—
Xylene	20.0	40.0
2-Ethylhexoic Acid (Catalyst)	1.0	1.0

Charge all ingredients to resin kettle and cook at reflux (between 285 and 300 F). Remove water azeotropically until the desired viscosity is obtained. Thin formula No. 2 with methyl isobutylketone. Cooking time is about 8½ hours for No. 1 and 12 hours for No. 2.

No. 3

Silicone-Epoxy Resin
(45% Silicone,
Solvent-Processed)

"Dow Corning"	
Z-6018	31.0
"Epi-Rez" 520	38.0
Xylene	31.0

Charge ingredients to resin kettle. Heat to reflux temperature (290°F) and hold until compatibility is reached (about 5 hours).

Red Lead, Zinc Chromate Primer

Formula No. 1

Pigment:

Red Lead (F.S. TT-R-191, Type I, Grade B)	380.0
Zinc Yellow (F.S. TT-Z-415)	70.0
Mica (Navy Spec 52M3)	65.0
Indian Red (F.S. TT-I-511, Type I)	10.0
Magnesium Silicate (Navy Spec 52M2, Type A)	160.0
Aluminum Stearate (Navy Spec 52A12)	6.0

Vehicle:

Alkyd Resin Solution (F.S. TT-R-266, Type I, Class A)	415.0
Dipentene (F.S. TT-D-376)	25.0
Paint Thinner (F.S. TT-T-291, Grade I)	190.0
Lead Drier (F.S. TT-D-643, Type I)	4.7
Cobalt Drier (F.S. TT-D-643, Type II)	4.7

No. 2

Pigment:

97% Grade Red Lead	570.0
Zinc Yellow	82.0
Graphitic Mica	163.0

Vehicle:

L.I. A. No. 27*	490.0
Mineral Spirits	137.0
24% Lead Naphthenate	7.7
6% Manganese Naphthenate	1.3

---

* L.I.A. Vehicle No. 27 Dehydrated Castor Oil Fatty Acids (Isoline)	31
Glycerol	15
Phthalic Anhydride	19
Mineral Spirits	35

No. 3

Pigment:

Red Lead (F.S.
TT-P-191*a*, Type
I, Grade C)   380.0
Zinc Yellow (F.S.
TT-Z-415, Type I) 70.0
Mica   65.0
Indian Red   10.0
Magnesium Silicate
(F.S. 52-MC-523,
Type I)   160.0
Aluminum Stearate   2.0

Vehicle:

Alkyd Resin (52-
MC-501, Type I) 415.0
Aromatic Naphtha 105.0
Petroleum Spirits
(F.S. TT-T-291*a*,
Grade I) and
Driers   110.0

Primer
(U.S. Maritime
Commission Specification
52-MC-201)

Pigment:

Red Lead   380
Zinc Yellow   70
Mica   65
Indian Red   10
Magnesium Silicate 160
Aluminum Stearate   2

Vehicle:

Alkyd Resin (52-
MC-501, Type I)   415
Aromatic Naptha   105
Petroleum Spirits   110

Litharge may be added as
drier, up to 6 lb/100 gal.

Applied to clean steel, this
primer will give excellent service when exposed to the atmosphere, particularly in marine
environments.

Red-Lead Primer

Pigment:

97% Grade Red
Lead   575.0
Basic Lead
Chromate   212.0
Zinc Oxide   105.0
Magnesium
Silicate   105.0
Diatomaceous
Silica   59.7

Vehicle:

Dehydrated Castor
Oil Alkyd   585.0

This is a general-purpose
primer that has performed
especially well in marine atmosphere. It should be applied

to thoroughly cleaned surfaces.

Metal-Wash Primer

Base Grind:

"Vinylite" XYHL	7.2
Basic Zinc Chromate	6.9
Talc	1.1
Isopropanol (99%)	48.7
Butyl Alcohol	16.1

Acid Diluent:

Phosphoric Acid (85%)	3.6
Water	3.2
Isopropanol (99%)	13.2

Primer, Chrome Wash

Chromic Acid Solution (33.3%)	1.40
Phosphoric Acid (85%)	1.10
Acetone	9.85

Mix and add slowly to the following which has been previously dissolved:

Vinyl Butyral	10.95
Ethanol (95%)	62.20

Keep at about 45°C for 20 minutes. Cool and add:

Butyl Alcohol	14.50

Zinc-Rich Primer

"Versamid" 401	20.5
"GenEpoxy" 525 or equivalent	12.3
"Solvesso" 150	26.1
Butanol	26.1
22 Zinc Dust Powder	240.1
	325.1

NOTE: The zinc dust powder should be stirred into the mixture of resins just before use because the zinc dust will react with both the "Versamid" and the epoxy resins.

Marine Architectural Primer

"Panarez" 6-70 Hydrocarbon Resin Solution	41.32
Red Lead	363.24
Zinc Chromate	62.54
Mica	59.63
Magnesium Silicate	146.00
Aluminum Stearate	6.05
"Krumbhaar" K-1111 Phenolic Resin (40% solution)*	72.67
Bodied Linseed Oil, Z-3 Viscosity	84.02
Tung Oil	29.32

Mineral Spirits	349.78
Cobalt Drier, 6%	4.92
Lead Drier, 24%	4.66
	1224.15

* 40 lb K-1111 Phenolic Resin in 54 lb mineral spirits and 6 lb isopropanol.

Premix the pigments, oils, and sufficient mineral spirits to prepare a suitable base for grinding. Grind to 4-5 Hegman on a three-roll mill; add remaining ingredients, and blend thoroughly.

Housepaint Primer

Rutile Titanium	
Dioxide	125
Basic Lead Silicate	350
Magnesium Silicate	250
"Linaqua"	300
"Kelecin" 1081	14
Acetic Acid	14
Cupric Acetate or	
Sodium Persulfate	14
6% Cobalt Naph-	
thenate	2.5
6% Manganese	
Naphthenate	1.0
24% Lead Naph-	
thenate	6.0
Water	325

1. To a heavy-duty pre-mixer charge 100 lb of water, 14 lb of 100% Acetic Acid, 14 lb of Cupric Acetate (or Sodium Persulfate) and 14 lb of "Kelecin" 1081. Mix until crystalline salts are dissolved (5 min).

2. Add 100 lb of "Linaqua" and mix until uniform (5 min).

3. Add the 725 lb of pigment and mix until smooth and uniform (15-30 min).

4. Reduce with 150 lb of "Linaqua" and 150 lb of water, to grinding consistency.

5. Pass paste through Morehouse Mill set for 3-4 grind.

6. To the letdown tank, charge the remaining 50 lb of "Linaqua" and 9.5 lb of driers. Thoroughly mix the driers into the "Linaqua" before the mill base or additional water is added.

7. Mix grinding paste with remainder of vehicle plus driers. Adjust to viscosity with remaining 75 lb of water, filter and package.

Zinc Yellow
Epoxy Metal Primer

Ball Mill Base:

Zinc Yellow Pigment	258
Titanium Oxide	72
Zinc Oxide	91

Raw Sienna	24
Magnesium Silicate	72
Aluminum Stearate	6
	523

Lecithin	2
Epoxy Resin (Epoxide equivalent 450-550)	226
Xylene	134
Methylisobutyl ketone	134
	496

Add when ready to use:

"Versamid" 115 resin	122
Xylene	73
Butanol	8
Or add "Versamid" 415 directly	174
Xylene	21
"Cellosolve"	8
Totals	1,222

Calculated weight of gallon is 11.70 pounds.

Viscosity of base is 80 K.U. fresh and does not change after seven days.

Viscosity of total paint is 75 K.U. when mixed.

Xylene (or toluene)-"Cellosolve," 9:1 may be used as a spraying reducer. The primer dries quickly to handle and cures hard and tough after a day at 70°F, or after 10 minutes at 300°F.

Steel Ship Paint

Red Lead (95% Grade)	55.0
Zinc Yellow	10.1
Mica (MIL-M-15176A)	9.4
Indian Red	1.5
Magnesium Silicate	23.1
Aluminum Stearate	0.9
Alkyd Resin Solution (Fed Spec TT-R-266, Type 1, Class A)	64.9
Dipentene	3.9
Petroleum Spirits	29.6
Lead Drier	0.8
Cobalt Drier	0.8

Anti-Corrosive Paint for Ship Bottom
U.S. Military Specification MIL-P-19450-Ships

Formula No. 1

lb/100 gal

Primer:

Zinc Chromate	297
Venetian Red	50
Mica	75
Diatomaceous Silica	75

Varnish	640
Lead Linoleate	
(Solid)	5
Xylene	20
Lampblack	10

Varnish:

Phenolic Resin	134
R-12-A "Neville"	
Resin	134
Tung Oil	84
Alkali Refined	
Linseed Oil	42
R-29 "Neville"	
Resin-10°	93
Xylene	243
Turpentine	81

Mix lead linoleate to a thin paste with a small amount of the varnish ingredients (about 1 gallon). Add the thin paste to about half of the total varnish and then add pigments. Mix and grind on roller mill, or charge all ingredients into steel-ball or pebble mill and rotate until required fineness of grind is obtained.

To prepare varnish: Heat the R-12-A "Neville" Resin, phenolic resin, tung oil, and linseed oil to 460°F in 45 to 50 minutes. Hold at 460°F for 20 minutes. Pull from fire and add R-29 "Neville" Resin-10°,

and stir. Cool to 275°F and thin. The varnish shall have a viscosity of B-C (Gardner-Holdt), nonvolatile 59-61%, and weight of not less than 8.1 pounds to gallon.

No. 2

Japanese Patent 3719 (1958)

Mercury Oxide, Red	3
Cuprous Oxide	25
2,3-Dichloro-1,4-	
Naphthoquinone	5
Rouge	12
Talc	4
Aluminum Stearate	1
Rosin	20 ½
Boiled Linseed Oil	9 ½
Solvent Naphtha	20

No. 3

U.S. Patent 2,738,283

Rosin	277
Fish Oil	118
Zinc Stearate	18
Copper Hydroxide	100
Zinc Oxide	161
Magnesium Silicate	56
Solvent Naphtha	241

No. 4

Pigment:

| Red Lead (97% | |
| Grade) | 47.0 |

Zinc Oxide	8.3
Zinc Yellow	8.8
Mica, 325 Mesh	8.3
Magnesium Silicate	27.2
Aluminum Stearate	0.4

Vehicle:

Coumarone-Indene Resin (M.R. 127-137°C)	25.4
Coumarone-Indene Resin (M.R. 5-15°C)	8.7
Liquid Coal Tar (MIL-T-15194)	8.9
Zinc Resinate	3.2
Chlorinated Rubber 20 Centipoise Viscosity	6.3
Hi-Flash Naptha (MIL-N-15178, Type A)	31.7
Aliphatic Petroleum Spirits (B.R. 205-260°F)	15.0
Pine Oil	0.8

Pigment by weight	51.9%
Vehicle by weight	48.1%
Wt/Gal	13.6 lb
PV	35

To prepare this vehicle heat the coumarone resins and liquid coal tar together to a temperature of 445°F, add the zinc resinate, and hold temperature until it is all melted in. All of the petroleum spirits and 80 gallons of the required Hi-Flash naptha may then be added. The chlorinated rubber can then be dissolved into the remaining Hi-Flash naphtha at room temperature. The chlorinated rubber should be blended in with the remainder of the ingredients only after the coumarone-coal tar-resinate mixture has cooled to 140°F or less. To avoid decomposition, the chlorinated rubber should never be heated above 140°F. Although this method is preferred, the vehicle may also be prepared by cold-cutting the indicated ingredients.

Coatings for Cans

Formula No. 1
Exterior,
Solvent-Resistant

"Neville" LX-685, 135	100.00
China Wood Oil, or Linseed Oil, Z-4 Viscosity	55.00
VM & P Naphtha	94.00
Mineral Spirits	95.00

6% Cobalt Naph-
thenate or
Tallate   0.23
24% Lead Naph-
thenate or
Tallate   0.69

Place the "Neville" LX-685, 135 and oil in kettle, run to 550°F, hold at 550°F for 30-40 minutes. Remove from fire, let cool, and reduce. Add drier last.

No. 2
Litho Coating

"Neville" LX-685,
135   100
China Wood Oil   118
Linseed Oil, Q
Viscosity   40
Solvent (Mineral
Spirits or
Aromatic)   258
Driers, as desired

Place the "Neville" LX-685, 135 and both oils in the kettle, and run to 500°F. Hold at this point for a sufficient interval so that upon reduction to desired solids a viscosity of around D-E (Gardner-Holdt) is realized.

No. 3
Fruit Cans

"Neville" LX-685,
135   75.00
Modified Phenolic
Resin   25.00
China Wood Oil   118.00
Bodied Linseed
Oil, Z-3 Vis-
cosity   81.00
Mineral Spirits   365.00
6% Cobalt
Naphthenate   0.84
6% Manganese
Naphthenate   0.42

Place both resins, china wood oil, and five gallons of bodied linseed oil in the kettle and run to 545-550°F. Hold at this point for approximately 30-45 minutes (depending on size of batch and desired final viscosity). Then check with remaining five gallons of bodied linseed oil and thin. Finally add driers.

NOTE: This vehicle contains 45% solids and has about a "C" Gardner-Holdt viscosity which is suggested for roller-coat application. Driers are approximate and may have to be increased. It might also be found necessary to use a small amount of iron naphthenate for right dryness and hardness of film.

Drum Enamels

Formula No. 1
(Light Color)

"Neville"	
LX-685, 135	100.00
China Wood Oil	78.00
Bodied Linseed Oil, Z-3 Viscosity	40.00
Mineral Spirits or VM & P Naphtha	179.00
6% Cobalt Naphthenate	1.00
24% Lead Naphthenate	2.00
6% Manganese Naphthenate	0.25

Place the "Neville" LX-685, 135 and china wood oil in the kettle and heat to 540-545°F. Hold at this temperature for around 30 minutes. Check with the bodied linseed oil. Reduce and add drier.

No. 2
(Weather Resistant)

"Neville" LX-782	70.00
Modified Phenolic Resin	30.00
China Wood Oil	94.00

Alkali Refined Linseed Oil	31.00
Bodied Linseed Oil, Z-2/Z-3 Viscosity	32.00
Mineral Spirits and/or VM & P Naphtha	257.00
6% Cobalt Naphthenate	1.30
24% Lead Naphthenate	2.60
6% Manganese Naphthenate	0.32

Place both resins, china wood oil, and the alkali refined linseed oil in the kettle and run to 540-545°F. Hold at this temperature approximately 45 minutes. Check with the bodied linseed oil. Reduce. Add driers after pigmentation.

No. 3
(Very Light Color)

"Neville"	
LX-685, 135	100.00
Bodied Linseed Oil, minimum Z-8 Viscosity	80.00
Mineral Spirits	60.00
VM & P Naphtha	60.00
6% Cobalt	

Naphthenate or	
Tallate	1.35
24% Lead	
Naphthenate or	
Tallate	1.00

Run the "Neville" LX-685, 135 and bodied linseed oil to 520-530°F, and hold for 15-20 minutes. (A small pill on glass should be clear.) Pull from fire and cool. Reduce and add drier.

No. 4
(Low Cost)

"Neville" LX-782	50.0
Gilsonite	50.0
Fish Oil	78.0
VM & P Naphtha	178.0
6% Cobalt	
Naphthenate	1.3
24% Lead	
Naphthenate	2.0

Charge the "Neville" LX-782, half of the gilsonite, and the fish oil, heating to 500°F in approximately 40 minutes. Hold at 500°F 45-60 minutes for body, then check with remaining gilsonite. Allow mixture to cool, reduce, and add

driers. Pigment with 2 to 4% carbon black if desired.

Automobile Undercoatings

Formula No. 1

"Neville" LX-782	90.0
FF Wood Rosin	10.0
Bodied Soya Oil,	
Z/Z-1 Viscosity	96.0
Mineral Spirits	196.0
6% Cobalt	
Naphthenate	1.5
24% Lead	
Naphthenate	2.5

Run the "Neville" LX-782, rosin, and oil to 550°F in 40-45 minutes and hold there for minimum of 1 hour (longer if possible). Cool, reduce and add drier. The formulation is completed by pigmenting with 3-5% carbon black in a steel ball mill overnight and adjusting to desired viscosity and solids with mineral spirits.

No. 2

"Neville" LX-782	100.00
Bodied Linseed	
Oil, Z-3	
Viscosity	81.00
Mineral Spirits	146.00

6% Cobalt
Naphthenate 1.35
24% Lead
Naphthenate 1.70

Place the "Neville" LX-782 and bodied linseed oil in kettle, run to 550°F and hold there for approximately 45 minutes to one hour. Then remove kettle from the fire, allow the product to cool below 500°F, reduce with the mineral spirits, and finally add the metallic driers. Pigment with carbon black to achieve the degree of coverage required.

No. 3

"Nebony" 60	95
FF Wood Rosin	5
"LX-767" Solvent	50
VM & P Naphtha	100

Melt the "Nebony" 60 and rosin at 350°F. Reduce first with the "LX-767" Solvent and finally with the VM & P Naphtha. This would yield a vehicle with 40% Solids. Pigments must be mixed in while the vehicle is still warm.

The usual pigments and fillers are short-fiber asbestos, asbestine, whiting, and small amounts of carbon black.

Heat-Resistant Coatings

Formula No. 1
(Resistance to 600°F)

R-16-A "Neville" Resin, 70% Solution in "Hi-Flash"	286
"Nevinol"	13
Mineral Spirits	110
6% Cobalt Naphthenate	5

Dissolve the "Nevinol" in the R-16-A "Neville" Resin Solution. Reduce with mineral spirits and add drier. Standard aluminum powder should be added to the above vehicle, 1 lb/gal.

Constants of Resulting Vehicle:

Viscosity, Gardner-Holdt	A-2
Solids Content	52%
Acid Number (on solids)	3

No. 2
(Resistance to 600°F)

R-16-A "Neville" Resin, 60% Solution in Mineral Spirits	167
Bodied Fish Oil, Extra Heavy	24
Xylene	60

Mineral Spirits
(40 K.B.) 60

Blend together at room temperature.

Constants of Resulting Vehicle:

Viscosity, Gardner-
Holdt A-3
Solids Content 40%
Acid Number
(on solids) 3

No. 3
(Resistance to 600°F)

"Neville" LX-685,
135 100
LP Menhaden Fish
Oil 156
Mineral Spirits 256
6% Cobalt
Naphthenate 3

Charge the "Neville" LX-685, 135 and the oil and heat to 540-550°F in about 45 min-

utes. Hold for body of A-B viscosity at 50% N.V. (approximately 45 minutes at 540-550°F) cool, reduce, and add drier. Pigment with extra-fine leafing grade of aluminum paste at approximately 2 pounds per gallon.

This coating is also resistant to water and to chemicals.

No. 4
(Resistance to 1000°F)

R-6 "Nevindene" 9
Silicone Resin, 60%
Solution in Toluene 5
Aluminum Lining
Paste 15
"Nevsolv" 30 71

Place the R-6 "Nevindene" in solution with the "Nevsolv" 30 and cold-blend the two resin solutions at room temperature. Finally add aluminum paste.

Chemical-Resistant Paint

	Primers		Top Coats			
Formula	No. 1[a]	No. 2[b]	No. 3	No. 4	No. 5[c]	No.6[d]
"Parlon" (20-cp)	100	100	100	100	100	100
Long-oil drying alkyd	—	—	—	—	—	44.5
"Aroclor" 1254	44	43	54.5	43.5	55.5	55.5
"Aroclor" 5460	27	27	27	27	33	—

Formula No.	Primers				Top Coats	
	No. 1[a]	No. 2[b]	No. 3	No. 4	No. 5[c]	No. 6[d]
Dioctyl Sebacate	—	—	—	11	—	—
Titanium Dioxide	—	—	116	116	89	89
Zinc Oxide	—	—	—	—	11	11
Zinc Dust	—	1015	—	—	—	—
Silica Flour	—	—	—	—	—	16.5
Carbon Black	—	—	1.2	1.2	3	1
Red Lead (97%)	555	—	—	—	—	—
Epichlorohydrin	0.6	0.45	0.5	0.5	—	—
Dibasic Lead Phosphite	4	4	4	4	—	—
"Bentone" 34 gelling agent	12	12	12	12	12	12
Xylene	—	—	—	—	—	239
"Solvesso" 140	104	75	79	79	—	—
"Solvesso" 150	—	—	—	—	237	—
Turpentine	33	25	21	21	26	—
"Amsco" D	280	200	200	200	—	—

[a] Good adhesion to steel; resists salt-fog test or water immersion at 100°F
[b] Outstanding resistance to salt-fog test or water immersion at 100°F
[c] Meets alkali-resistance requirements of Federal Specification TT-P-91
[d] For basement floors

Chemical-Resistant Tank Lining

"Vistanex" MM L-100	100
EPC Black	120
MT Black	80
Zinc Stearate	10
Paraffinic Process Oil	7
Polyethylene or "Piccolastic" D-75	40

Corrosion Resistant Primers

Formula No. 1
Orange Primer

Pigment:

Basic Lead Silico-Chromate	1000.0
"Bentone" 34	6.0

Vehicle:

Raw Linseed Oil	309.0
Alkyd Resin (70% Solids)	112.0
Mineral Spirits, Heavy	87.0
Lead Naphthenate (24%)	4.0
Manganese Naph- thenate (6%)	1.9

No. 2
Maroon Intermediate

Pigment:

Basic Lead Silico- Chromate	650.0
Red Iron Oxide (Siliceous, 85% Fe$_2$O$_3$ min)	350.0
"Bentone" 34	6.0

Vehicle:

Raw Linseed Oil	290.0
Alkyd Resin (70% Solids)	210.0
Mineral Spirits, Heavy	91.8
Lead Naphthenate (24%)	4.5
Manganese Naph- thenate (6%)	2.2
Anti-Skinning Agent	1.0

No. 3
Dark Green Finish

Pigment:

Basic Lead Silico- Chromate	850.0
C.P. Chromium Oxide Green	93.6
Phthalocyanine Green	6.7
"Bentone" 34	6.1

Vehicle:

Raw Linseed Oil	294.0
Alkyd Resin (70% Solids)	256.0
Mineral Spirits, Heavy	81.0
Lead Naphthenate (24%)	5.0
Manganese Naph- thenate (6%)	2.3
Anti-Skinning Agent	1.0

Paints for Alkaline Surfaces

	Fed. Spec. TT-P-91	Concrete Swimming Pool	Maximum Acid- and Alkali- Resistance	White Marine Paint
"*Parlon*" (20 cp)	18.00	10.2	18.00	20
"Aroclor" 1254	10.00	5.1	8.00	6
"Aroclor" 1260	—	—	6.00	—
"Aroclor" 5460	6.00	—	—	—
Long-Oil, Oxidizing, Alkyd Resin	—	—	—	6
Medium-Oil, Drying, Alkyd Resin	—	5.1	—	—
Carbon Black	0.50	—	—	—
Iron Oxide	—	—	18.00	—
Titanium Dioxide	16.00	11.4	—	25
Zinc Oxide	2.00	—	—	—
"*Asbestine*" 3X	—	11.7	—	—
"Bentone"34 gelling agent	—	1.3	—	—
Epichlorohydrin	—	0.1	0.09	—
Stabilizer	—	—	0.90	—
"*Solvesso*" 150	42.75	—	—	—
"*Solvesso*" 100	—	9.1	45.00	—
Turpentine	4.75	—	5.00	—
Xylene	—	45.7	—	23
Hi-Flash Naptha	—	—	—	20

Latex Exterior Paint

Pigment Dispersion
(Three Roll Mill)

	lb/100 gal
Titanium Dioxide (Rutile)	263.0

Non-Chalking

Zinc Oxide (low reactive type)	43.0
Mica, Water-ground	38.2
#680 "Multicel"	28.7

Pigment Dispersant	
(25% N.V.)	1.6
4000 cps Methyl-	
cellulose, 2.5%	
Solution (high	
gel)	153.0
Water	47.8
Monoethanolamine	
Oleate, 50% in	
Ethanol	23.9

Paint Let Down:

Water	95.6
Styrene-Butadiene	
Latex (48%	
N.V.) (60-40	
type)	392.0
Ortho phenyl	
Phenol, 20%	
Soln KOH-Cut	16.3
Antifoamer	1.4

Deep Tone Tint Base

Water	128
Potassium	
Tripolyphosphate	1
"Advawet" #33	6
"Ti-Pure"R-610 (1)	50
"Alsilate" W (2)	100
"Gold Bond" "R"	
Silica	50
"Micromite"	95
"Methocel" 400 cps	
(2% soln) (3)	125
"Foamicide" 581-B	1

Charge all into mixer with agitation; disperse in Morehouse or other high speed mill, then add slowly:

"Resyn" 12K51	290
"Methocel" 400 cps	
(2% soln) (3)	60
Water	105
Hexylene Glycol	15
Ethylene Glycol	30

Interior White Emulsion Paint

Paint Base:

a	Dibutyl Phthalate	19.4
	Hexylene Glycol	15.0
	"Neville" R7	
	(108-112°C)	18.6
	"Tergitol" NPX	2.0
	"Aerosol" OT	2.0
b	"Methocel"	
	4000 cps,2%	55.0
c	Water	134.0
d	"Gelva"	
	Emulsion TS 22	211.0

Pigment Slurry:

Water	155
"Tamol" 731, 1%	20
"Methocel"	
4000 cps, 2%	100
"Titanox" RA 50	135
Wollastonite PL	100

"Gold Bond" "R"
Silica	50
"ASP" 400	50
"Albalith" 14	45

PV Ac Latex
Primer-Sealer
PV Ac Base

White Pigment-Extender Dispersion:

Water	106.61
"Vicatmide" Solution (25% solids)	5.12
Defoamer	6.40
Ethylene Glycol	15.99
"Ti-Pure" R-510	58.62
Barytes #1 White	74.63
"Mineralite" 3X	53.31
#400 "Silver Bond" "B" Silica	53.31

Stir all these together well for 5-10 minutes; then add

#7 Guar Gum Solution (2% solids)	106.61

Pass once through mill; then add

B-3 PV Ac Latex vehicle	223.47

Then add:

#7 Guar Gum Solution (2% solids)	319.83

Red Paint for Roof and Barn

"Panarez" 7-70 Resin Solution	104.00
Red Oxide Pigment	164.00
Magnesium Silicate	123.50
Aluminum Stearate	1.26
Bodied Fish Oil, Z-6	306.00
Mineral Spirits	253.00
Cobalt Drier, 6%	1.92
Lead Drier, 24%	4.80
Calcium Drier, 4%	7.60
	966.08

Charge pigments, aluminum stearate, 132 pounds of the oil, and 132 pounds of mineral spirits to a steel ball mill. Grind to 5-6 Hegman grind. Add remaining ingredients and blend.

White Non-Penetrating Flat Paint

"Panarez" 6-60 Resin Solution	43.8

"Ti-Cal" R-25
Pigment 533.0
"Surfex" Calcium
Carbonate 45.3
"Asbestine"
Pigment 90.7
Aluminum Stearate
Gel, 10% 20.9
Bodied Linseed
Oil, Z-9 128.5
Mineral Spirits 316.5
Lead Naph-
thenate, 24% 4.2
Cobalt Naph-
thenate, 6% 2.6
"ASA" Anti-
Skinning Agent 1.0
———
1,186.5

Non-Volatile Matter
(Vehicle), wt % 33.0
Pigment, wt % 56.4

The normal plant practice in preparing this paint would involve grinding in either a ball mill or a stone mill. In the laboratory preparation, one-half of the mineral spirits, all of the driers, and the anti-skinning agent were withheld, and a paste was prepared with remaining components. This paste was given one pass through a 3-roll mill. The remaining mineral spirits, anti-skinning agent and driers were then added.

———

Gray Porch and Deck Paint

This very inexpensive formula is an excellent interior-exterior finish for wood, and has satisfactory hardness for exterior flooring. Its high resistance to caustic makes it suitable for concrete floors.

	lb/100 gal
30% Titanium- Calcium Pigment	300.0
Lampblack	1.0
"Bentone" 38	3.0
Ethanol Premix just before use	1.0
Calcium Carbonate	50.0
"Neville NevcoVar"	285.0
Mineral Spirits	40.0

Grind on 3-roll mill to 7 N.S. or better, then add

"Neville NevcoVar"	200.0
Mineral Spirits	104.0
Cobalt (6%)	

Naphthenate	2.0
Manganese (6%)	
Naphthenate	2.0
Lead (24%)	
Naphthenate	4.8

Low-Cost, Satin Finish Floor Varnish

Flatting Paste:

"Celite" 165-S	800
"Aroplaz" 1241 M-70 (70% N.V.)	572
Mineral Spirits	614
"R-Lecin" TS (Lecithin)	12

Grind in pebble mill overnight.

Satin Floor Varnish:

Flatting Paste (Above)	300
"Aroplaz" 1081-M-50 (50% N.V.)	460
Mineral Spirits	186
6% Cobalt Naphthenate	2
6% Manganese Naphthenate	1
24% Lead Naphthenate	5
Anti-Skinning Agent ("ASA")	1

Floor Sealer, Varnish Type (For Wood and Cork) Federal Specification IT-S-176a

Formula No. 1

"Neville" LX-685, 135	75.00
"BR"-254 Resin	25.00
China Wood Oil	117.50
"Drisoy" (Y-Z Viscosity)	280.00
Xylene	150.00
Mineral Spirits	602.00
Cobalt Naphthenate, 6%	4.70
Lead Naphthenate, 24%	11.50
"ASA" Anti-Skin	1.57

Place "BR"-254 Resin and "Drisoy" in the kettle, heat to 550-555°F, and hold there for 20 to 25 minutes. Then add china wood oil and "Neville" LX-685, 135, holding at 520°F for proper viscosity (20 to 30 minutes). Let cool and reduce. Finally add driers and anti-skinning agent.

No. 2

"Neville" LX-685, 135	75.0

"BR"-254	25.0
China Wood Oil	117.5
"Drisoy" (Y-Z Viscosity)	280.0
2-50-W "Hi-Flash" Solvent	300.0
Mineral Spirits	1194.0
6% Cobalt Naphthenate	4.7
24% Lead Naphthenate	11.5

Place "BR"-254 Resin and "Drisoy" in the kettle, heat to 550-555°F, and hold there for about 30 minutes. Then add china wood oil and "Neville" LX-685, 135, holding at 520°F (20 to 25 minutes). Let cool and reduce. Finally add driers.

Multicolor (Speckled or Polka-Dot) Coatings
For Enamels

Ingredients lb/100 gal	Red	Green	Yellow	Black	White	Blue
Indian Red Iron Oxide	108.7					
CP Chrome Oxide Green	—	106.7				
Hansa Yellow	—	—	59.1			
Lamp Black	—	—	—	43.8		
Titanium Dioxide Rutile	—	—	—	—	151.5	156.1
Phthalocyanine Blue	—	—	—	—	—	17.4
China Clay	96.5	106.6	177.4	175.2	61.5	34.7
Pliolite VT	149.1	125.4	167.6	193.6	132.4	136.4
Chlorinated Paraffin (40%)	14.9	12.6	16.8	19.4	13.2	13.6
China Wood Oil	7.5	6.3	8.4	9.7	6.6	6.8
Soya Lecithin	—	—	—	—	—	2.1
Mineral Spirits (36KB)	480.0	497.8	419.0	414.3	493.6	487.2
	856.7	855.4	848.3	856.0	858.8	854.3

These enamels were ground in a pebble mill for 16 hours.

For Paints

Mix in the order given

Ingredients lb/100 gal	Red	Green	Yellow	Black	White	Blue
Methanol	8.5	8.5	6.2	8.5	7.6	8.5
Colloidal Clay (Organic Dispersible)	8.5	8.5	6.2	8.5	7.6	8.5
Enamel (above)	425.0	425.0	425.0	425.0	425.0	425.0

Add these to following solution while stirring

	Red	Green	Yellow	Black	White	Blue
Water	425.0	425.0	425.0	425.0	425.0	425.0
Hydroxyethyl Cellulose	4.3	4.3	—	—	—	—
Polyvinyl Alcohol	—	—	1.0	1.0	—	—
Methyl Cellulose	—	—	—	—	4.3	4.3
Anionic Surfactant	—	—	0.2	—	—	—
Colloidal Silica (25%)	—	—	—	4.0	—	—
Preservative	1.0	1.0	—	—	—	—
Colloidal Clay (water disp)	2.5	2.5	—	—	2.5	2.5

When desired particle size is obtained, add

	Red	Green	Yellow	Black
Glyoxal	8.5	8.5	—	—
Borax	—	—	0.4	0.4

Algacidal and Fungicidal Coating

U.S. Patent 3,034,953

Isobutyl Acetate	30-50 cc
Ethanol-Butanol (2:1)	8-12 cc
Toluene	20-30 cc
Xylene	20-30 cc
Cellulose Butyrate (½ Sec)	5-10 g
2,4,5-Trichlorophenol	5-8 g

Fungicidal Coating
for Books

Bleached Shellac	2 lb
"Shirlan" or "G4"	1.5 oz
Denatured Alcohol or Methanol	1 gal

Flash Bulb Coating

U.S. Patent 3,022,653

Cellulose Acetate (20 sec)	7.2
Cellulose Acetate (7 sec)	16.8
Acetone	81.0
Ethanol	9.0
Ethyl Lactate	5.0
Diacetone Alcohol	5.0

Traffic Paint

Formula No. 1
(Yellow)

CP Medium Chrome Yellow	302
"Celite" 281 (4)	35
"Asbestine" 3X (5)	348
"Pliolite" S-5 (6)	70
"Chlorowax" 40 (7)	53
"Piccodiene" 2215	70
Lecithin	8
VM & P Naphtha	293
Toluene	64
	1243

Yield	100 gal
P.V.C.	50%
Non-Vol Solids	71%
Viscosity	70-80 K.U.

Charge all the "Pliolite" S-5, toluene, "Piccodiene," and 75 lb of VM & P naphtha into a mixing tank, and agitate until solution is complete. Charge this solution and all remaining ingredients to ball mill and grind.

No. 2
(White)

"Panarez" 6-210 Hydrocarbon Resin	91.50
Titanium Dioxide	145.00
Lithopone	350.00
Zinc Oxide	226.00
Magnesium Silicate	91.60
Diatomaceous Silica	90.30
Bakelite "CKR 2400" Resin	30.40
Tung Oil	71.50
Bodied Linseed Oil, Z-6	71.60
VM & P Naphtha	250.00
Cobalt Naphthenate, 6%	3.60

Lead Naphthenate, 24%	8.06
Manganese Naphthenate, 6%	3.64
	1433.20

Charge in a pebble mill and grind for 24-30 hours; then add driers.

No. 3
Plastic Composition
Swiss Patent 301,822

Rosin, water-white	155
Calcium Carbonate	130
Quartz Sand	600
Titanium Dioxide	80
Mineral Oil	35

Heat and mix until uniform; spread at about 280°C.

If a yellow color is desired add

Chrome Yellow	50

No. 4
Reflecting Paint
U.S. Patent 2,879,171

Chrome Yellow	28

Urea Resin (50% in Butanol)	14
Soybean Alkyd Resin (50% in Xylene)	58

To one gallon of this solution add

Glass Beads (3-10 mils diameter)	3-10 lb

Anti-Static Coating

A special formula was developed for use as an anti-static or conductive coating. This particular formulation has been successfully used in flour mills to discharge static electricity on the polyethylene ducts used to transport flour and wheat products.

Anti-Static Duct Coating

"Versamid" 940	25.5
"Versamid" 100	6.5
Isopropanol	40.0
"Skellysolve" B	25.0
"Norite" A	3.0

The carbon ("Norite A") provides the basis for the conductive effect of the coating.

Plastisol Wrinkle Finish

	Formula No. 1	No. 2	No. 3
"Bakelite" Vinyl Resin QYNV	100.0	100.00	100.0
Monomer MG-1	70.0	70.00	70.0
"Flexol" Plasticizer DOP	10.0	10.00	10.0
Stabilizer A-5	1.0	1.00	1.0
Stabilizer D-22	1.0	1.00	1.0
Cobalt Octoate	1.2	1.20	1.2
CW-2015 Catalyst	3.0	3.00	3.0
White Paste (1/1: R110/DOP)	4.5	10.00	—
Blue Paste (1/1: Phthalocyanine Blue/DOP)	—	0.33	—
Black Paste (1/3 Lampblack Pigment/DOP)	0.5	—	4.0

Mixing Procedure

Using "Hobart" or similar mixer, premix dry ingredients; then premix MG-1 and plasticizer. Combine liquids with dry ingredients to make a 10:4 plastisol; mix until a paste is formed. Add cobalt drier and pigment paste. Add remainder of liquids 5 parts at a time mixing 2 minutes after each addition. Add pigment paste and mix until completely dispersed.

Artists' Water-Color Paint

Formula No. 1

U.S. Patent 2,662,031

"Carbowax" 6000	30
Stearyl Alcohol	1.25
Sorbitol Solution	2-4
"Lomar" PW	0.5
Extender (e.g., Kaolin)	0-25
Color (e.g., Carbon Black)	0-15

No. 2

U.S. Patent 2,594,273

Triethanolamine	10
Oleic Acid	30
Starch	20
Dextrin	25
Glycerol	5

Formaldehyde	3	Calcium Chloride	2.50
Water	100	Copper Sulfate	0.25

Mix until uniform.

Alkali-proof pigments must be used for coloring.

Magnesium Sulfate 0.50
Calcined Magnesite 6.00
Feldspar 4.00
Talc 0.25

Fireproof Water Paint

U.S. Patent 2,532,099

Magnesium
Chloride 7.50

Add water to suit.

Heat-Indicating Paints

Pigments	*Color Produced*
1. Cobaltous Hexamethyl-enetetramine Iodide·10 H_2O	Dull brownish pink to green at 50°C.
2. Cobaltous Pyridine Arsenate·10 H_2O	Brown to light blue-green at 50°C.
3. Nickelous Hexamethyl-enetetramine Bromide· 10 H_2O	Green to blue at 62°C.
4. Cobaltous Hexamethyl-enetetramine Thiocyan-ate·10 H_2O	Orange-pink to blue at 74°C.
5. Cobaltous Acetate	Pink to purple at 82°C.
6. Cobaltous Fluoride	Orange-pink to light lavender pink at 84°C.
7. Cobaltous Borate	Drab pink to purple at 85°C.
8. Cobaltous Pyridine Thio-cyanate·10 H_2O	Lavender to blue at 93°C.
9. Cobaltous Silicofluoride	Dull orange-pink to bright pink at 99°C.
10. Cobaltous Phosphate	Pink to blue at 112°C.

APPENDIX

TABLES

Weights and Measures
Troy Weight

24 grains = 1 pwt.
20 pwts. = 1 ounce
12 ounces = 1 pound

Apothecaries' Weight

20 grains = 1 scruple
3 scruples = 1 dram
8 drams = 1 ounce
12 ounces = 1 pound
The ounce and pound are the same as in Troy Weight.

Avoirdupois Weight

$27\frac{11}{32}$ grains = 1 dram
16 drams = 1 ounce
16 ounces = 1 pound
2000 lbs. = 1 short ton
2240 lbs. = 1 long ton

Dry Measure

2 pints = 1 quart
8 quarts = 1 peck
4 pecks = 1 bushel
36 bushels = 1 chaldron

Liquid Measure

4 gills = 1 pint
2 pints = 1 quart
4 quarts = 1 gallon
$31\frac{1}{2}$ gals. = 1 barrel
2 barrels = 1 hogshead
1 teaspoonful = $\frac{1}{8}$ oz.
1 tablespoonful = $\frac{1}{2}$ oz.
16 fluid oz. = 1 pint

Circular Measure

60 seconds = 1 minute
60 minutes = 1 degree
360 degrees = 1 circle

Long Measure

12 inches = 1 foot
3 feet = 1 yard
$5\frac{1}{2}$ yards = 1 rod
5280 feet = 1 stat. mile
320 rods = 1 stat. mile

Square Measure

144 sq. in. = 1 sq. ft.
9 sq. ft. = 1 sq. yard
$30\frac{1}{4}$ sq. yds. = 1 sq. rod
43,560 sq. ft. = 1 acre
40 sq. rods = 1 rood
4 roods = 1 acre
640 acres = 1 sq. mile

Metric Equivalents
Length

1 inch = 2.54 centimeters
1 foot = 0.305 meter
1 yard = 0.914 meter
1 mile = 1.609 kilometers
1 centimeter = 0.394 in.
1 meter = 3.281 ft.
1 meter = 1.094 yd.
1 kilometer = 0.621 mile

Capacity

1 U. S. fluid oz. = 29.573 milliliters
1 U. S. Liquid qt. = 0.946 liter
1 U. S. dry qt. = 1.101 liters
1 U. S. gallon = 3.785 liters
1 U. S. bushel = 0.3524 hectoliter
1 cu. in. = 16.4 cu. centimeters
1 milliliter = 0.034 U. S. fluid ounce
1 liter = 1.057 U. S. liquid qt.
1 liter = 0.908 U. S. dry qt.
1 liter = 0.264 U. S. gallon
1 hectoliter = 2.838 U. S. bu.
1 cu. centimeter = .061 cu. in.
1 liter = 1000 milliliters or 100 cu. c.

Weight

1 grain = 0.065 gram
1 apoth. scruple = 1.296 grams
1 av. oz. = 28.350 grams
1 troy oz. = 31.103 grams
1 av. lb. = 0.454 kilogram
1 troy lb. = 0.373 kilogram
1 gram = 15.432 grains
1 gram = 0.772 apoth. scruple
1 gram = 0.035 av. oz.
1 gram = 0.032 troy oz.
1 kilogram = 2.205 av. lbs.
1 kilogram = 2.679 troy lbs.

483

Trade-Mark Chemicals

The practice of marketing raw materials under names which in themselves are chemically not descriptive of the products they represent has become very widespread. No modern book of formulae could justify its claims either to completeness or modernity without including numerous formulae containing trade-mark chemicals.

Without wishing to enter into a discussion of the justification of trade-marks, the Editor recognizes the tremendous service rendered to commercial chemistry by manufacturers of trade-mark products, both in the physical data supplied and the formulations suggested.

Deprived of the protection afforded their products by this system of nomenclature, these manufacturers would have been forced to stand by helplessly while the fruits of their labor were being filched from them by competitors who, unhampered by expenses of research, experimentation and promotion, would be able to produce something "just as good" at prices far below those of the original producers.

That these competitive products were "just as good" solely in the minds of the imitators would only be evidenced in costly experimental work on the part of the purchaser and, in the meantime, irreparable damage would have been done to the truly ethical product. It is obvious, of course, that under these circumstances, there would be no incentive for manufacturers to develop new materials.

Because of this, and also because the CHEMICAL FORMULARY is primarily concerned with the physical results of compounding rather than with the chemistry involved, the Editor felt that the inclusion of formulae containing various trade-mark chemicals would be of definite value to the manufacturer of chemical products. If they had been left out, many ideas and processes would have been automatically eliminated. Trade-marks are in quotes throughout the book for easy distinction.

As a further service, the trade-mark products used in this volume are included in the list of chemicals and supplies.

CHEMICALS AND SUPPLIES
Where to Buy Them

Numbers on right refer to list of suppliers on pages directly following this list. Chemicals not listed here may be located by consulting *Chemical Week*, 330 West 42nd Street, New York 10036; or their *Annual Buyers' Guide;* or *Oil, Paint and Drug Reporter*, 100 Church Street, New York 10007.

A		Product	No.
Acetulan	30	Areskap	500
Acintol	75	Arlex	90
ACL—	500	Armac	80
Acrawax C	355	Arochem	500
Acryloid	665	Aroclor	500
Actamer	500	Aroplaz	65
Adipol	305	Arquad	80
Advance—	5	ASA	330
Advawet	190	Asbestine	420
Aerosil	190	Atlas	90
Aerosol OT	35	Atomite	795
Agent RE-610	325	Avitex	235
Age Rite	825		
Alathon	235	**B**	
Albalith	540	Baker's	100
Albusoy	145	Bayol	270
Alcolan	660	Beckacite	650
Aldo	355	Beetle	35
Aldrin	705	Bentone	525
Alipal	60	Biopal	325
Aliquat	335	Blancophor	325
Alkanol	235	Breadlac	115
Alox	20	Brij	90
Alro Amine	320	BR-254 Resin	820
Alrosol	320	BTC	565
Amberol	665		
Ambropur	230	**C**	
Amerchol	30	Cab-O-Sil	120
Ammonyx	565	Calcene	165
Amsco	45	Calcotone	125
Anodex	25	Calgon	380
Antarate	60	Carbitol	815
Antaron	60	Carbopol	365
Antarox	60	Carbowax	815
Apco thinner	50	Carnea Oil	710
Aquarex	235	Catalyst AC	500

Product	No.	Product	No.
Cedrenen	652	Durmont	240
Celite	430	Duroxon	240
Cellosize	815	Dylex	445
Cellosolve	815	Dymerex	395
Celluflex	140	Dyphos	525
Celogen	530		
Centifiol	230	**E**	
Centrophil	145	Elvax	235
Cerelose	170	Emcol	850
Ceroxin	195	Emersol	260
Cetiol	275	Emulphor	60
Cetonia	345	Epi-Rez	435
Cheelox	60	Epolene	792
Chemocide	150	Epon	705
Chlorhydrol	645	Esso	290
Chlorothene	220	Estane	365
Chlorowax	205	Ethofat	80
Ciba Resin	155	Ethomeen	80
Cinacoa	805	Ethylan	660
Citroflex	610	Eutanol	275
Citronal-S	375	Exkin	555
Citroviol	220	Exon	290
CMC	395	Extrapone	230
Colloids—	160		
Comperlan	195	**F**	
Crag	815	Fastolux	760
Crown Wax	605	Filtrosol	690
Cumar	110	Fixative 404	295
Cumaryl	375	Flexol Plasticizor	815
Cumate	845	Flexo Wax	355
		Fluilan	185
D		Foamicide 581-B	160
Dantoin	355	Fractol	290
Darvan	200	Franklin Clay	822
DC 200	200	Freon	235
Dehydag	195	Frigene	405
Deltyl	345		
Deriphat	335	**G**	
Dieldrin	705	G-11	715
Distabex	740	Gafac	325
Dixie Clay	825	Gelva	700
Dowanol	220	Gen-Epoxy	335
Dowicide	220	Geon	365
Drago—	230	Glim	95
Dresinate	395	Glucarine	355
Drinox	580	Glyco—	355
Drisoy	730		
Duomeen	80	**H**	
Duponol	235	Halane	860
Duraplex	665	Halowax	385
Durez	245	Harflex	835

Product	No.	Product	No.
Harshaw—	390	Linaqua	730
Hartolan	185	Lomar	855
HB-40	500	Lopor	270
Heliocrete	315	Loramin	250
Heroflex	395	Lorol	235
Hercolyn	395	LPB	565
Hetrofoam	410	Lupersol	460
Hoechst Wax	405	Lustrex	500
Hyamine	665		
Hycar	365	**M**	
Hylene	235	Macisol	510
Hyonic	550	Magnesol	305
Hypalon	235	Makon	765
		Mapro	565
I		Maprofix	565
Idonyx	565	Marasperse	470
Igepal	60	Marcol	270
Igepon	60	Mark—	70
Indoflor	375	Methocel	220
Indopol	50	Methylparaben	310
Insular	675	Michelene	475
Intracaine	735	Micro-Cel	430
Iso-Bergamot	230	Micromite	340
Isocreme	185	Mineralite	480
Iso-lan	360	Miranol	485
Isothan	565	Modulan	30
Ivory	630	Monamid	495
		Monapon	495
J		Monastral	235
Jacinthol	375	Mondur	490
Jonon	375	Monoplex	665
		Moskene	345
K		Mousse-S	117
Karathane	665	Multi-Cel	790
Katapol	325	Multifex	205
Kelecin	440	Multranil	490
Kessco	442	Myrj	90
KP—	305		
Krumbhaar	450	**N**	
		Nacconol	515
L		Nadic	515
Lactoscaton	230	Nebony	535
Laminac	35	Nekal	325
Lan-Aqua-Sol	280	Neocol	210
Lanette Wax	195	Neo Cryl	625
Lanogel	660	Neo-Fat	80
Lantox	455	Neophax	750
Lanwax	280	Neoprene	235
Lavandozon	375	Neo-Spectra	167
Lethane	665	Neutronyx	565
Lilial	345	Nevastain	535

Product	No.	Product	No.
Robane	655	Synprowax	780
Rogepel	220	Synvar	785
Rosottone	220		
Ross Wax	670	**T**	
		Tamol	665
S		Tecsol	255
Santicizer	500	Teepol	415
Santocel	500	Teflon	235
Santolite	500	Tegin	360
Santomerse	500	Tegosept	360
Santonox	500	Telloy	825
Santophen	500	Telura Oil	755
Santowax	500	Tergitol	815
Sarkosyl	320	Tetrine	355
Saturn Yellow	775	Texapon	275
Selectro foam	615	Thanite	665
Seqlene	610	Thermax	825
Sequestrene	320	Thiovanic Acid	274
Shanco	695	Thixcin	100
Shellsol	710	Ti-Cal	235
Shirlan	415	Ti-Pure	235
Silene	165	Titanox	800
Silicone LE-46	815	Tolusol	710
Sipon	10	Toncarine	345
Siponol	10	Toxaphene	395
Skellysolve	720	Tribase	525
Snow-Floss	430	Trimal	525
Solulan	30	Triton	665
Solvesso	270	Truodor	425
Sorbo	90	Tuads	825
Soyalose	350	Tween	90
Span	90		
Spiceolate	215	**U**	
Stabelan	550	Ubatol	810
Stabilizer # 1	715	Ucon	815
Sta-Sol	745	Uformite	665
Staybelite	395	Ultrawet	85
Stepan	765	Unads	825
Stepanol	765	Univul	60
Sterox	500	Unox	815
Sucaryl	1	Uvitex	155
Sunwax	770		
Super-Amide	565	**V**	
Super Beckacite	650	Vandex	825
Supermultifex	205	Varsol	270
Surfex	205	Veegum	825
Sweetose	745	Velvatex	635
Switzer	775	Velvetine	220
Synalyl	220	Versalide	345
Synfleur	300	Versamid	335
Synpron	780	Versene	220

SELLERS OF CHEMICALS AND SUPPLIES

No.	Name	Address
1.	Abbott Laboratories	North Chicago, Ill.
5.	Advance Solvents & Chemical Div., Carlisle Chemical Works	New Brunswick, N. J.
10.	Alcolac Chemical Corp.	Baltimore, Md.
15.	Allied Chemical Corp.	New York, N. Y.
20.	Alox Corp.	Niagara Falls, N. Y.
25.	American Anode Co.	Akron, Ohio
30.	American Cholesterol Products Inc.	Edison, N. J.
35.	American Cyanamid Company	Wayne, N. J.
40.	American Maize Products Co.	New York, N. Y.
45.	American Mineral Spirits Co.	Chicago, Ill.
50.	Amoco Chemical Corp.	Chicago, Ill.
55.	Anderson-Prichard Oil Corp.	Chicago, Ill.
60.	Antara Chemicals Division General Aniline & Film Corp.	New York, N. Y.
65.	Archer-Daniels-Midland Co.	Minneapolis, Minn.
70.	Argus Chemical Laboratory	Brooklyn, N. Y.
75.	Arizona Chemical Co.	New York, N. Y.
80.	Armour Industrial Chemical Co.	Chicago, Ill.
85.	Atlantic Refining Co.	Philadelphia, Pa.
90.	Atlas Chemical Industries, Inc.	Wilmington, Del.
95.	Babbitt, B. T., Inc.	New York, N. Y.
100.	Baker Castor Oil Co.	Jersey City, N. J.
105.	Bareco Wax Co.	Kilgore, Tex.
110.	Barrett Div., Allied Chemical Corp.	New York, N. Y.
115.	Borden Co.	New York, N. Y.
117.	Bruno-Court	Grasse, France
120.	Cabot Corp.	Boston, Mass.
130.	Cambridge Industries Co.	Cambridge, Mass.
135.	Carwin Polymer Products, Inc.	North Haven, Conn.
140.	Celanese Corp. of America	New York, N. Y.
145.	Central Soya Co.	Ft. Wayne, Ind.
150.	Chemo Puro Mfg. Corp.	Newark, N. J.
155.	Ciba Corp.	New York, N. Y.
160.	Colloids Inc.	Newark, N. J.
165.	Columbia-Southern Chemical Corp.	Pittsburgh, Pa.
175.	Columbian Carbon Co.	New York, N. Y.
170.	Corn Products Co.	New York, N. Y.
180.	Court, Bruno-	Grasse, France
185.	Croda Ltd.	Goole, England
190.	Degussa	Frankfort, Germany
195.	Dehydag	Dusseldorf, Germany
200.	Dewey & Almy Chemical Co.	Cambridge, Mass.
205.	Diamond Alkali Co.	Painesville, Ohio

No.	Name	Address
210.	Dispergent Co.	Guilford, Conn.
215.	Dodge & Olcott, Inc.	New York, N. Y.
220.	Dow Chemical Co.	Midland, Mich.
225.	Dow Corning Corp.	Midland, Mich.
230.	Dragoco, Inc.	Totowa, N. J.
235.	du Pont de Nemours, E. I. & Co.	Wilmington, Del.
240.	Dura Commodities Corp.	New York, N. Y.
245.	Durez Plastics Div. Hooker Chemical Corp.	North Tonawanda, N. Y.
250.	Dutton & Reinisch, Ltd.	London, England
255.	Eastman Chemical Products, Inc.	New York, N. Y.
260.	Emery Industries	Cincinnati, Ohio
265.	Emulsol Corp.	Chicago, Ill.
270.	Esso Standard Div. Humble Oil & Refining Co.	New York, N. Y.
274.	Evans Chemetics, Inc.	New York, N. Y.
275.	Fallek Chemical Corp.	New York, N. Y.
280.	Fanning Chemical Corp.	Newark, N. J.
285.	Fine Organics, Inc.	Lodi, N. J.
290.	Firestone Plastics Co.	Pottstown, Pa.
295.	Firmenich & Co.	New York, N. Y.
300.	Florasynth Laboratories, Inc.	New York, N. Y.
305.	FMC Corp.	San Jose, Cal.
310.	Fries & Bro.	New York, N. Y.
315.	Fritzsche Bros., Inc.	New York, N. Y.
320.	Geigy Chemical Corp.	Ardsley, N. Y.
325.	General Aniline & Film Corp.	New York, N. Y.
330.	General Chem. Div. Allied Chemical Corp.	New York, N. Y.
335.	General Mills, Inc.	Minneapolis, Minn.
340.	Georgia Marble Co.	Tate, Ga.
345.	Givaudan-Delawanna, Inc.	New York, N. Y.
350.	Glidden Co.	Cleveland, Ohio
355.	Glyco Chemicals, Inc.	New York, N. Y.
360.	Goldschmidt Chemical Corp.	New York, N. Y.
365.	Goodrich, B. F., Chemical Co.	Cleveland, Ohio
370.	Goodyear Tire & Rubber Co.	Akron, Ohio
375.	Haarmann & Reimer	Holzminden, Germany
380.	Hagan Chemicals & Controls, Inc.	Pittsburgh, Pa.
385.	Halowax Div. Union Carbide Corp.	New York, N. Y.
390.	Harshaw Chemical Co.	Cleveland, Ohio
395.	Hercules Powder Co., Inc.	Wilmington, Del.
400.	Heyden Newport Chemical Corp.	New York, N. Y.
405.	Hoechst Chemicals	New York, N. Y.
410.	Hooker Chemical Corp.	Niagara Falls, N. Y.
415.	Imperial Chemical Industries, Ltd.	London, England
420.	International Talc Co., Inc.	New York, N. Y.
430.	Johns-Manville Corp.	New York, N. Y.
435.	Jones Dabney Co.	Louisville, Ky.
440.	Kelco Co.	Clark, N. J.
442.	Kessler Chemical Co., Inc.	Philadelphia, Pa.
445.	Koppers Co., Inc.	Pittsburgh, Pa.
450.	Krumbhaar Chemical Resin Div. Lawter Chemicals, Inc.	Kearny, N. J.
455.	Lanatex Products Co.	Elizabeth, N. J.

No.	Name	Address
460.	Lucidol Corp.	Buffalo, N. Y.
465.	Malmstrom Chemical Corp.	Linden, N. J.
470.	Marathon Div. American Can Co.	Neenah, Wisc.
505.	McLaughlin Gormley King Co.	Minneapolis, Minn.
475.	Michel, M. Co., Inc.	New York, N. Y.
480.	Mineralite Sales Corp.	Mineola, N. Y.
485.	Miranol Chemical Co., Inc.	Irvington, N. J.
490.	Mobay Chemical Co.	Pittsburgh, Pa.
495.	Mona Industries, Inc.	Paterson, N. J.
500.	Monsanto Chemical Co.	St. Louis, Mo.
510.	Naarden, N. V., Chem. Fabrik	Naarden, Holland
515.	National Aniline Div. Allied Chemical Corp.	New York, N. Y.
520.	National Lead Co.	New York, N. Y.
525.	National Starch and Chemical Corp.	New York, N. Y.
530.	Naugatuck Chemical Div. U. S. Rubber Co.	New York, N. Y.
535.	Neville Chemical Co.	Pittsburgh, Pa.
540.	New Jersey Zinc Co.	New York, N. Y.
545.	Nipa Laboratories, Ltd.	London, England
550.	Nopco Chemical Co.	Newark, N. J.
555.	Nuodex Products Co. Div. Heyden Newport Chemical Corp.	Elizabeth, N. J.
560.	Olin-Mathieson Chemical Corp.	New York, N. Y.
565.	Onyx Chemical Corp.	Jersey City, N. J.
570.	Oronite Chemical Co.	San Francisco, Cal.
580.	Panogen Inc.	Ringwood, Ill.
585.	Penick, S. B. & Co.	New York, N. Y.
590.	Pennotex Oil Corp.	Sparta, N. J.
595.	Pennsylvania Industrial Chemical Corp.	Clairton, Pa.
605.	Petrolite Corp.	New York, N. Y.
610.	Pfizer, Chas. & Co., Inc.	New York, N. Y.
600.	Pilot Chemical Co. of California	Santa Fe Springs, Calif.
615.	Pittsburgh Plate Glass Co.	Pittsburgh, Pa.
620.	Polak's Frutal Works	Middletown, N. Y.
625.	Polyvinyl Chemicals	Peabody, Mass.
630.	Procter & Gamble Co.	Cincinnati, Ohio
635.	Proctor Chemical Co.	Salisbury, N. Car.
640.	Raybo Chemical Co.	Huntington, W. Va.
645.	Reheis Co., Div. Armour Pharmaceutical Co.	Berkeley Heights, N. J.
650.	Reichhold Chemicals, Inc.	White Plains, N. Y.
652.	Riedel-De Haen, A. C.	Hanover, Germany
655.	Robeco Chemicals, Inc.	New York, N. Y.
660.	Robinson, Wagner Co., Inc.	Mamaroneck, N. Y.
665.	Rohm & Haas Co.	Philadelphia, Pa.
670.	Ross, F. B. Co.	Jersey City, N. J.
675.	Rubber Corporation of America	Hicksville, N. Y.
680.	Rumford Chemical Works	Rumford, R. I.
740.	St. Denis Ind. Chimie	Paris, France
685.	Salomon, L. A. & Bro., Inc.	New York, N. Y.
690.	Schimmel & Co., Inc.	Newburgh, N. Y.
695.	Shanco Plastics & Chemical Co., Inc.	Tonawanda, N. Y.
700.	Shawinigan Products Corp.	New York, N. Y.

No.	Name	Address
705.	Shell Chemical Co., Div. Shell Oil Co.	New York, N. Y.
715.	Sindar Corp.	New York, N. Y.
720.	Skelly Oil Co.	Kansas City, Mo.
725.	Sonneborn Chemical & Refining Corp.	New York, N. Y.
730.	Spencer Kellogg & Sons, Inc.	Buffalo, N. Y.
735.	Squibb, E. R. & Sons	New York, N. Y.
745.	Staley, A. E. Mfg. Co.	Decatur, Ill.
750.	Stamford Rubber Supply Co.	Stamford, Conn.
755.	Standard Oil Co. (N. J.)	New York, N. Y.
760.	Stein, Hall & Co., Inc.	New York, N. Y.
765.	Stepan Chemical Co.	Chicago, Ill.
770.	Sun Oil Co.	Philadelphia, Pa.
775.	Switzer Bros.	Cleveland, Ohio
780.	Synthetic Products Co.	Cleveland, Ohio
785.	Synvar Corp.	Wilmington, Del.
790.	Tamms Industries Inc.	Chicago, Ill.
792.	Tennessee Eastman Co.	Kingsport, Tenn.
795.	Thompson-Weinman & Co.	New York, N. Y.
800.	Titanium Pigment Corp.	New York, N. Y.
805.	Traders Oil Mill Co.	Ft. Worth, Tex.
810.	Union Bay State Chemical Corp.	Cambridge, Mass.
815.	Union Carbide Corp.	New York, N. Y.
820.	Union Carbide Plastics Co.	New York, N. Y.
822.	United Clay Mines Corp.	Trenton, N. J.
825.	Vanderbilt, R. T. Co., Inc.	New York, N. Y.
830.	Victor Chemical Works Div. Stauffer Chemical Co.	Chicago, Ill.
835.	Wallace & Tiernan, Inc.	Newark, N. J.
840.	Warwick Chemical Co. Div. Sun Chemical Corp.	Wood River Junction, R. I.
845.	Waverly Petroleum Products Co.	Philadelphia, Pa.
850.	Witco Chemical Co., Inc.	New York, N. Y.
855.	Wolf, Jacques & Co. Nopco Chemical Co.	Newark, N. J.
860.	Wyandotte Chemicals Corp.	Wyandotte, Mich.

INDEX

495

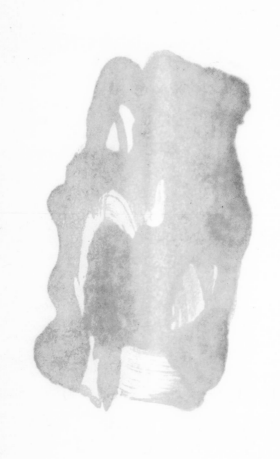